AF594992

Real-Time Sensor Networks and Systems for the Industrial IoT

Real-Time Sensor Networks and Systems for the Industrial IoT

Editors

Christos Koulamas
Mihai T. Lazarescu

MDPI • Basel • Beijing • Wuhan • Barcelona • Belgrade • Manchester • Tokyo • Cluj • Tianjin

Editors

Christos Koulamas
Industrial Systems Institute/
R.C. "Athena"
Greece

Mihai T. Lazarescu
Politecnico di Torino
Italy

Editorial Office
MDPI
St. Alban-Anlage 66
4052 Basel, Switzerland

This is a reprint of articles from the Special Issue published online in the open access journal *Sensors* (ISSN 1424-8220) (available at: https://www.mdpi.com/journal/sensors/special_issues/industrial_IoT).

For citation purposes, cite each article independently as indicated on the article page online and as indicated below:

LastName, A.A.; LastName, B.B.; LastName, C.C. Article Title. *Journal Name* **Year**, *Article Number*, Page Range.

ISBN 978-3-03943-430-5 (Hbk)
ISBN 978-3-03943-431-2 (PDF)

Contents

About the Editors

Christos Koulamas PhD. He is currently a Research Director at the Industrial Systems Institute (ISI), ATHENA Research and Innovation Centre in Patras, Greece. He has more than 25 years of experience in R&D, in the areas of real-time distributed embedded systems, industrial networks, wireless sensor networks and their applications in industrial and building automation and the transportation and energy sectors. He is the author or co-author of more than 70 publications in international journals, conferences and books, and he has participated in numerous European and national research projects with multiple roles. He has served as a member of the board of ERCIM, as Guest Editor for MDPI journals and as PC/TPC or panel member of many IEEE and other conferences and workshops. He is an IEEE Senior Member and a member of the Technical Chamber of Greece.

Mihai T. Lazarescu (M'92) received his PhD in 1998 from the Politecnico di Torino, Italy, where he serves now as Assistant Professor. He was Senior Engineer at Cadence Design Systems, working on high-level synthesis (HLS) of embedded systems; he founded several start-ups, working on real-time embedded Linux and WSN for long-term environmental monitoring; and he has participated in numerous European- and national-founded research projects on topics related to the design of ASIC, EDA for HLS and WSN hardware, software and development tools. He has authored and co-authored more than 60 scientific publications at international conferences and journals, several books and international patents. He is an IEEE Senior Member and served as Guest Editor for ACM and MDPI journals and as PC Chair for international conferences. His research interests include real-time embedded systems, low-power sensing and data processing for IoT and WSN platforms, high-level hardware/software co-design and high-level synthesis.

Editorial

Real-Time Sensor Networks and Systems for the Industrial IoT: What Next?

Christos Koulamas [1,*] and Mihai T. Lazarescu [2]

[1] Industrial Systems Institute, Athena Research Center, PSP Bldg, Stadiou Strt, 26504 Patras, Greece
[2] Dipartimento di Elettronica e Telecomunicazioni, Politecnico di Torino, I-10129 Turin, Italy; mihai.lazarescu@polito.it
* Correspondence: koulamas@isi.gr; Tel.: +30-2610-911-597

Received: 20 August 2020; Accepted: 2 September 2020; Published: 4 September 2020

1. Introduction

The Industrial Internet of Things (Industrial IoT—IIoT) is the emerging core backbone construct for the various cyber-physical systems constituting one of the principal dimensions of the 4th Industrial Revolution. While initially born as a concept inside specific industrial applications of the generic IoT technologies, for the optimization of the operational efficiency in automation and control, it quickly became the vehicle for the achievement of the total convergence of Operational Technologies (OT) with Information Technologies (IT). Today, it is already breaking the traditional borders of automation and control functions in the process and manufacturing industry, towards a wider domain of functions and industries, embraced under the dominant, global initiatives and architectural frameworks of Industry 4.0 (or Industrie 4.0) in Germany, Industrial Internet in the US, Society 5.0 in Japan, and Made-in-China 2025 in China. As real-time embedded systems are quickly achieving ubiquity in people's everyday life and in industrial environments, and many processes already depend on real-time cyber-physical systems and embedded sensors, IoT integration with cognitive computing and real-time data exchanges is essential for real-time analytics and realization of digital twins in smart environments and services under the various frameworks' provisions [1,2]. In this context, real-time sensor networks and systems for the Industrial IoT encompass multiple technologies and raise significant design, optimization, integration, and exploitation challenges.

2. The Current Issue

The ten articles in this Special Issue provide advances of real-time sensor networks and systems that are significant enablers of the Industrial IoT paradigm. In the relevant landscape, the wireless networking technologies domain is centrally positioned, as expected. Seferagić et al. comparatively discuss various alternatives for the engineering and deployment of industrial wireless sensor and actuator networks (IWSAN) in [3]. In this work, LoRa, IEEE 802.11ah (Wi-Fi HaLow), Narrowband-IoT (NB-IoT), WirelessHART, ISA100.11a, Bluetooth Low Energy (BLE), and IEEE 802.15.4g with IEEE 802.15.4e TSCH are reviewed and compared over multiple axes, including communication range, latency, reliability, data rates, energy consumption, scalability, and spectrum regulations. WSAN reliability and real-time performance are key aspects for effective and stable wireless control, and nearly half of the articles of this Special Issue focus on the optimization of the quality of service offered by the wireless stack. In this view, Park et al. propose and evaluate in [4] three different wireless transmission scheduling schemes serving multiple and heterogeneous control systems. The reported results suggest that the centralized Lyapunov-based scheduling approach stands closer to the ideal solution, and that the distributed random access is a good candidate for control systems with a small number of control loops. Furthermore, multipath retransmission in multiple channel deterministic wireless networks, such as WirelessHART, are discussed by Wang et al. in [5], proposing the CEM-RM resource-scheduling algorithm. Through simulations as well as data collected from real experimental setups, the authors demonstrate improved performance in terms of schedulability, end-to-end delay, and resource usage efficiency compared to the M-LLF and M-RM scheduling policies. The WirelessHART protocol also provides the context for Wu et al.'s research in [6], and especially the priority assignment for real-time data flows mapped over the TDMA operation at the MAC layer. Cheng et al. address soft real-time industrial applications over IEEE 802.11ah in [7]. In this work, the authors propose the channel-aware contention window adaption (CA-CWA) algorithm, which improves the packet loss rate and average delay by adapting the contention window based on the channel status. Along the wireless network technologies, legacy wired real-time networks can play significant roles in the modern IIoT landscape, as shown by Lee et al. in [8]. The authors propose a lightweight CAN virtualization technology for virtual controllers containerized on top of an OS-based virtualization technology that provides virtual CAN interfaces and buses at the device driver level.

Several articles in this Special Issue address important industrial IoT security and infrastructure protection issues. Tedeschi et al. analyze the integration of modern IoT and legacy industrial equipment for real-time machine condition monitoring applications, and the resulting security challenges in [9]. The authors propose a security-by-design approach that introduces real-time adaptation features for device security through subsystem isolation and lightweight authentication. Similarly, the protection of legacy devices in critical infrastructures is addressed by Fournaris et al. in [10]. The proposed dedicated hardware security token (HST) supports a secure event log and real-time monitoring mechanism for anomaly detection, thus isolating the typically insecure legacy OS logging mechanisms. In connection to this, Wielgosz et al. introduce in [11] AI concepts in real-time smart sensors to detect anomalies for the protection of superconducting industrial machinery. In particular, an embedded Recurrent Neural Network (RNN) has been designed for a device protecting the main subsystem of CERN's Large Hadron Collider (LHC) accelerator. The proposed scheme demonstrates a low memory footprint and architectural uniformity, allowing an efficient hardware implementation and a distributed edge-computing cluster of sensors that reduce resource consumption, latency, and throughput. AI is also the core technological domain exploited by Ntalianis et al. in [12], which is an excellent example of the extended scope of today's Industrial IoT. Specifically, the authors propose a deep CNN sparse coding used for the classification of inhaler sounds in real-time directly on the time-domain, thus avoiding computationally expensive feature extraction techniques and allowing for an efficient implementation on constrained hardware and the integration with real-time IIoT technologies.

3. What Next?

The real-time performance of the evolving IIoT networks and systems is increasingly considered among the major IIoT challenges, together with energy efficiency, security, and interoperability [13]. Coming as a challenging requirement from various application domains, real-time operation is also recognized as a mitigator of other horizontal challenges, such as energy efficiency or reliability. For example, the TSCH operation of IEEE 802.15.4e is quite often proposed either as an energy efficiency measure, since nodes can sleep until their precise communication slot time, or as a reliability enhancement, through its inherent frequency hopping capability, which still requires perfect time synchronization [14]. Furthermore, timeliness and security emerge as two major driving forces behind the increasing shift of processing towards the edge, tightly coupled with AI capabilities in constrained embedded devices. The components needed to build real-time wireless sensor network segments are already widely available. However, there is still open space for tools and mechanisms that will master the complexity of configuring the scheduling functions and other relevant specificities, including secure bootstrapping of extremely constrained structures in an automatic or semi-automatic way, enabling both specialists and non-specialists to deploy and exploit these new technologies. Most importantly though, the challenge consists in providing end-to-end timeliness guarantees in complex paths of heterogeneous sensor networks and systems, that may also include heavier real-time data analytics services over virtualized cloud infrastructures.

For all these reasons, we expect that the pressure for solutions will be increased with the upcoming penetration of 5G [15] and the deployment of more and more complex infrastructures for critical monitoring and control, either inside the traditional process and manufacturing industry, or outside it, in modern IIoT verticals, such as autonomous vehicles and V2V/V2I interactions, smart agriculture, smart energy, smart buildings and cities, and many others.

Author Contributions: C.K. and M.T.L. worked together during the whole editorial process of the Special Issue, "Real-Time Sensor Networks and Systems for the Industrial IoT", published in the MDPI journal *Sensors*. C.K. drafted this editorial summary. C.K. and M.T.L. reviewed, edited, and finalized the manuscript. All authors have read and agreed to the published version of the manuscript.

Funding: This research received no external funding.

Acknowledgments: We thank all authors who submitted excellent research work to this Special Issue. We are grateful to all reviewers who contributed evaluations of the scientific merits and quality of the manuscripts and provided countless valuable suggestions to improve their quality and the overall value for the scientific community. Our special thanks go to the editorial board of MDPI's *Sensors* journal for the opportunity to guest edit this Special Issue, and to the *Sensors* Editorial Office staff for the hard and precise work to keep to a rigorous peer review schedule and timely publication.

Conflicts of Interest: The authors declare no conflict of interest.

References

1. Koulamas, C.; Lazarescu, M.T. Real-Time Embedded Systems: Present and Future. *Electronics* **2018**, *7*, 205. [CrossRef]
2. Koulamas, C.; Kalogeras, A. Cyber-Physical Systems and Digital Twins in the Industrial Internet of Things, CPS in Control. *IEEE Comput.* **2018**, *51*, 95–98. [CrossRef]
3. Seferagić, A.; Famaey, J.; De Poorter, E.; Hoebeke, J. Survey on Wireless Technology Trade-Offs for the Industrial Internet of Things. *Sensors* **2020**, *20*, 488. [CrossRef] [PubMed]
4. Park, B.; Nah, J.; Choi, J.-Y.; Yoon, I.-J.; Park, P. Transmission Scheduling Schemes of Industrial Wireless Sensors for Heterogeneous Multiple Control Systems. *Sensors* **2018**, *18*, 4284. [CrossRef] [PubMed]
5. Wang, H.; Ma, J.; Yang, D.; Gidlund, M. Efficient Resource Scheduling for Multipath Retransmission over Industrial WSAN Systems. *Sensors* **2019**, *19*, 3927. [CrossRef] [PubMed]
6. Wu, Y.; Zhang, W.; He, H.; Liu, Y. A New Method of Priority Assignment for Real-Time Flows in the WirelessHART Network by the TDMA Protocol. *Sensors* **2018**, *18*, 4242. [CrossRef] [PubMed]
7. Cheng, Y.; Zhou, H.; Yang, D. CA-CWA: Channel-Aware Contention Window Adaption in IEEE 802.11ah for Soft Real-Time Industrial Applications. *Sensors* **2019**, *19*, 3002. [CrossRef] [PubMed]
8. Lee, S.-H.; Kim, J.-S.; Seok, J.-S.; Jin, H.-W. Virtualization of Industrial Real-Time Networks for Containerized Controllers. *Sensors* **2019**, *19*, 4405. [CrossRef] [PubMed]
9. Tedeschi, S.; Emmanouilidis, C.; Mehnen, J.; Roy, R. A Design Approach to IoT Endpoint Security for Production Machinery Monitoring. *Sensors* **2019**, *19*, 2355. [CrossRef] [PubMed]
10. Fournaris, A.P.; Dimopoulos, C.; Lampropoulos, K.; Koufopavlou, O. Anomaly Detection Trusted Hardware Sensors for Critical Infrastructure Legacy Devices. *Sensors* **2020**, *20*, 3092. [CrossRef] [PubMed]
11. Wielgosz, M.; Skoczeń, A.; De Matteis, E. Protection of Superconducting Industrial Machinery Using RNN-Based Anomaly Detection for Implementation in Smart Sensor. *Sensors* **2018**, *18*, 3933. [CrossRef] [PubMed]
12. Ntalianis, V.; Fakotakis, N.D.; Nousias, S.; Lalos, A.S.; Birbas, M.; Zacharaki, E.I.; Moustakas, K. Deep CNN Sparse Coding for Real Time Inhaler Sounds Classification. *Sensors* **2020**, *20*, 2363. [CrossRef] [PubMed]
13. Sisinni, E.; Saifullah, A.; Han, S.; Jennehag, U.; Gidlund, M. Industrial Internet of Things: Challenges, Opportunities, and Directions. *IEEE Trans. Ind. Inform.* **2018**, *14*, 4724–4734. [CrossRef]
14. Kim, H.-S.; Kumar, S.; David, E.C. Thread/OpenThread: A Compromise in Low-Power Wireless Multihop Network Architecture for the Internet of Things. *IEEE Commun. Mag.* **2019**, *57*, 55–61. [CrossRef]
15. Varga, P.; Peto, J.; Franko, A.; Balla, D.; Haja, D.; Janky, F.; Soos, G.; Ficzere, D.; Maliosz, M.; Toka, L. 5G support for Industrial IoT Applications—Challenges, Solutions, and Research gaps. *Sensors* **2020**, *20*, 828. [CrossRef] [PubMed]

Review

Survey on Wireless Technology Trade-Offs for the Industrial Internet of Things

Amina Seferagić [1,*], Jeroen Famaey [2], Eli De Poorter [1] and Jeroen Hoebeke [1]

1 IDLab, Department of Information Technology, Ghent University—imec, 9000 Ghent, Belgium; Eli.DePoorter@UGent.be (E.D.P.); Jeroen.Hoebeke@ugent.be (J.H.)

2 IDLab, Department of Computer Science, University of Antwerp—imec, 2000 Antwerp, Belgium; jeroen.famaey@uantwerpen.be

* Correspondence: amina.seferagic@ugent.be

Received: 7 December 2019; Accepted: 13 January 2020; Published: 15 January 2020

Abstract: Aside from vast deployment cost reduction, Industrial Wireless Sensor and Actuator Networks (IWSAN) introduce a new level of industrial connectivity. Wireless connection of sensors and actuators in industrial environments not only enables wireless monitoring and actuation, it also enables coordination of production stages, connecting mobile robots and autonomous transport vehicles, as well as localization and tracking of assets. All these opportunities already inspired the development of many wireless technologies in an effort to fully enable Industry 4.0. However, different technologies significantly differ in performance and capabilities, none being capable of supporting all industrial use cases. When designing a network solution, one must be aware of the capabilities and the trade-offs that prospective technologies have. This paper evaluates the technologies potentially suitable for IWSAN solutions covering an entire industrial site with limited infrastructure cost and discusses their trade-offs in an effort to provide information for choosing the most suitable technology for the use case of interest. The comparative discussion presented in this paper aims to enable engineers to choose the most suitable wireless technology for their specific IWSAN deployment.

Keywords: Industrial Internet of Things (IIoT); LoRa; IEEE 802.11ah; WiFi HaLow; Time Slotted Channel Hopping (TSCH); Narrowband IoT (NB-IoT); Bluetooth Low Energy (BLE); BLE Long Range; WirelessHART; ISA100.11a

---

1. Introduction

Industrial networks for process automation are deployed in sites which can be hundreds of meters wide, hosting very dense networks consisted of hundreds or thousands of nodes. Harsh industrial environments impose a number of challenges for wireless communications: reliability, fault-tolerance and low latency being the biggest ones. Unpredictable variations in temperature, humidity, vibrations and pressure make the industrial environments harsh, as well as the presence of highly reflective (metal) objects and electromagnetic noise. Even though not much data needs to be communicated in an industrial application, reliability and latency are critical, that is, delivery of all data must be guaranteed in real-time. Wired networks have met these requirements and are being used in spite of the high cost of wiring and the often present installation difficulties (see Figure 1) because wireless solutions are not as robust as their wired counterparts. Industrial automation systems in chemical industry, power plants, oil refineries or underground water supply systems implement complex monitoring and control processes. Thousands of devices send measured values (i.e., temperature, pressure, flow, position) to the actuators that control processes and to the servers that coordinate the production phases. Wiring is generally both challenging and costly (cca. 20 \$/m): flammable, explosive and hot environments have to be avoided (e.g., in the presence of flammable gases in an oil refinery), remote or unavailable locations are hard to reach and mobile nodes can hardly be connected at all. Although wired networks at this time cannot fully be replaced by wireless networks in this domain, supervision and non-critical control with loose enough requirements could be realized over wireless. In addition, significant constrains that limit the practical deployments of wireless networks in such scenarios are battery capacity and power consumption of the devices. Ideally, communication and power cables can be mitigated to enable a fully wireless solution. For that, the devices should be energy efficient and able to power from a battery for years. Moreover, wireless networks introduce logical benefits that could be used in maintenance and commissioning, such as "plug-n-play" automation architectures to reduce downtime and speed-up tests and "hot-swapping" faulty modules. In addition to control and supervision, global wireless plant coverage could enable localization and tracking of parts in production, coordination of autonomous transport vehicles and mobile robots [1].

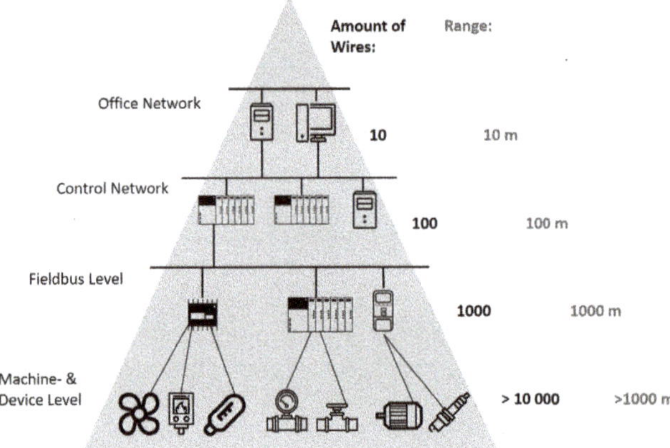

Figure 1. Node density exponentially increases from the top (office network/Internet/Intranet) to the bottom (machine- and device-level) in a typical automation system network hierarchy.

Industrial Wireless Sensor and Actuator Network (IWSAN) are gaining popularity in process industries due to their advantage in lowering infrastructure cost and deployment effort. The advent of Industry 4.0 already resulted in the successful use of IWSANs for monitoring applications and non-critical open-loop control in factory automation. A few new wireless technologies, such as WirelessHP [2], OFDMA wirelesscontrol [3], Real-Time-WiFi [4], Wireless network for Industrial Automation and Process Automation (WIA-PA) [5], can replace extensive wiring on industrial machinery, providing connectivity between machine parts with μs order of magnitude latency. Even though they enable reliable and fast communication, the range of such networks is limited to only a few meters, making them unsuitable for broad usage across an entire industrial site in process automation or for reaching remote areas if infrastructure cost has to be kept low. Ranges up to a few hundred meters are feasible with 802.15.4-based technologies such as WirelessHART [6], ISA 100.11a [7], 802.15.4g [8] with Time Slotted Channel Hopping (TSCH) [9] and WIA-PA [10], at the cost of other performance metrics. Sub-GHz wireless technologies, such as LoRa and SigFox, further extend the coverage due to the better signal propagation characteristics (up to 15 km and 50 km respectively) but are not suitable for frequent critical traffic given their low data rates (up to 50 kbps and 0.1 kbps respectively) which lead to very long transmission times in both uplink and downlink. In downlink, long transmission times also limit the gateway to serve many nodes, more so considering the duty cycle limitations [11]. Moreover, LoraWAN Class A and Sigfox only allow downlink transmissions that immediately follow uplink, resulting in substantial downlink delays due to buffering. NB-IoT experiences downlink delays due to buffering as well, when using the Power Saving Mode (PSM). The existing trade-off between the range and latency varies across different technologies (cf. Figure 2), aiming to cover a variety of use cases. This paper explores the aforementioned trade-offs and the conditions that enable the use of particular Internet of Things (IoT) wireless technologies in heterogeneous sensor-actuator networks for mid-range communication able to cover an industrial site ranging up to more than one kilometer in diameter.

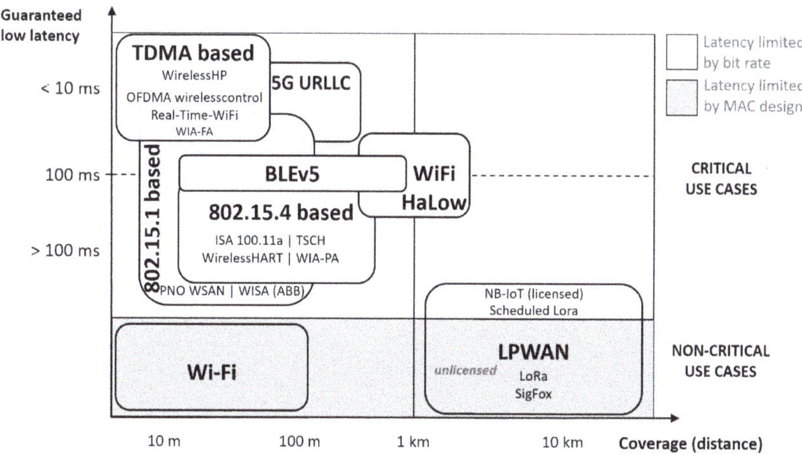

Figure 2. Different wireless technologies have different range/latency capabilities. This article discusses the trade-offs in mid-range technologies that could provide coverage of an entire industrial site (black boxes) with sufficiently low latency.

Wireless Sensor Networks (WSNs) have been evaluated from different perspectives in the state of the art literature. However, significantly smaller amount of research is conducted in the context of IWSANs that have much more strict application requirements. An overview of key issues and challenges of wireless technologies in industrial networks is surveyed in References [1,12–19]. Communication requirements and a general profile of a wireless fieldbus for low level short-range factory automation systems are discussed in Reference [20]. References [13,21] discuss security and Quality of Service (QoS) perspectives of IWSAN in industrial automation. Furthermore, an in-depth review of recent advances in real-time IWSANs for industrial control systems is given in Reference [22], with a focus on WirelessHART. Reference [22] reviews real-time scheduling and analytic techniques for achieving real-time performance in Reference IWSANs. An extensive survey on wireless network design for control systems is presented in Reference [17], briefly reviewing a few of the existing wireless technologies in that context but mostly focusing on the joint design considerations of both control systems and wireless networks. A comparative examination of ZigBee, WirelessHART, ISA100.11a and WIA-PA in terms of network architecture and protocol design in the context of IWSAN is given in Reference [19]. State of the art in Low-Power Wide-Area Network (LPWAN) solutions for Industrial Internet of Things (IIoT) services is explored in Reference [23].

This paper takes a different approach and interprets the wireless standards from the practical standpoint, offering the readers concrete values on achievable sampling rates, energy consumption, scalability and coverage in practice. Both standard specifications and product datasheets provide extensive low level data, and are insufficient on their own without additional empirical research. This paper quantifies the existing trade-offs in wireless technologies for wireless sensor and actuator networks with coverage of at least a couple of hundred meters, able to cover a production site or at least a large part of it. Range is crucial in process automation for all slave nodes to be able to reach a master node, considering that control is typically done by one or few master nodes (controllers) and a large number of slave nodes (sensors and actuators) that take part in bidirectional communication with the controller and are spatially distributed over the entire site. This paper presents cost, scalability, latency, reliability, range and energy consumption evaluation of wireless technologies with promising range and latency potential, including LoRa, IEEE 802.11ah (Wi-Fi HaLow), Narrowband-IoT (NB-IoT), WirelessHART, ISA100.11a, Bluetooth Low Energy (BLE) and 802.15.4g physical layer with 802.15.4e TSCH on data-link layer. These technologies offer the possibility of a dense heterogeneous wireless network deployment able to serve both actuators and sensors in critical applications, as well as provide an infrastructure for supervisory traffic. Along with the practical limitations of each technology with respect to the existing trade-offs between latency, throughput, coverage and scalability, a direct projection of the aforementioned wireless technologies to their key performance indicators is made, aiming to enable adequate network design in particular industrial applications.

The remainder of this article is organized as follows. The requirements and challenges that industrial networks must comply with are summarized in Section 2. Section 3 presents a general discussion of the key trade-offs in wireless network design in the context of the requirements, while in Section 4 the discussed trade-offs are quantified for each particular technology. Overall discussion and the experimental evaluation of energy consumption is presented in Section 5. Finally, Section 6 presents the conclusions.

2. Requirements and Challenges

The International Society of Automation (ISA) classified industrial systems into six classes [13] on the basis of data urgency and operational requirements. These classes range from critical control systems to monitoring systems, from the strictest requirements to the most relaxed ones respectively:

1. Safety systems—require immediate actions on events (usually in the order of tens or hundreds of μs or a few ms).
2. Closed loop regulatory systems - control the system via feedback loops operating either periodically or based on events. They may or may not have stricter timing requirements than safety systems.
3. Closed loop supervisory systems—similar to regulatory systems with the difference that the feedbacks are usually non-critical and event-based, for example, collecting statistical data and reacting only when a certain trend is observed by issuing a notification or alarm.
4. Open loop control systems—where sensors collect data and store it to the central database. An operator (human) analyzes the data and acts upon it if needed.
5. Alerting systems—send periodical or event-based alerts indicating different stages, for example, heating up the boiler and alerting every once in a while to indicate the progress.
6. Information gathering systems—collect the data (logging) and forward the logs to a server. These systems have no immediate operational consequence.

Wireless coverage of the entire industrial site may benefit classes 2–6, whereas class 1 requires a solution combining both ultra-high reliability, redundancy and ultra-low latency, which is infeasible with long range wireless considering the trade-offs. Performance requirements of different classes are depicted in Figure 3. For site-wide coverage, a range of at least a few hundred meters is needed. Site-wide coverage would enable multicasting measurements to several destinations, for example, actuators, supervision systems, databases, enabling the support of different services for several classes of industrial systems, making the network heterogeneous. Thus, site-wide IWSANs need to be scalable enough to accommodate new nodes and provide QoS, considering that they are expected to run for several decades. Different applications require different performance and services, as examples in Table 1 show.

Table 1. Cycle time and communication range requirements broadly vary over industrial automation use-cases [12,24–26].

Application	Range [m]	Cycle Time
Building automation	10–200	100 ms—seconds
Monitoring and supervision	100–1000	seconds—days
Process control	50–500	10–1000 ms
Factory automation	10–50	0.5–100 ms
Automotive	1–10	1–100 ms
Interlocking and control	50–100	10–250 ms
Power-system protection	100–10 k	0.01 μs–50 ms
Event-based control	10–100	1–100 ms

Besides the key performance requirements illustrated in Figure 3, deployment cost, energy consumption, interoperability, QoS and service differentiation come to focus especially when considering heterogeneous networks. A wired fieldbus network is very expensive to deploy because of tens of kilometers of cables needed to connect devices to their master nodes, the time needed for deployment and the maintenance of such deployment. Lower deployment and implementation costs are the prime motivations for the transition from wired to wireless solutions wherever possible. Among wireless solutions, subscription fees for operator based networks also vary. Operator based solutions are generally not ideal for industrial purposes as the dependence on the operator in case of failure increases the repair time. However, reliable full-duplex operation of wired industrial networks is a large advantage over wireless technologies that are the subject of this article. Namely, reliability inherently suffers in the full-duplex wireless solutions because of the self-interference and increased interference from the neighbours [27]. Opting for half-duplex instead causes the inability to send and receive at the same time on the same channel, which in turn largely increases the latency in wireless networks. IWSANs must operate in real time to serve class two systems. Specifically, closed loop regulatory systems require IWSANs to sample, process and exchange the data between a sensor and an actuator in a time frame that is less than the cycle time of the loop, with typical values ranging from microseconds to hundreds of milliseconds (depending on the concrete process being controlled). Critical applications (classes 1 and 2) also require redundancy, resistance to noise and robustness against failure as they must ensure timely and successful delivery. In addition, the failure of one or a few nodes must not compromise the operation of the network as whole.

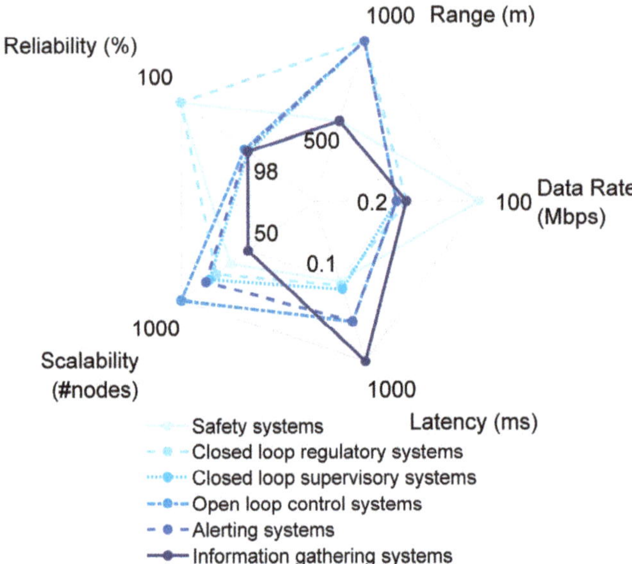

Figure 3. Different classes of industrial systems have significantly different performance requirements.

Many of the stated requirements are interconnected and there is no single technology that covers all of them simultaneously. The inevitable trade-offs, their causes and consequences are elaborated in the next section.

3. Trade-offs in Wireless Network Design

Providing wireless communication to heterogeneous applications, including the time-critical ones, over a wide industrial site is a conflicting task. For example, l All three are partly determined by the choice of frequency band and bandwidth but also with other design choices that create additional interlocks between the performance parameters. These trade-offs, illustrated in Figure 4, complicate the design of wireless solutions.

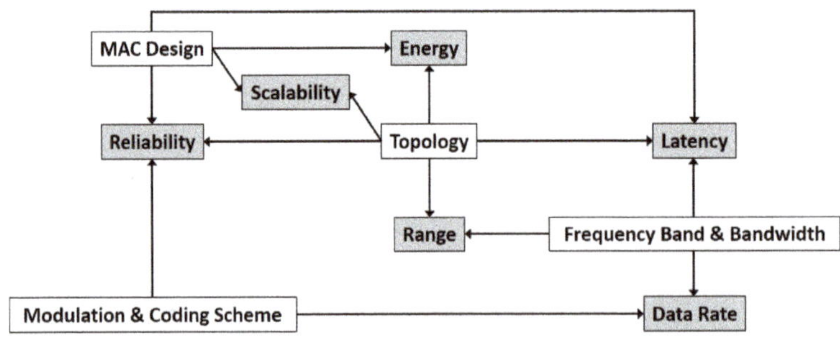

Figure 4. Network design choices (white rectangles) influence several performance properties simultaneously (grey rectangles), thus creating the trade-offs between them.

3.1. The Transmission Range

The transmission range is mainly determined by the transmission power, typically limited by regulations [11], the radio and propagation properties, as well as coding and modulation complexity. If a radio transmits at a constant power, lowering this complexity rate permits the correct decoding of a weaker or more distorted signal by a receiver, thus extending the transmission range. Also, higher frequency bands with more bandwidth available enable higher data rates and faster data transmission but they also have worse penetration capabilities which reduces the range in an industrial site full of obstacles. Range is largely determined by topology as well. Multi-hop topologies extend the range at the expense of latency, design complexity and energy consumption because of the need for synchronization of nodes, routing and so forth. In conclusion, low data rates at low frequencies and multi-hop topologies are prolonging the range but they all increase latency.

3.2. Latency

Latency is reduced by increasing the data rate, in turn enabled by more complex codings and larger bandwidths. Furthermore, multi-hop topologies increase the latency considering that forwarding and routing introduce additional delays. In addition, computing a new route when a link fails also introduces delay which can render multi-hop topologies useless in low-latency time-critical applications. Medium Access Control (MAC) design has a significant impact on latency as well, especially in IoT technologies where devices aim to sleep as long as possible to save energy, therefore delaying transmissions and receptions. MAC protocols can be classified into four classes: (1) Fixed Assignment Protocols where resources are divided among the nodes for a defined time duration, (2) Demand Assignment Protocols where resources are provided to a node on demand, (3) Random Access Protocols where resources are divided randomly and (4) Hybrid Protocols that combine fixed or demand assignment with random access. Fixed Assignment Protocols such as Time Division Multiple Access (TDMA) introduce determinism and achieve lower latency than random access protocols under very high load, but under low load they waste resources by inefficient usage of the channel time, where random access protocols achieve lower latency. Demand-based protocols are not suitable for low-latency time-critical communications given that explicitly asking for resources every time takes up bandwidth and adds up to latency. For heterogeneous industry applications, hybrid approaches are the most promising given that they aim to combine the benefits of both fixed assignment and random access protocols, while surpassing their limits at the same time and adapting to the network conditions. In addition, retransmissions need to be kept a minimum as they also increase the latency.

3.3. Reliability

Reliability is determined by topology, MAC design and Modulation and Coding Scheme (MCS). One of the major setbacks of wireless technologies in terms of reliability, in comparison to their wired counterparts, is the inter- and intra-technology interference on air which can cause collisions and increase packet loss. Technologies that operate in licensed bands reserve a part of the spectrum for themselves, mitigating the issue. However, spectrum is a scarce resource and it comes at a high price. Private deployments are not possible in reserved spectrum, disabling the possibility of local control over a network. Shared spectrum, on the other hand, can be shared by any number of technologies which can try and mitigate interference by channel hopping or using some MAC layer mechanisms such as Listen Before Talk (LBT). Furthermore, in a single-hop networks, the success probability is entirely dependent on a single link, opposed to multiple links in multi-hop networks. Reliability can be improved by employing both retransmissions and repetitions at the MAC layer, which also add to the latency. To reduce the number of retransmissions, error control techniques such as Forward Error Correction (FEC) can be used. Coding schemes and modulation largely define reliability. Coding rates create extra error checking bits that make modulation more reliable. Modulation schemes are more reliable as they have fewer points on the constellation diagram but also slower. That makes Binary Phase Shift Keying (BPSK) the slowest and the most reliable modulation compared to Quadrature Amplitude Modulation (QAM) and Quadrature Phase Shift Keying (QPSK), given that it only accommodates two points (one bit per burst). QPSK uses four constellations, whereas QAM can have any number of points. Any increase in the number of points on the constellation diagram reduces the space between them, leaving fewer margins for error. This makes QAM the fastest modulation scheme but more unreliable over longer distances.

3.4. Data Rate

Data rate is directly correlated with the available bandwidth and thus frequency band, on one hand, and with modulation and coding scheme on the other. More bandwidth enables higher data rates, while modulation techniques and coding schemes can further contribute to the achievable data rate by encoding more data into the signal. Unlicensed wireless technologies operate either in sub-GHz frequency bands (400 MHz, 800–900 MHz), in 2.4 GHz or in 5 GHz. Sub-GHz technologies generally (although not universally) use narrower channels (few hundred kHZ) than those in GHz frequency bands (22 MHz Wi-Fi, 2 MHz 802.15.4) and thus have more limited data rates than the GHz ones.

3.5. Energy Consumption

Energy consumption depends on data rate, topology and MAC design, as well as the hardware design of course. Low data rates result in long transmission times, which increases the energy consumption of the node and reduces the battery lifetime. Topology wise, nodes in multi-hop networks consume more energy than in single-hop networks given that, besides their own transmissions and receptions, they also need to forward other nodes' packets. Energy-efficiency of data forwarding paths give the routing protocols a strong influence over energy consumption as well. Complex coding and decoding operations also contribute to energy consumption. For example, FEC has been omitted in commercial 802.15.4 based networks due to the energy consumption of the decoding operation. Nevertheless, employing FEC could reduce the overall energy consumption as less energy would be spent on retransmissions and rescheduling [28]. Besides, MAC design has a significant impact on energy consumption as it defines scheduling and hence the radio on and off times.

3.6. Scalability

Scalability is primarily determined by MAC design. Scheduling, contention resolution and other MAC mechanisms work together to provide maximum network capacity. In single-hop networks, the network capacity upon reaching the upper limit can only be extended by deploying more base stations. However, in practice the density of such base stations is limited. Multi-hop networks address this issue by allowing for wireless data forwarding, at the expense of overall throughput. In TDMA-based protocols, network density is limited by the need for synchronization and time division in combination with QoS requirements.

3.7. Spectrum Regulations

Another tackling design choice is the one between unlicensed Industrial, Scientific and Medical (ISM) and licensed bands. On the one hand, worldwide permitted unlicensed operation reduces the runtime costs but has no regulatory protection against interference by other wireless networks operating in the same frequency band. On the other hand, even though licensed bands prevent interference, they typically depend on an external operator. Therefore, network issues cannot be immediately resolved on site, the external operator needs to resolve them. This introduces administrative delays which are unaffordable in time-critical industrial applications. Communication technologies operating in the unlicensed spectrum are maintained and managed locally. However, several co-located or overlapping wireless networks operating in the same frequency band will interfere with each other and can experience decreased QoS and extensive packet loss [14]. In an effort to alleviate coexistence issues in unlicensed spectrum, regulatory bodies have issued a number of norms such as a Clear Channel Assessment (CCA) check before each transmission by all devices, that is, a device has to sense if the channel is free by energy detection or other types of Detect And Avoid (DAA) mechanisms [14]. Although the DAA mechanisms improve the coexistence between the contending wireless nodes and networks, collisions can still occur. Apart from collisions, medium sensing adds to the latency and introduces non-determinism due to the medium congestion. Aforementioned facts significantly limit the use of wireless solutions in closed loop control applications in automation industry. A limiting regulation is present in unlicensed sub-GHz spectrum as well. Devices with an operating range of 863–868 MHz in Europe, 916.5–927.5 MHz in Japan and 902–928 MHz in the US must comply with the maximum duty cycle limit of 2.8% and 10% for the, Access Point (AP) provided that they support LBT and Adaptive Frequency Agility (AFA) features, 1% otherwise [11].

4. Wireless Technologies for Industrial Applications

To support cyclic communication between sensors, actuators and controllers, sufficient throughput, latency and range is needed. This paper only considers wireless technologies that have the potential to enable real-time cyclic communication over a range comparable to the size of an industrial site, thus larger than a hundred meters. In line with that, we consider the promising IIoT technologies to be LoRa, IEEE 802.11ah and NB-IoT out of single-hop long range networks, WirelessHART, ISA100.11a, BLE and 802.15.4g/e physical (PHY) with 802.15.4e TSCH MAC out of long range multi-hop networks. The performance of each individual technology in terms of requirements presented in Section 2 and trade-offs presented in Section 3 is discussed below.

4.1. Long Range Networks

Single-hop long range networks that have the potential to enable real-time cyclic communication over a range comparable to the size of an industrial site are introduced in the remainder of this section.

4.1.1. LoRa

LoRa is a proprietary wireless data communication technology which specifies a PHY layer only. A popular MAC for use with LoRa is the open LoRaWAN specification. LoRa PHY uses Semtech's proprietary Chirp Spread Spectrum (CSS) radio modulation to reduce receiver complexity while achieving long range. CSS is resistant to Doppler effects and multipath fading. A LoRa receiver can decode transmissions ~20 dB below the noise floor, enabling very long communication distances while using very limited power. CSS is a spread spectrum technique where the signal is modulated by chirp pulses whose frequency linearly varies, parametrized by the orthogonal Spreading Factors (SFs), which can take values 7–12. The higher the SF, the longer packet transmission time and the more reliable its reception. Therefore, high SFs improve robustness against interference and counteract heavy multipath fading characteristic for indoor propagation and urban environments. This comes at the cost of low data rates and much higher energy consumption. Considering 125 kHz channels, the data rates range from 0.25 kbps to 5.47 kbps. These very low rates result in long transmission times (and medium usage) even for small packets. For example, a 17-byte sensor reading would take over 1.5 s to transmit at SF12 and cca. 70 ms at SF7. Combining the LoRaWAN MAC and LoRa PHY data rates results in an rather low network capacity per gateway of less than 0.02 MB per hour [29].

An experimental study on the range of LoRaWAN [30] showed that it can achieve ranges up to 7.5 km using SF10 and packets with 10 bytes of payload, resulting in 0% Packet Error Rate (PER) and −126 dBm Received Signal Strength Indication (RSSI). When using the specification of the Wireless M-Bus according to EM13757-4, 50 bytes of payload, Frequency Shift Keying (FSK) modulation with an FSK deviation value of 50 kHz and a data rate of 100 kbps, LoRa achieves up to 1.35 km of range with 0% PER and up to 3.6 km of range for <10% PER.

LoRaWAN is based on pure ALOHA, thus collisions pose the biggest issue in such networks given the long air-times. According to LoRaWAN, edge-devices' downloading needs determine their class:

- Class A devices have a single Receive Window (RW) scheduled immediately after a corresponding uplink connection,
- Class B devices can schedule additional RWs,

- Class C devices continuously listen and can receive almost anytime.

The more RWs, the more energy devices consume, that is, the power consumption increases over the classes A through C.

LoRaWAN supports both confirmed and unconfirmed messages. However, downlink capability of LoRaWAN networks is highly limited. With an increasing traffic load, RWs for sending acknowledgements (ACKs) to confirmed messages are more frequently missed as the gateway cannot transmit at the start of a RW due to the duty cycle restrictions. When a gateway sends an ACK in either RW1 or RW2, it aborts all ongoing receptions further decreasing the Packet Delivery Ratio (PDR) [31]. Currently, various scheduling solutions are being investigated that aim to improve LoRaWAN capacity and reliability, properties that are of interest to industrial IWSANs [32]. At this moment, even with the adaptation of LoRa/LoRaWAN to TSCH mechanism, this technology can only fulfill requirements for those industrial application scenarios classified to classes 3–6 that need cycle time in the order of seconds in the best case scenario, or even minutes for very dense deployments [33]. Reference [34] analyzed end-to-end latency of LoRaWAN in IIoT use cases and showed that LoRaWAN may be applicable for a subset of IIoT use cases where lower SFs can achieve end-to-end delay below 400 ms. Long distance transmissions ranging up to 15 km require a high SF and result in the end-to-end latency well above one second, which is far beyond real-time availability of sensor data.

In conclusion, the LoRa PHY allows long range, but the low network capacity and current class A MAC design restricts use cases to those with a combination of low data requirements, a majority of uplink traffic and overall traffic volumes that remain far below the theoretical capacity.

4.1.2. IEEE 802.11ah/Wi-Fi HaLow

IEEE 802.11ah, marketed as Wi-Fi HaLow, can serve up to 8192 stations per AP, a much higher value than the previous 802.11 iterations. In addition to the 1 km coverage, its relay functionality can further increase both the network size and coverage. To limit interference between stations and collisions, it introduces the Restricted Access Window (RAW) mechanism which combines the deterministic and the stochastic channel access. RAW restricts the channel access for a specified group of stations assigned to the time slots within the RAW. Stations assigned to specific slots within a RAW contend for the medium in their corresponding slots employing Enhanced Distributed Channel Access (EDCA)/Distributed Coordination Function (DCF). The stations are not allowed to contend for the medium in RAW slots they are not assigned to. RAW configuration and assignment is configurable and can change every beacon interval. RAW can reserve channel time for any group of stations. It improves throughput in dense IoT networks where many stations contend for the medium simultaneously. RAWs can contribute to introducing determinism in a network, if configured accordingly. However, an arbitrary RAW configuration can also degrade network performance if RAW is not configured with respect to the traffic patterns of the end devices. RAW duration can be configured to any value between 500 μs and the beacon interval.

IEEE 802.11ah demonstrates robustness in industrial environments given that it has better penetration capabilities than 2.4 GHz technologies due to sub-GHz frequency bands. It employs Wi-Fi WPA3 for security and can be used for over-the-air software updates [35], besides monitoring and control.

The minimal feasible cycle time in a 802.11ah network equals 2 or 4 beacon intervals, given that 2 to 4 hops are needed to complete a single cycle in a control loop in a star network: two if the controller is wired to the AP and four otherwise. Each hop can only be executed in a single beacon interval. Beacon intervals can be reduced in order to support low-latency loops but reducing the beacon intervals makes the spectrum usage less efficient [36]. The minimal cycle time for a single 99.99% reliable control loop with a wired controller is 32 ms and it can operate alongside 75 sensors reporting 64-byte measurements every 1 s. Longer cycles enable the reliable support of more control loops, for example, 4 control loops with 64 ms cycle time and 70 sensors reporting every 1 s. IEEE 802.11ah can offer differentiated QoS to different types of end nodes in a dense deployments [37,38]. In addition to classes

3–6, this technology may be capable of providing QoS to class 2 as well, supporting low latency closed loops (<100 ms) along with other non-critical traffic. However, the lower the latency requirement, the fewer nodes the network can support due to the used up bandwidth. Hence, there is a trade-off between the scalability and the latency that can be supported in practice. Given the lack of available hardware at the market at this time, it is yet to be seen how does this technology perform in the real world, outside of simulation studies.

4.1.3. Narrowband-IoT

NB-IoT aims to offer deployment flexibility allowing an operator to use only a small part of the available spectrum for the technology. It targets ultra-low-end IoT applications. NB-IoT supports three coverage classes, namely (1) normal, (2) robust and (3) extreme coverage class. Those correspond to the Minimum Coupling Losses (MCLs) of 144 dB, 158 dB and 164 dB respectively. In NB-IoT, the uplink latency consists of the system synchronization, broadcast information reading, random access, resource allocation, data transmission and feedback response [39]. In uplink, NB-IoT makes use of single- (20 kbps) and multi-tone (250 kbps) channels. A single-tone technology implies either 12 or 48 continuous sub-carriers with sub-carrier spacing of either 15 kHz or 3.75 kHz, respectively. Multi-tone technology implies 12 continuous sub-carriers, with 15 kHz spacing, grouped in 3, 6 or 12 continuous sub-carriers. Spacing of 3.75 kHz results in higher coverage than 15 kHz spacing due to the higher power spectral density, which makes the cell capacity 8% larger for 3.75 kHz-spacing [40]. In downlink, Orthogonal Frequency-Division Multiple-Access (OFDMA) is employed with sub-carrier spacing of 15 kHz.

Recent measurements on an actual public NB-IoT network showed that the achievable application layer throughput is significantly lower, at around 10 and 15 kbps for uplink and downlink, respectively [41]. In terms of latency, a one-way latency of about 50 ms could be achieved under perfect channel conditions for small packets of 8 bytes. However, for deep indoor scenarios, latency increased to around 16 s for uplink and 8 s for downlink transmissions. Another experimental evaluation of NB-IoT using two different commercial platforms in a public NB-IoT network has observed the inconsistency in performance metrics, namely the energy consumption and latency [42]. The variability of energy consumption results in an imprecise predictability of battery life and causes a difference in performance of similar devices. Although NB-IoT is designed for delay-insensitive applications, some cases cannot tolerate the variability of latency amounting to tens of seconds or even minutes. Guaranteed reliability in NB-IoT comes at a cost of variability [42].

NB-IoT is superior than most of the competition in terms of range, security and availability. However, its unpredictable latencies that can be in the order of seconds make it applicable for latency insensitive processes only. Its random resource reservation procedures make the connection latency high in dense networks. When connected, its throughput and latency can significantly vary depending on channel conditions due to the repetitions and changes in MCSes. Even though the hardware is inexpensive, NB-IoT comes with a subscription fee, while all other technologies listed in Tables 2 and 3 can be deployed privately. In conclusion, it is significantly more robust than competition which makes it suitable for class 2 latency insensitive applications. It can also be used in classes 3–6.

4.2. Long Range Multi-hop Networks

Multi-hop networks not only extend the range but also substantially contribute to the reliability as in mesh networks there are (typically) 2 or more paths from the source to the destination, meaning that the loss of a link will not result in a communication failure like in single-hop networks. However, nodes in multi-hop networks consume more energy because aside from their own packets, they also must transmit and receive (forward) other nodes' packets. Routing the packets adds up to latency and complexity. Thus, even though the scalability of multi-hop networks is not explicitly limited, deployments typically do not include more than 5 hops. The theoretical limit on the maximal network density is based on the addressing space, which for most of the analyzed technologies is 2^{32}, thus considered unlimited. However, the practical capacity limit is determined by the network latency and individual device's energy consumption. They both increase with the number of the communication links in the network. In addition, bottlenecks can occur at one or more devices that communicate directly with the gateway in a mesh network. Because of this, data traffic can significantly increase resulting in the increase of the devices' energy consumption. Also, the maximum achievable update rate is greatly influenced by the network density given that high update rates generate more traffic than low update rates.

4.2.1. WirelessHART

Along with the PHY, MAC, network and transport layer, the WirelessHART protocol stack also includes the application layer. The application layer is HART, which is compatible with existing wired HART solutions.

Both WirelessHART and ISA100.11a combine TDMA and frequency hopping at the MAC layer. To access the channel, WirelessHART uses a two-dimensional matrix consisting of time slots and 15 available channels. Time slots are grouped in superframes which periodically repeat throughout the entire network lifetime. Variable-length superframes are also supported and at least one superframe must be enabled. Superframes can be added or removed during the operation of the network and they are managed by the network/system manager. Time slot duration is fixed to 10 ms in WirelessHART. This scheme allows multiple devices to transmit data at the same time using different channels. However, a single device can only make use of a single channel per time slot. WirelessHART's (single) time slotted channel hopping mechanism also identifies noisy channels and blacklists them, thereby greatly improving the network reliability. In addition, this technology can achieve up to 99.999% reliability by supporting mesh topologies [43]. Security in WirelessHART communication is provided by the 128-bit Advanced Encryption Standard (AES) authentication.

WirelessHART is the oldest and the most experimentally evaluated wireless solution for IIoT [44–47]. A study has demonstrated control over WirelessHART with the update rate of 20 ms [45]. However, this and similar studies evaluate the simplest variant of a WirelessHART network comprising of a sensor, an actuator, an AP, a gateway (GW) and a network manager. Such a low update rate is achievable only in star topologies and with very short superframes that disable scaling the network up. Even though not more than 5 hops are advised in industrial networks, scaling WirelessHART is possible (1) using multiple APs and (2) using several WirelessHART gateways connected to a HART-over-IP backbone. This is very much needed in dense industrial environments as only 8 devices with 500 ms reporting period can be supported by a single AP [48]. In practice, a single WirelessHART AP can support 25, 50, 80 or 100 devices at 1 s, 2 s, 4 s and 8 s reporting period respectively. Both bidirectional (cyclic) traffic and an increment in hops doubles the latency or halves the number of nodes. Hence, WirelessHART is eligible for use in classes 2–6 of industrial systems, although it is limited either by latency or by scale.

4.2.2. ISA100.11a

The ISA100.11a protocol is designed for secure and reliable wireless operation for noncritical monitoring, alerting, supervisory control, open- and closed-loop control applications (classes 2–6). Although it is in many aspects similar to WirelessHART, unlike WirelessHART it provides flexibility for customizing the operation of a system. It is based on 802.15.4 PHY and achieves a single-hop range up to 100 m. The ISA100.11a utilizes TDMA and Carrier Sense Multiple Access with Collision Avoidance (CSMA/CA) on MAC layer, as well as graph routing and channel hopping which improves reliability by avoiding busy channels. The CSMA/CA mode is commonly used for retries, association requests, exception reporting and burst traffic. The use of any single channel is reduced by the time-synchronized slots and channel hopping, resulting in the improvement of ISA100.11a's coexistence with other networks in the shared spectrum. Communication takes place in time-slots with configurable constant lengths, varying from 10 ms to 12 ms. Time-slots are grouped in a superframe that periodically repeats in time. The length of the superframe is configurable and can differ for each node. Generally, long-period superframes result in high latency and low bandwidth. However, they conserve energy and contribute to the outspread allocation of digital bandwidth. Three methods are employed for limiting the use of undesirable radio channels and reducing the interference with other wireless networks: (1) CCA, (2) spectrum management and (3) and adaptive channel hopping. There are 15 (16 if the optional channel 26 is enabled) available channels to chose from in each time-slot. Three channel hopping sequences are defined, along with five hopping patterns.

In dense deployments, ISA100.11a devices can join the network faster than WirelessHART devices, as shown in Reference [49]. The WirelessHART network manager allows the initiation of the join process for each transmission opportunity (TXOP) and allocates the resources for publishing data during the join process. On the other hand, ISA100.11a employs dedicated advertisement links for join process initiation. At the network startup, the ISA100.11a gateway acts as a join proxy. However, each joined device may act as a join proxy, thus allowing more join opportunities. This results in faster joining of ISA100.11a devices in a dense deployment but not when there are only few devices. Reference [49] observed ISA100.11a to be slightly more reliable than WirelessHART, but also to have larger latency than WirelessHART in a cyclic communication.

Unlike WirelessHART, not all devices in ISA100.11a network must have routing capability. Without it, devices must be within one hop of a routing-capable device or the gateway. This significantly constrains the network's re-routing capability in case of link fading due to changes or movements in the plant, making the network less adaptable to changes. In addition, even though the standard does not impose a limit on the number of nodes, in practice it is only possible to connect up to 10 devices reporting every 0.5 s to a single AP, 25 devices reporting every 1 s, 50 at 2 s, 80 at 4 s or 100 at 10 s [48]. Both bidirectional (cyclic) traffic and an increment in hops doubles the latency or halves the number of nodes. This makes ISA100.11a suitable for application to classes 3–6, as well as to class 2 assuming the specific requirements of a class 2 application are in line with the latency and scale constrains of this technology.

4.2.3. Bluetooth Low Energy

Unlike classic Bluetooth which was designed as a point to point wire replacement, BLE provides increased range with 1 Mbps data rates and recently introduced meshing. An experimental study [50] observed the range of around 50 m indoors. The observed outdoors range for non line of sight scenarios was 123 m and 165 m for 0 dBm and 9 dBm transmission gain respectively, versus 490 m and 790 m in line of sight scenarios. Instead of Bluetooth's 1-MHz wide 79 channels, BLE uses 40 channels of 2 MHz. It employs frequency hopping over 37 channels for (bidirectional) communication and 3 for (unidirectional) advertising. Two MAC layer roles are defined for the devices in BLE, master and slave. The master coordinates the medium access using a TDMA-based polling mechanism, in which it periodically polls the slaves in their corresponding *Connection Events (CEs)* which are uniformly spaced within a configurable periodic interval called a *Connection Interval (CI)*. Frequency hopping is employed in a way that each CE uses a different channel. The standard defines the CI duration as a multiple of 1.25 ms in the range from 7.5 ms to 4 s. Therefore, the minimal cycle time equals the CI in a BLE control loop. Considering that the sample/update rate in an industrial system should be 3–4 times faster than the process time constant for condition monitoring and open loop control and 4–10 times faster for regulatory closed loop control, the minimal cycle times achievable with BLE increase to 22.5 ms–30 ms for non-critical traffic and 30 ms–75 ms for regulatory loops. BLE is theoretically capable of supporting 6 control loops exchanging 47-byte (max. size) packets with the minimal CI. Extending the CI for n time units of 1.25 ms results in S control loops, as follows:

$$S = \left\lfloor \frac{7.5\text{ ms} + n \cdot 1.25\text{ ms} + T_{GB}}{2 \cdot 0.376\text{ ms} + 0.15\text{ ms} + T_{GB}} \right\rfloor \tag{1}$$

where 0.376 ms is the transmission duration of a 47-byte packet at 1 Mbps, 0.15 ms stands for an Inter-Frame Spacing (IFS) and T_{GB} (in ms) for implementation-specific guard band between two successive CEs. This reasoning is based on the one-to-one mapping between control loops and CEs in a single CI. Besides $2S$ transmissions in a CI, we must fit S IFSs between successive transmissions in CEs and $S-1$ guard bands between CEs as well.

Bluetooth 5 provides the options to double the speed to 2 Mbit/s at the expense of range or up to fourfold the range at the expense of data rate and eightfold the data broadcasting capacity of transmissions by increasing the packet lengths. However, BLE's limited range makes it unsuitable for large scale IIoT deployments, as over 1000 basestations would have to be deployed to cover the same area as a single Wi-Fi HaLow basestation. To overcome this issue, Bluetooth introduced meshing. Communication in the Bluetooth Mesh standard in implemented using BLE advertising and scanning on the 3 advertisement channels. The standard uses a flooding mechanism where each node in the network repeats incoming messages, relaying them until the destination. Unlike in normal BLE advertising, Bluetooth Mesh transmissions are not scheduled based on an advertising interval. Instead, the nodes transmit after a random backoff time. In a Bluetooth Mesh network, increasing the number of relaying neighbors from 1 to 10 reduces the 2-hop round-trip time (RTT) from 47 ms to 33 ms for 41-byte packets and 10 ms scanning interval [51]. Increasing the number of hops from 1 to 4 increases the RTT from 22 ms 88 ms. This is because the latency is influenced the most by the random backoff mechanism. However, the absence of a random backoff would increase the collision probability, decreasing the reliability and scalability of Bluetooth Mesh networks which makes it suitable only for classes 4–6 of industrial applications.

4.2.4. Time Slotted Channel Hopping

A MAC layer with TSCH can operate on top of both 2.4 GHz- and sub-GHz 802.15.4-based PHY. Assuming up to 4 hops, 802.15.4 networks that operate in unlicensed 2.4 GHz frequency bands can achieve a range of up to 200 m. On the other hand, their sub-GHz counterparts based on 802.15.4 g can reach up to 1 km single-hop range [52].

In TSCH mode, time is slotted and time slots are grouped in slot frames. Both the time-slot duration and the number of slots in the slot frame are configurable. Slot duration is determined by the time needed to transmit a packet and receive an ACK and typically amounts 6 ms–10 ms. The length of the slot frame is based on the trade-off between energy consumption and throughput. Decreasing the slot frames increases both the energy consumption and the throughput and vice versa. For cyclic communication in a closed loop, a node would need two slots in a slot frame. The cycle could repeat every slot frame, thus the cycle time trades off with the number of nodes in the network. A network of two nodes could exchange packets with ~20 ms cycle time but 25 closed loops would increase the minimal cycle time to 500 ms. Considering the best practices regarding the sampling in industrial communications, the minimal cycle times achievable with TSCH in practice increase to 60 ms–80 ms for non-critical traffic and 80 ms–200 ms for a single regulatory loop (2 s–5 s for 25 loops).

Besides time slotting, TSCH also utilizes channel hopping to diminish the impact of interference and channel fading. Nodes communicate with each other over 16 orthogonal channels and retransmit packets on different channels if initial attempts fail, thereby increasing reliability. Channel hopping scheme could benefit the scalability and the aggregate throughput of the network if all channels were used in a same time slot. However, this benefit is largely limited in industrial networks given that all slave nodes communicate with the same master node in the network, as illustrated in Figure 1. Unless a master node is able to send and receive on multiple channels simultaneously, which is not the default capability, nodes will be forced not to transmit simultaneously, regardless of the channel hopping scheme.

Furthermore, reliability of 2.4 GHz TSCH networks may be significantly reduced in the presence of interference of co-located Wi-Fi networks, as shown in Reference [53]. When only a few channels in the TSCH matrix are affected by interference, only an increase of the transmission latency due to retransmissions could be expected. However in reality, the probability of eventual packet dropout is no longer negligible because of the quite low default retry limit (4) and the fact that one Wi-Fi channel overlaps many IEEE 802.15.4 channels (four to five). As Reference [53] shows, this results in an increase of packet losses, in turn reducing the reliability. Reliability could be increased by increasing the retry limit, hence trading off the latency and power consumption.

In conclusion, TSCH networks utilize the schedule on each node to achieve determinism and robustness against channel fading, as well as to conserve energy. That robustness and determinism makes TSCH suitable for class 2–6 industrial applications, assuming it can comply with the range requirement. However, there is a strong trade-off between latency and scalability, given that multiple channels cannot be used in parallel by different slave nodes due to the centralized master node.

5. Discussions

Key specifications of each technology introduced in Section 4 are listed in Tables 2 and 3. Note that not all numbers in Tables 2 and 3 can be taken for granted as they do not depict the trade-offs, but only present theoretical limits. NB-IoT operates in licensed spectrum. It can reach as far as its cellular infrastructure goes, and it is inherently more reliable than technologies that operate in shared spectrum. LoRa, IEEE 802.11ah and 802.15.4g operate in unlicensed sub-GHz frequency bands, which makes their long range a feature of their physical layers. LoRa operates in 863–870 MHz frequency band in Europe, offering three sub-bands at 864 MHz, 867 MHz and 868 MHz. Five 125 kHz channels are defined in 867 MHz band and three in 864 MHz and 868 MHz band [54,55]. The three default channels in 868 MHz band are to be implemented by every node, whereas the rest are optional. Wi-Fi HaLow operates in 863–868 MHz band in Europe and defines five 1-MHz channels and two 2-MHz channels [56]. Considering this, it is clear that Wi-Fi HaLow and LoRa's optional channels overlap in 3 out of 7 Wi-Fi HaLow channels and do not interfere with each other otherwise, which makes their parallel deployments possible. However, Wi-Fi HaLow can severely interfere with 802.15.4g [57], given that they operate in the same bands. The 2.4 GHz unlicensed frequency band is widely utilized today and the technologies that operate in this band must pay special attention to coexistence. WirelessHART, ISA100.11a and 802.15.4e TSCH are all based on the 802.15.4 2.4 GHz PHY layer that operates in sixteen 2 MHz-wide and 5 MHz-spaced channels in 2.4 GHz frequency band. They all employ frequency-hopping to improve the reliability of their transmissions, as they make use of exactly the same spectrum. Given their shorter single-hop ranges, these technologies employ multi-hop topologies to extend their range.

Table 2. Technical properties (top) and key performance indicators (bottom) of sub-GHz wireless technologies for the IIoT domain.

	LoRa	IEEE 802.11ah	NB-IoT	802.15.4g TSCH
Band	unlicensed sub-GHz	unlicensed sub-GHz	licensed (LTE band)	unlicensed sub-GHz
Bandwidth	125 kHz/250 kHz	1/2/4/8/16 MHz	180 kHz	200 kHz–1.25 MHz
Topology	star-of-stars	star/tree	cellular	star, p2p mesh
Deployment	private/operator-based	private	operator-based	private
MAC	LoRaWAN	hybrid EDCA/DCF	LTE based OFDMA (DL) & SC-FDMA (UL)	TSCH
Retransmissions	yes	yes	yes	yes
Reliability mechanisms	orthogonal SFs 32-bit MIC	FEC, WPA1 (MIC), WPA2 (CCM) WPA3 (BIP-GMAC-256)	FEC, ARQ	FSK/O-QPSK/OFDM
Range	15 km	1 km	20 km	1 km
Nodes per network	unlimited	8192	52,247 per cell	6000
Data rate	250 bps–5.5 kbps/11 kbps/50 kbps	150 kbps–78 Mbps	<250 kbps	6.25 kbps–800 kbps
Min. cycle time	>1 s	>20 ms	>1.6 s	> 20 ms

Orthogonal Frequency-Division Multiplexing (OFDM); Orthogonal Frequency-Division Multiple Access (OFDMA); Single Carrier FDMA (SC-FDMA); Uplink (UL); Downlink (DL); Automatic Repeat reQuest (ARQ); Offset QPSK (OQPSK); Counter Mode Cipher Block Chaining Message Authentication Code (CCM); Broadcast/Multicast Integrity Protocol (BIP); Galois Message Authentication Code (GMAC); Message Integrity Code (MIC).

Table 3. Technical properties (top) and key performance indicators (bottom) of IIoT wireless technologies based on 802.15.4 2.4 GHz PHY layer and BLE.

	WirelessHART	ISA100.11a	BLE	802.15.4e TSCH
Band	2.4 GHz ISM	2.4 GHz ISM	2.4/5 GHz ISM	2.4 GHz ISM
Bandwidth	200 kHz–1.2 MHz	2 MHz	2 MHz	2 MHz/5 MHz
Topology	mesh	star/mesh/star-mesh	p2p/star/mesh	star, tree, mesh
Deployment	private	private	private	private
MAC	time sync., freq. hopping TSMP (TDMA, 10ms)	TDMA / CSMA/CA (10–12 ms)	TDMA	TSCH (TDMA/CSMA/CA)
Retransmissions	yes	yes	yes	yes
Reliability mechanisms	ARQ, FHSS DSSS, 32-bit MIC	ARQ, FHSS, DSSS 32-to-128-bit MIC	FHSS, 24-bit CRC, 32-bit MIC, FEC	DSSS/OQPSK
Range	<1.5 km (225 m)	<1.5 km (100 m)	<100 m/<1000 m	<200 m
Nodes per network	30,000/hundreds per AP	unlimited/thousands per GW	unlimited	unlimited
Data rate	<250 kbps	<250 kbps	125 kbps/1 Mbps/2 Mbps	250 kbps
Min. cycle time	500 ms	500 ms	50 ms	20 ms

Automatic Repeat reQuest (ARQ); Frequency-Hopping Spread Spectrum (FHSS); Direct Sequence Spread Spectrum (DSSS); Gateway (GW); Offset QPSK (OQPSK); Message Integrity Code (MIC).

Wireless networks have the benefit of mitigating cabling for communication but the question of power cables still remains. Ideally, power cables could also be mitigated when the devices can live long enough drawing the power from the batteries available today. This is feasible for sufficiently large cycle times. Therefore, it is important to also consider the energy consumption. To evaluate the possibility of entirely wireless deployments that do not need power cables, an experimental energy efficiency comparison between LoRa, NB-IoT, IEEE 802.11ah and IEEE 802.15.4g (Wi-SUN with TSCH) was performed. The experiments combine energy consumption values of state-of-the-art off-the-shelf radios obtained from their data sheet (cf. Table 4), with simulated performance analysis. At the time of this study, no off-the-shelf radio was available for Wi-Fi HaLow and the same radio was assumed as for 802.15.4g (Atmel AT86RF215), as it supports the required modulation and coding schemes. The simulation experiments were performed using the ns-3 network simulator for LoRa [58], NB-IoT [59] and Wi-Fi HaLow [60]. Using the 6TiSCH simulator, 802.15.4g was evaluated [61]. A single device and a single gateway/AP is considered in the simulations. Hence, the results do not take into account scalability or contention. To maintain simplicity, predictability and comparability of the results, the experiments did not take into account any packet loss due to propagation errors, interference or collisions. As such, the PHY and channel contribution to the energy consumption is limited to the time-on-air of the radio for the reported packet sizes (including preamble and PHY/MAC header). The main contribution of the results pertains to the energy consumption of the MAC layer protocols of the considered technologies. For this reason, we used a simplified linear battery discharging model.

Table 4. Technology-specific parameters for energy consumption simulations.

	IEEE 802.15.4g	IEEE 802.11ah	LoRaWAN	NB-IoT
Radio module	Atmel AT86RF215		SEMTECH SX1272	uBlox SARA N210
TX power (dBm)	14		23	20
Power (mA) [RX/TX/idle/sleep]	28/62/6.28/0.03		11.2/125/0.0015/0.0001	46/220/6/0.003
Technology-specific parameters	113 slots @ 40 ms per frame 2-FSK—100 kHz (50 kbps) 0.33% DIO and EB probability	4096 ms beacon interval MCS 10 —1 MHz (150 kbps) 1 RAW group, 1 slot	no repetitions no ACK	RRC: 10 s, DRX: 0 s, PSM: TI s no repetitions
Payload size	104 bytes		12 bytes	
Microcontroller	ARM Cortex M3 @ 32 MHz (3.38 mA power consumption)			

Frequency Shift Keying (FSK); Enhanced Beacon (EB); DODAG Information Object (DIO); Destination Oriented Directed Acyclic Graph (DODAG); Modulation and Coding Scheme (MCS); Restricted Access Window (RAW); Radio Resource Control (RRC); Power Saving Mode (PSM); Transmission Interval (TI); Discontinuous Reception (DRX)

Wi-Fi HaLow is significantly more energy efficient than 802.15.4g with TSCH. This improvement is due to the higher supported data rate that leads to shorter transmission times. This result is also reflected in the predicted battery lifetime shown in Figure 5. Given a sufficiently large battery of 2000 mAh, the lifetime of Wi-Fi HaLow is expected to be above 10 years for a 10-min transmission interval, without battery replacement. For 802.15.4g, battery life expectancy is under 3 years in the same scenario. For the long-range contenders, it is under a year, as shown in Figure 6b, except the LoRaWAN SF7 that can live up to 7.5 years (cf. Figure 6a). The compared configurations of 802.15.4g and 802.11ah do not represent the best case scenario regarding energy consumption, as both technologies are configured to use low data rates. However, the chosen settings result in similar coverage range. In a line-of-sight scenario, 802.15.4g FSK-50 reaches a good PDR up to 700 m, while FSK-200 goes up to 420 m [62]. For 802.11ah [63], MCS10 (150 kbps) has a range of 700 m to more than 1000 m. For a higher data rate, 802.15.4g will achieve better energy efficiency but it's achievable range would be very different from that of MCS10 of 802.11ah. Similarly, 802.11ah will achieve better energy efficiency for higher data rates (MCS10 represents the lowest possible data rate and thus the worst case).

Figure 6 illustrates the benefits of low power modes in the design of a technology. As shown in the Figure, the lifetime of LoRaWAN devices is much higher than that of NB-IoT. Power consumption of LoRaWAN devices in transmission (TX)/reception (RX) mode is the dominant factor of the overall consumption, considering the very low power consumption in the idle/sleep mode (cf. Table 4). This results in a large difference in lifetime between LoRaWAN devices that use SF7 and SF12. The difference between SF7 and SF12 is especially large when the transmissions are frequent. The rarer the transmissions, the smaller the difference between SF7 and SF12 as idle/sleep time prevails and the TX/RX time becomes less significant in comparison to the idle/sleep time.

The larger difference in battery lifetime between NB-IoT MCS9 and MCS4 occurs due to the better efficiency of MCS9 which reflects not only in data TX/RX but also in signaling between the transmissions which adds up to the difference. The difference between MCS9 and MCS4 also reduces as the traffic interval increases, similarly as the difference between spreading factors of LoRaWAN. NB-IoT has higher data rate than LoRaWAN, thus it is more efficient in terms of TX/RX. However, the NB-IoT hardware is less efficient in terms of energy consumption in all four modes, so LoRaWAN becomes more energy efficient relative to NB-IoT. Higher data rates of NB-IoT cannot compensate for the energy efficiency of LoRaWAN in the evaluated scenarios.

It is important to note that the shelf life also affects the battery life. The lifetimes illustrated in Figures 5 and 6 represent the ideal cases which only take into account the consumption of a radio and a microcontroller. Hence, the presented results do not take into account energy consumption of peripherals, self-discharge of battery nor other power drains. For example, when considering only radio and microcontroller, NB-IoT platform using MCS4 and MCS9 would need around 230 mAh and 155 mAh respectively to run for three years. For the same time period, LoRaWAN radio and microcontroller using SF7 and SF12 would consume 8 mAh and 47 mAh, when ignoring other power drains and employing sleep mode (45 mAh and 83 mAh with idle mode). However, the above mentioned factors need to be taken into account when estimating very long battery lives as they make the largest part of energy consumption over a long time.

This paper focused mostly on PHY and MAC layer features of the considered wireless technologies, even though some technologies also address commissioning, security, roaming and other higher layer features. In addition to PHY and MAC, LoRaWAN defines a complete network architecture, various device types, commissioning and security (network and application). Parameter configurations and network management are generally implementer specific for all the considered technologies. LoRaWAN is configurable in terms of many parameters such as:

- spreading factor (and thus data rate): fixed choice or adaptive,
- reliability: ACKs or multiple transmissions of the same packet without downlink ACKs,
- higher layer logic: raw payload or Internet Protocol (IP) compliant stack.

Choosing the network setup for a certain scenario is not always straightforward, which motivated using optimization techniques to choose the optimal parameters [64]. Moreover, LoRaWAN does not impose the coordination of transmissions of class A devices, which might impact scalability in real deployments. Similarly, Wi-Fi HaLow defines PHY and MAC layer, as well as layer 2 security upon connecting. On top, IP compliant stack is to be used. Wi-Fi HaLow is highly configurable, defining a multitude of different parameters to be configured such as MCS, RAW, Traffic Indication Map (TIM) and others [37,65,66]. The configurable parameters can significantly influence the performance and the lifetime of the network. Higher layers of NB-IoT can be both IP- and non-IP compliant. NB-IoT also defines some configurable parameters such as extended Discontinuous Reception (eDRX) and PSM timers. However, other settings are under control of the network operator. The standard IEEE 802.15.4g/e TSCH defines PHY and MAC, with TSCH MAC layer specifying how to execute the schedule but not how to define it. Given that a schedule significantly impacts the network performance, there are standardization efforts in Internet Engineering Task Force (IETF) on scheduling functions but there is also a choice between centralized and decentralized scheduling. WirelessHART and ISA100.11a share a similar concept as TSCH. They have self-configuration capabilities greatly simplifying the deployment. They define the full stack and are centrally managed by the network manager that makes use of commands defined by the standard for network management [67]. They are out there for a long while and are quite well understood. BLE also defines the entire stack as well as other aspects such as commissioning. BLE has configurable parameters such as advertisement interval and connection interval that influence its performance. Different combinations of the configurable parameters can result in a different performance of any technology in various scenarios, thus the interdependence of the configurable parameters needs to be empirically examined.

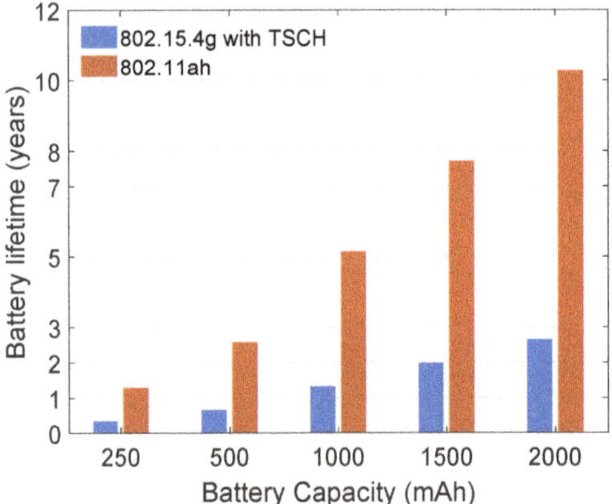

Figure 5. Battery lifetime of a device (both radio and microcontroller) transmitting once every 10 min for mid-range technologies (>1 km).

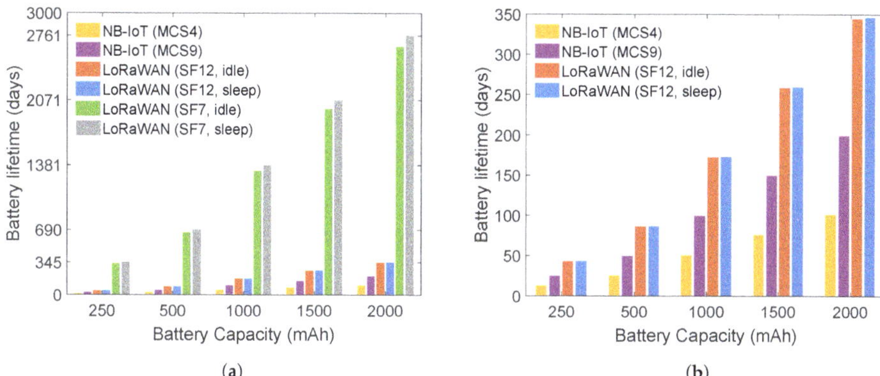

Figure 6. Battery lifetime of a device (both radio and microcontroller) transmitting once every 10 min for long-range technologies (<10 km). (**b**) is a subset of (**a**) where LoRaWAN SF7 is omitted for clarity.

6. Conclusions

LoRa, IEEE 802.11ah, NB-IoT, BLE, WirelessHART, ISA100.11a and TSCH can provide site-wide network coverage if configured and deployed accordingly, at the cost of data rate and/or latency and/or scalability. The trade-offs of the evaluated technologies are basically equivalent qualitatively but greatly vary in quantity, as elaborated in Section 4. NB-IoT and LoRa have potential to serve non-critical class 4–6 systems. On top, NB-IoT is more suitable for frequent downlink traffic because of LoRaWAN's downlink limitations, which makes it eligible for class 3 systems, as well as latency insensitive class class 2 systems (regulating very slow processes). However, NB-IoT typically comes with a subscription fee and due to the operator involvement, maintenance is slower. Thus, both have pros and cons, as well as similarities. BLE could serve supervising non-critical systems given its inherent unreliability in mesh topology and range limitations in star. TSCH, on the other hand, provides necessary reliability for critical services but its scale and latency are limited by the typical industrial fieldbus topology. WirelessHART, 802.15.4e TSCH and ISA100.11a are similar in terms of both reliability and latency. They all operate in the same frequency band and they all employ channel hopping to improve coexistence. Although they provide high data rates in comparison to other technologies, mesh topology increases the latency and limits the scalability. Still, these technologies are being deployed in class 2–6 systems. Finally, Wi-Fi HaLow trades data rate for range and latency but can still support the lowest latencies compared to reviewed technologies in networks of limited scale. Its star topology and novel MAC design contribute to high energy efficiency. Other technologies exhibit various trade-offs as well, ergo the choice of appropriate technology for a certain application is very challenging, given the number of options. This paper quantitatively evaluates the aforementioned technologies, outlining their main advantages and disadvantages and enabling the appropriate choice of the technology for an IWSAN of interest, considering the specific latency, reliability, scalability, energy consumption, data rate and coverage requirements.

Author Contributions: Conceptualization, A.S., E.D.P. and J.H.; validation, A.S., J.F., E.D.P. and J.H.; investigation, A.S.; resources, A.S., J.F., E.D.P. and J.H.; writing–original draft preparation, A.S.; writing–review and editing, A.S., J.F., E.D.P. and J.H.; visualization, A.S., J.F.; supervision, E.D.P. and J.H.; project administration, E.D.P. and J.H.; funding acquisition, E.D.P. and J.H. All authors have read and agreed to the published version of the manuscript.

Funding: This research was funded by Fund for Scientific Research-Flanders (FWO-V) grant number S004017N in the context of Strategic Basic Research (SBO) project IDEAL-IoT (Intelligent DEnse And Longe range IoT networks) project.

Acknowledgments: The authors would like to thank Serena Santi, Ashish Kumar Sultania, Glenn Daneels and Jetmir Haxhibeqiri for assisting in setting up the energy consumption measurement experiments.

Conflicts of Interest: The authors declare no conflict of interest.

References

1. Willig, A.; Matheus, K.; Wolisz, A. Wireless technology in industrial networks. *Proc. IEEE* **2005**, *93*, 1130–1151. [CrossRef]
2. Luvisotto, M.; Pang, Z.; Dzung, D. High-performance wireless networks for industrial control applications: New targets and feasibility. *Proc. IEEE* **2019**, *107*, 1074–1093. [CrossRef]
3. Weiner, M.; Jorgovanovic, M.; Sahai, A.; Nikolié, B. Design of a low-latency, high-reliability wireless communication system for control applications. In Proceedings of the IEEE International Conference on Communications (ICC), Sydney, Australia, 10–14 June 2014. [CrossRef]
4. Wei, Y.H.; Leng, Q.; Han, S.; Mok, A.K.; Zhang, W.; Tomizuka, M.; Li, T.; Malone, D.; Leith, D. RT-WiFi: Real-time high speed communication protocol for wireless control systems. *ACM SIGBED Rev.* **2013**, *10*, 28. [CrossRef]
5. Das Neves Valadão, Y.; Künzel, G.; Müller, I.; Pereira, C.E. Industrial Wireless Automation: Overview and Evolution of WIA-PA. *IFAC-PapersOnLine* **2018**, *51*, 175–180. [CrossRef]
6. HART Communications Foundation. *HART Field Communication Protocol Specification*; HFC_SPEC-12, Revision 6.0; HART Communications Foundation: Austin, TX, USA, 2001.
7. International Society of Automation (ISA) Standard. *Wireless Systems for Industrial Automation: Process Control and Related Applications*; ISA-100.11 a-2009; International Society of Automation (ISA) Standard: Research Triangle Park, NC, USA, 2009.
8. Harada, H.; Mizutani, K.; Fujiwara, J.; Mochizuki, K.; Obata, K.; Okumura, R. IEEE 802.15. 4g based Wi-SUN communication systems. *IEICE Trans. Commun.* **2017**, *100*, 1032–1043. [CrossRef]
9. Hermeto, R.T.; Gallais, A.; Theoleyre, F. Scheduling for IEEE802. 15.4-TSCH and slow channel hopping MAC in low power industrial wireless networks: A survey. *Comput. Commun.* **2017**, *114*, 84–105. [CrossRef]
10. IEC-International Electrotechnical Commission. *Industrial Networks: Wireless Communication Network and Communication Profiles*; IEC PAS 62601; IEC-International Electrotechnical Commission: Geneva, Switzerland, 2015.
11. Saelens, M.; Hoebeke, J.; Shahid, A.; De Poorter, E. Impact of EU duty cycle and transmission power limitations for sub-GHz LPWAN SRDs: An overview and future challenges. *EURASIP J. Wirel. Commun. Netw.* **2019**, *2019*, 219. [CrossRef]
12. Åkerberg, J.; Gidlund, M.; Björkman, M. Future research challenges in wireless sensor and actuator networks targeting industrial automation. In Proceedings of the 2011 9th IEEE International Conference on Industrial Informatics (INDIN), Lisbon, Portugal, 26–29 July 2011; pp. 410–415.
13. Ovsthus, K.; Kristensen, L.M. An industrial perspective on wireless sensor networks–a survey of requirements, protocols, and challenges. *IEEE Commun. Surv. Tutor.* **2014**, *16*, 1391–1412.

14. Frotzscher, A.; Wetzker, U.; Bauer, M.; Rentschler, M.; Beyer, M.; Elspass, S.; Klessig, H. Requirements and current solutions of wireless communication in industrial automation. In Proceedings of the 2014 IEEE International Conference on Communications Workshops (ICC), Sydney, Australia, 10–14 June 2014; pp. 67–72.
15. Li, X.; Li, D.; Wan, J.; Vasilakos, A.V.; Lai, C.F.; Wang, S. A review of industrial wireless networks in the context of industry 4.0. *Wirel. Netw.* **2017**, *23*, 23–41. [CrossRef]
16. Wollschlaeger, M.; Sauter, T.; Jasperneite, J. The future of industrial communication: Automation networks in the era of the internet of things and industry 4.0. *IEEE Ind. Electron. Mag.* **2017**, *11*, 17–27. [CrossRef]
17. Park, P.; Ergen, S.C.; Fischione, C.; Lu, C.; Johansson, K.H. Wireless network design for control systems: A survey. *IEEE Commun. Surv. Tutor.* **2018**, *20*, 978–1013. [CrossRef]
18. Gunatilaka, D. High Performance Wireless Sensor-Actuator Networks for Industrial Internet of Things. Ph.D. Thesis, Washington University in St. Louis, St. Louis, MO, USA, 2019.
19. Wang, Q.; Jiang, J. Comparative examination on architecture and protocol of industrial wireless sensor network standards. *IEEE Commun. Surv. Tutor.* **2016**, *18*, 2197–2219. [CrossRef]
20. De Pellegrini, F.; Miorandi, D.; Vitturi, S.; Zanella, A. On the use of wireless networks at low level of factory automation systems. *IEEE Trans. Ind. Inf.* **2006**, *2*, 129–143. [CrossRef]
21. Christin, D.; Mogre, P.S.; Hollick, M. Survey on wireless sensor network technologies for industrial automation: The security and quality of service perspectives. *Future Internet* **2010**, *2*, 96–125. [CrossRef]
22. Lu, C.; Saifullah, A.; Li, B.; Sha, M.; Gonzalez, H.; Gunatilaka, D.; Wu, C.; Nie, L.; Chen, Y. Real-time wireless sensor-actuator networks for industrial cyber-physical systems. *Proc. IEEE* **2016**, *104*, 1013–1024. [CrossRef]
23. Sanchez-Iborra, R.; Cano, M.D. State of the art in LP-WAN solutions for industrial IoT services. *Sensors* **2016**, *16*, 708. [CrossRef]
24. Varghese, A.; Tandur, D. Wireless requirements and challenges in Industry 4.0. In Proceedings of the 2014 International Conference on Contemporary Computing and Informatics (IC3I), Mysore, India, 27–29 November 2014; pp. 634–638.
25. Zurawski, R. *Industrial Communication Technology Handbook*; CRC Press: Boca Raton, FL, USA, 2014.
26. Tindell, K.; Burns, A. Guaranteeing message latencies on control area network (CAN). In Proceedings of the 1st International CAN Conference, 1994. Available online: https://citeseer.ist.psu.edu/viewdoc/download?doi=10.1.1.43.3251&rep=rep1&type=pdf (accessed on 15 January 2020).
27. Luvisotto, M.; Tramarin, F.; Vitturi, S. Assessing the impact of full-duplex wireless in real-time industrial networks. In Proceedings of the IECON 2018—44th Annual Conference of the IEEE Industrial Electronics Society, Washington, DC, USA, 21–23 October 2018; pp. 4119–4124.
28. Vuran, M.C.; Akyildiz, I.F. Error control in wireless sensor networks: A cross layer analysis. *IEEE/ACM Trans. Netw. (TON)* **2009**, *17*, 1186–1199. [CrossRef]
29. Bor, M.C.; Roedig, U.; Voigt, T.; Alonso, J.M. Do LoRa low-power wide-area networks scale? In Proceedings of the 19th ACM International Conference on Modeling, Analysis and Simulation of Wireless and Mobile Systems, Malta, 13–17 November 2016; pp. 59–67.
30. Aref, M.; Sikora, A. Free space range measurements with Semtech LoRa™ technology. In Proceedings of the 2014 2nd International Symposium on Wireless Systems within the Conferences on Intelligent Data Acquisition and Advanced Computing Systems, Offenburg, Germany, 11–12 September 2014; pp. 19–23.
31. Van den Abeele, F.; Haxhibeqiri, J.; Moerman, I.; Hoebeke, J. Scalability analysis of large-scale LoRaWAN networks in ns-3. *IEEE Internet Things J.* **2017**, *4*, 2186–2198. [CrossRef]
32. Haxhibeqiri, J.; De Poorter, E.; Moerman, I.; Hoebeke, J. A survey of lorawan for iot: From technology to application. *Sensors* **2018**, *18*, 3995. [CrossRef]
33. Rizzi, M.; Ferrari, P.; Flammini, A.; Sisinni, E.; Gidlund, M. Using LoRa for industrial wireless networks. In Proceedings of the 2017 IEEE 13th International Workshop on Factory Communication Systems (WFCS), Trondheim, Norway, 31 May–2 June 2017; pp. 1–4.
34. Pötsch, A.; Hammer, F. Towards end-to-end latency of LoRaWAN: Experimental analysis and IIoT applicability. In Proceedings of the 2019 15th IEEE international workshop on factory communication systems (WFCS), Sundsvall, Sweden, 27–29 May 2019; pp. 1–4.
35. Šljivo, A.; Kerkhove, D.; Tian, L.; Famaey, J.; Munteanu, A.; Moerman, I.; Hoebeke, J.; De Poorter, E. Performance Evaluation of IEEE 802.11ah Networks With High-Throughput Bidirectional Traffic. *Sensors* **2018**, *18*, 325. [CrossRef]

36. Seferagić, A.; Moerman, I.; De Poorter, E.; Hoebeke, J. Evaluating the suitability of IEEE 802.11 ah for low-latency time-critical control loops. *IEEE Internet Things J.* **2019**, *6*, 7839–7848. [CrossRef]
37. Santi, S.; Šljivo, A.; Tian, L.; De Poorter, E.; Hoebeke, J.; Famaey, J. Supporting heterogeneous IoT traffic using the IEEE 802.11 Ah restricted access window. In Proceedings of the 15th ACM Conference on Embedded Network Sensor Systems, Delft, The Netherlands, 6–8 November 2017; p. 32.
38. Šljivo, A.; Kerkhove, D.; Moerman, I.; De Poorter, E.; Hoebeke, J. Reliability and Scalability Evaluation with TCP/IP of IEEE802. 11ah Networks. In Proceedings of the Workshop on ns3 (WNS3), Porto, Portugal, 13–14 June 2017; pp. 1–4.
39. Chen, J.; Hu, K.; Wang, Q.; Sun, Y.; Shi, Z.; He, S. Narrowband internet of things: Implementations and applications. *IEEE Internet Things J.* **2017**, *4*, 2309–2314. [CrossRef]
40. Wang, Y.P.E.; Lin, X.; Adhikary, A.; Grovlen, A.; Sui, Y.; Blankenship, Y.; Bergman, J.; Razaghi, H.S. A primer on 3GPP narrowband Internet of Things. *IEEE Commun. Mag.* **2017**, *55*, 117–123. [CrossRef]
41. Basu, S.S.; Sultania, A.K.; Famaey, J.; Hoebeke, J. Experimental Performance Evaluation of NB-IoT. In Proceedings of the International Conference on Wireless and Mobile Computing, Networking and Communications (WiMob), Barcelona, Spain, 21–23 October 2019.
42. Martinez, B.; Adelantado, F.; Bartoli, A.; Vilajosana, X. Exploring the Performance Boundaries of NB-IoT. *IEEE Internet Things J.* **2019**, *6*, 5702–5712 [CrossRef]
43. Martinez, B.; Vilajosana, X.; Chraim, F.; Vilajosana, I.; Pister, K.S. When scavengers meet industrial wireless. *IEEE Trans. Ind. Electron.* **2015**, *62*, 2994–3003. [CrossRef]
44. Petersen, S.; Carlsen, S. Performance evaluation of WirelessHART for factory automation. In Proceedings of the 2009 IEEE Conference on Emerging Technologies & Factory Automation, Palma de Mallorca, Spain, 22–25 September 2009; pp. 1–9.
45. Han, S.; Zhu, X.; Aloysius, K.M.; Nixon, M.; Blevins, T.; Chen, D. Control over WirelessHART network. In Proceedings of the IECON 2010-36th Annual Conference on IEEE Industrial Electronics Society, Glendale, AZ, USA, 7–10 November 2010; pp. 2114–2119.
46. Åkerberg, J.; Gidlund, M.; Reichenbach, F.; Björkman, M. Measurements on an Industrial Wireless Hart Network Supporting Profisafe: A Case Study. In Proceedings of the International Conference on Emerging Technologies and Factory Automation (ETFA), Toulouse, France, 5–9 September 2011; IEEE, pp. 1–8.
47. Hassan, S.M.; Bingi, K.; Ibrahim, R.; Chein, L.J.; Supramaniam, T. Implementation of flow control over wirelessHART sensor network using wirelessHART adaptors. *Indones. J. Electr. Eng. Comput. Sci.* **2019**, *15*, 910–919.
48. One Wireless Field Device Access Point Specification Release 310 OW03-650-310 Kernel Description. 2018. Available online: https://www.honeywellprocess.com/library/marketing/tech-specs/onewireless-fdap-specification.pdf (accessed on 14 December 2018).
49. Jecan, E.; Pop, C.; Padrah, Z.; Ratiu, O.; Puschita, E. A dual-standard solution for industrial Wireless Sensor Network deployment: Experimental testbed and performance evaluation. In Proceedings of the 14th IEEE International Workshop on Factory Communication Systems (WFCS), Imperia, Italy, 13–15 June 2018.
50. Karvonen, H.; Pomalaza-Ráez, C.; Mikhaylov, K.; Hämäläinen, M.; Iinatti, J. *Experimental Performance Evaluation of BLE 4 Versus BLE 5 in Indoors and Outdoors Scenarios;* Advances in Body Area Networks I; Fortino, G.; Wang, Z., Eds.; Springer: Cham, Switzerland, 2019; pp. 235–251.
51. Baert, M.; Rossey, J.; Shahid, A.; Hoebeke, J. The Bluetooth mesh standard: An overview and experimental evaluation. *Sensors* **2018**, *18*, 2409. [CrossRef]
52. Mochizuki, K.; Obata, K.; Mizutani, K.; Harada, H. Development and field experiment of wide area Wi-SUN system based on IEEE 802.15. 4g. In Proceedings of the 2016 IEEE 3rd World Forum on Internet of Things (WF-IoT), Reston, VA, USA, 12–14 December 2016; pp. 76–81.
53. Cena, G.; Scanzio, S.; Valenzano, A.; Zunino, C. Experimental Analysis and Comparison of Industrial IoT Devices based on TSCH. In Proceedings of the 24th IEEE International Conference on Emerging Technologies and Factory Automation (ETFA), Zaragoza, Spain, 10–13 September 2019. [CrossRef]
54. ETSI. System Reference document (SRdoc). In *Technical characteristics for Low Power Wide Area Networks Chirp Spread Spectrum (LPWAN-CSS) operating in the UHF spectrum below 1 GHz;* ETSI TR 103 526 V1.1.1; ETSI: Sophia Antipolis Cedex, France, 2018.

55. LoRa Alliance. *LoRaWAN™ Specification*; LoRa Alliance Inc., Beaverton, OR, USA, 2015. Available online: https://lora-alliance.org/sites/default/files/2018-04/lorawantm_specification_-v1.1.pdf (accessed on 11 October 2017).
56. *IEEE Standard for Information Technology–Telecommunications and Information Exchange between Systems—Local and Metropolitan Area Networks–Specific Requirements—Part 11: Wireless LAN Medium Access Control (MAC) and Physical Layer (PHY) Specifications Amendment 2: Sub 1 GHz License Exempt Operation*; IEEE Std 802.11ah-2016 (Amendment to IEEE Std 802.11-2016, as amended by IEEE Std 802.11ai-2016); IEEE: New York, NY, USA 2017; pp. 1–594. [CrossRef]
57. Liu, Y.; Guo, J.; Orlik, P.; Nagai, Y.; Watanabe, K.; Sumi, T. Coexistence of 802.11 ah and 802.15. 4g networks. In Proceedings of the 2018 IEEE Wireless Communications and Networking Conference (WCNC), Barcelona, Spain, 15–18 April 2018; pp. 1–6.
58. Haxhibeqiri, J.; Van den Abeele, F.; Moerman, I.; Hoebeke, J. LoRa scalability: A simulation model based on interference measurements. *Sensors* **2017**, *17*, 1193. [CrossRef]
59. Sultania, A.K.; Delgado, C.; Famaey, J. Implementation of NB-IoT Power Saving Schemes in ns-3. In Proceedings of the 2019 Workshop on Next-Generation Wireless with ns-3, Florence, Italy, 21 June 2019; pp. 5–8.
60. Tian, L.; Šljivo, A.; Santi, S.; De Poorter, E.; Hoebeke, J.; Famaey, J. Extension of the IEEE 802.11 ah ns-3 Simulation Module. In Proceedings of the 10th Workshop on ns-3, Mangalore, India, 13–14 June 2018; pp. 53–60.
61. Municio, E.; Daneels, G.; Vučinić, M.; Latré, S.; Famaey, J.; Tanaka, Y.; Brun, K.; Muraoka, K.; Vilajosana, X.; Watteyne, T. Simulating 6TiSCH networks. *Trans. Emerg. Telecommun. Technol.* **2019**, *30*, e3494. [CrossRef]
62. Muñoz, J.; Chang, T.; Vilajosana, X.; Watteyne, T. Evaluation of IEEE802. 15.4 g for Environmental Observations. *Sensors* **2018**, *18*, 3468. [CrossRef]
63. Bellekens, B.; Tian, L.; Boer, P.; Weyn, M.; Famaey, J. Outdoor IEEE 802.11 ah range characterization using validated propagation models. In Proceedings of the GLOBECOM 2017—2017 IEEE Global Communications Conference, Singapore, 4–8 December 2017; pp. 1–6.
64. Ousat, B.; Ghaderi, M. LoRa Network Planning: Gateway Placement and Device Configuration. In Proceedings of 2019 IEEE International Congress on Internet of Things (ICIOT), Milan, Italy, 8–13 July 2019; pp. 25–32.
65. Tian, L.; Khorov, E.; Latré, S.; Famaey, J. Real-Time Station Grouping under Dynamic Traffic for IEEE 802.11ah. *Sensors* **2017**, *17*, 1559. [CrossRef] [PubMed]
66. Khorov, E.; Lyakhov, A.; Krotov, A.; Guschin, A. A survey on IEEE 802.11ah: An enabling networking technology for smart cities. *Comput. Commun.* **2015**, *58*, 53–69. [CrossRef]
67. Shamsi, M. WirelessHART Field Device Installation and Configuration. 2016. Available online: https://www.emerson.com/documents/automation/article-wirelesshart-field-device-installation-configuration-en-38282.pdf (accessed on 8 January 2019).

Article

Transmission Scheduling Schemes of Industrial Wireless Sensors for Heterogeneous Multiple Control Systems

Bongsang Park [1], Junghyo Nah [2], Jang-Young Choi [2], Ick-Jae Yoon [2] and Pangun Park [1,*]

1 Department of Radio and Information Communications Engineering, Chungnam National University, Daejeon 34134, Korea; bong4ang@cnu.ac.kr

2 Department of Electrical Engineering, Chungnam National University, Daejeon 34134, Korea; jnah@cnu.ac.kr (J.N.); choi_jy@cnu.ac.kr (J.-Y.C.); ijyoon@cnu.ac.kr (I.-J.Y.)

* Correspondence: pgpark@cnu.ac.kr; Tel.: +82-42-821-6862

Received: 14 November 2018; Accepted: 1 December 2018 ; Published: 5 December 2018

Abstract: The transmission scheduling scheme of wireless networks for industrial control systems is a crucial design component since it directly affects the stability of networked control systems. In this paper, we propose a novel transmission scheduling framework to guarantee the stability of heterogeneous multiple control systems over unreliable wireless channels. Based on the explicit control stability conditions, a constrained optimization problem is proposed to maximize the minimum slack of the stability constraint for the heterogeneous control systems. We propose three transmission scheduling schemes, namely centralized stationary random access, distributed random access, and Lyapunov-based scheduling scheme, to solve the constrained optimization problem with a low computation cost. The three proposed transmission scheduling schemes were evaluated on heterogeneous multiple control systems with different link conditions. One interesting finding is that the proposed centralized Lyapunov-based approach provides almost ideal performance in the context of control stability. Furthermore, the distributed random access is still useful for the small number of links since it also reduces the operational overhead without significantly sacrificing the control performance.

Keywords: transmission scheduling scheme; industrial internet of things; wireless networks; industrial control systems; wireless networked control systems

---

1. Introduction

Industrial internet of things (IIoT) through wireless sensors and actuators have tremendous potential to improve the efficiency of various industrial control systems in both process automation and factory automation [1–5]. IIoTs are integrated systems of computation, wireless networking, and physical systems, in which embedded devices are connected to sense, actuate, and control the physical plants. In wireless networked control systems (WNCSs), the wireless sensors basically measure the physical plants and transmit its information to the controllers. The controllers then compute the control signal based on the received sensing information in order to manipulate the physical plants through the actuators. The wireless networks of IIoTs provide many benefits such as simple deployment, flexible installation and maintenance, and increased modularity in many practical control systems with the low cost [1].

Unfortunately, the network constraints such as delays and losses can significantly degrade the control performance and can even lead to unstable control systems [1,6]. In the control community, extensive research has been conducted to analyze the communication effects on control systems and design the control algorithm to handle its effects on the control performance [7]. Comparatively, much less work of wireless network design for control systems has been proposed. In fact, current wireless networks do not offer guaranteed stability of heterogeneous multiple control systems over lossy channels [1]. The main reason is that the numerous parameters of the control system and the communication system influence each other due to the complex interactions among different layers. It is important to understand how these communication constraints affect the control stability and performance properties in a quantitative manner. The quantitative result is an important factor to bridge the gap between control and communication layers for the efficient and stable control operations using IIoTs. Furthermore, the communication protocol must guarantee the stability of all control-loops since each industrial process affects the performance of overall connected control systems [7].

The transmission scheduling policies of sensors and controllers must efficiently optimize the traffic generation instance and transmit slot allocation since it directly affects the network delay and loss, and eventually leading to the stability issue of control systems. Decreasing the traffic generation interval of sensors and controllers generally improves the performance of the control system at the cost of lossy and delayed control feedback due to the increasing network congestion. Moreover, increasing the traffic generation rate may not satisfy the schedulability constraint of the communication system. Thus, the transmission scheduling must adjust its operations dependent on the control system requirements and the link conditions.

The main contribution of the paper is to propose three transmission scheduling policies, namely centralized stationary random access, distributed random access, and Lyapunov-based scheduling scheme, of wireless networks to guarantee the stability of heterogeneous multiple control systems over different lossy links. The three transmission scheduling policies are based on the max-min optimization problem where the objective is to maximize the minimum slack of the stability constraint of the control systems. We show the performance of the proposed scheduling schemes in terms of the stability region of heterogeneous multiple control systems over different link conditions.

The paper is organized as follows. Section 2 discusses the related works on both control and communication aspects. In Section 3, we present a general WNCS modeling framework to include both communication constraints such as varying delays, packet losses, and sampling intervals. In Section 4, an illustrative WNCS example is used to present the fundamental performance issues of general communication protocols in terms of the control stability. Based on the fundamental observation, we formulate a novel optimization problem for the transmission scheduling of wireless networks in Section 5. We present three different scheduling schemes, namely centralized stationary random access, distributed random access, and centralized Lyapunov-based schedule schemes, to solve the proposed optimization problem in Sections 6–8, respectively. In Section 9, we analyze the robust performance of the proposed three transmission scheduling schemes to guarantee the stability of heterogeneous control systems. Finally, we summarize the contributions of the paper in Section 10.

Notations: $\mathbb{Z}^+$ denotes all nonnegative integers. Normal font x, bold font $\mathbf{x}$, and calligraphic font $\mathcal{X}$ denote scalar, vector, and set, respectively.

2. Related Works

In the control community, extensive research has been conducted to analyze control stability and to design control algorithms by considering the communication constraints [1,7,8]. The control community generally considers network imperfections and constraints such as the varying packet dropouts, network delays, and traffic generation intervals [1]. Note that all these network factors are highly correlated dependent on the assumptions of the NCS literature. However, most NCS studies only consider some of the network effects, while ignoring the other factors, due to the high complexity of the stability analysis. For instance, the network effects of packet dropouts are investigated in [9,10], of time-varying sampling intervals in [11,12], and of delays in [6,13,14]. By considering the structure of NCSs, previous works analyze the stability of control systems by using either only wireless sensor–controller channel (e.g., [15,16]) or both sensor–controller and controller–actuator (e.g., [8,17–19]).

In [19], the explicit bounds on the maximum allowable transfer interval and the maximally allowable delay are derived to guarantee the control stability of NCSs, by considering time-varying sampling period and time-varying packet delays. If there are packet losses for the time-triggered sampling, its effect is modeled as a time-varying sampling period from receiver point-of-view. The maximum allowable transfer interval is the upper bound on the transmission interval for which stability can be guaranteed. If the network performance exceeds the given requirements, then the stability of the overall system could not be guaranteed. The developed results lead to tradeoff curves between maximum allowable transfer interval and maximally allowable delay. These tradeoff curves provide effective quantitative information to the network designer when selecting the requirements to guarantee stability and a desirable level of control performance.

In the communication community, most existing approaches design the scheduling algorithms to meet the delay constraint of each packet for a given traffic demand [20–23]. An interesting design framework is proposed to minimize the energy consumption of the network, while meeting reliability and delay requirements from the control application [24,25]. In [24], modeling of the slotted random access scheme of the IEEE 802.15.4 medium access control (MAC) is developed by using a Markov chain model. The proposed Markov chain model is used to derive the analytical expressions of reliability, delay, and energy consumption. By using this model, an adaptive IEEE 802.15.4 MAC protocol is proposed. The protocol design is based on a constrained optimization problem where the objective function is the energy consumption of the network, subject to constraints on reliability and packet delay. The protocol is implemented and experimentally evaluated on a testbed. Experimental results show that the proposed algorithm satisfies reliability and delay requirements while ensuring a longer network lifetime under both stationary and transient network conditions.

The cross-layer protocol solution, called Breath, is designed for industrial control applications where source nodes attached to the plant must transmit information via multihop routing to a sink [25]. The protocol is based on randomized routing, MAC, and duty-cycling to minimize the energy consumption, while meeting reliability and packet delay constraints. Analytical and experimental results show that Breath meets reliability and delay requirements while exhibiting a nearly uniform distribution of the work load.

Since the joint design of controller and wireless networks necessitates the derivation of the required packet loss probability and packet delay to achieve the desired control cost, we provided the formulation of the control cost function as a function of the sampling period, packet loss probability, and packet delay [26]. We first presented how the wireless network affects the performance of NCSs by showing the feasible region of the control performance. By considering these results, the joint design between communication and control application layers is proposed for multiple control systems over the IEEE 802.15.4 wireless network. In particular, a constrained optimization problem is studied, where the objective function is the energy consumption of the network and the constraints are the packet loss probability and delay, which are derived from the desired control cost. We clearly observe the tradeoff between the control cost and power consumption of the network.

Recently, a novel framework of communication system design is proposed by efficiently abstracting the control system in the form of maximum allowable transmission interval and maximum allowable delay constraints [27,28]. The transmission interval is the traffic generation interval of successfully received information. The objective of the optimization is to minimize the total energy consumption of the network while guaranteeing interval and delay requirements of the control system and schedulability constraints of the wireless communication system. A schedulability constraint is introduced as the sum of the utilization of the nodes, namely the ratio of the delay to the sampling periods. The decision variables are the set of transmission rate and sampling period. The proposed mixed-integer programming problem is converted to an integer programming problem based on the analysis of optimality and relaxation [27]. Then, the centralized resource allocation algorithm gives the suboptimal solution for the specific case of M-ary quadrature amplitude modulation and earliest deadline first scheduling. For a fixed sampling period, the formulation is also extended for any non-decreasing function of the power consumption as the objective function, any modulation scheme, and any scheduling algorithm in [28]. All related works [27,28] propose the centralized algorithm to adapt the communication parameters for the homogeneous control systems over the equal link condition.

In [29], the cross-layer optimized control protocol is proposed to minimize the worst-case performance loss of multiple control systems over a multihop mesh network. The design approach relies on a constrained max–min optimization problem, where the objective is to maximize the minimum resource redundancy of the network and the constraints are the stability of the closed-loop control systems and the schedulability of the communication resources. The stability condition of the control system has been formulated in the form of stochastic transmission interval constraint [19]. The centralized algorithm gives the optimal solutions of the protocol operation in terms of the sampling period, slot scheduling, and routing.

In comparison to these works [27–29], the transmission control policies of IIoTs must optimize both traffic generation instance and transmit slot allocations to guarantee the stability of heterogeneous control systems over different link conditions. The earliest deadline first scheduling only guarantees the optimal performance for the homogeneous requirements [30]. Furthermore, the centralized approach generally provides the optimal performance but at the cost of the monitoring and network control overhead and single point failure problem. In this paper, we focus on the robust performance guarantee of the control stability rather than the energy efficiency issue of wireless networks. In fact, researchers have recently applied the IEEE 802.11 standards for the real-time control applications instead of low-power wireless standards such as IEEE 802.15.4 and 802.15.3 [1,31,32].

3. System Model

Figure 1 illustrates the general structure of multiple control processes over wireless networks. It consists of a number of plants and controllers, which are connected through wireless networks. When the plant and controller are connected over wireless networks, it leads to the following operation aspects of WNCSs in Figure 1.

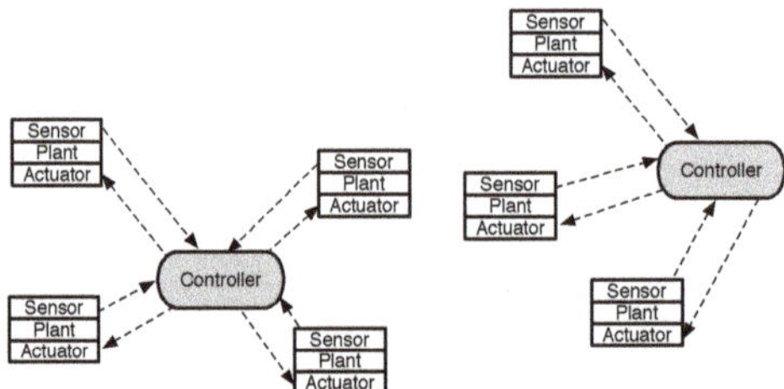

Figure 1. General structure of wireless networked control systems.

- While each sensor operates in a time-driven fashion, both the controller and actuator operate in an event-driven fashion. In general, wireless sensors transmit data in each assigned time slot dependent on the transmission scheduling scheme. However, both the controller and actuator only respond to newly received data over unreliable wireless links. We assume that the controllers are collocated with the actuators since the control signal is more critical than the sensing information in many practical NCSs [7].
- In [1], we defined three major metrics of WNCSs, namely sampling interval, packet dropout, and packet delay. Two main reasons of packet dropouts are packet discard due to the control algorithm and packet loss due to the wireless network itself. Most works of control systems model the dropouts as prolongations of the sampling interval [6,19]. The reason is that a new packet is transmitted at the next transmission time with new data if a packet is dropped. Hence, both the controller and actuator observe the time-varying sampling interval even if the

sensing and actuating links operate in a fixed time interval. The time-varying sampling interval of successfully received information called transmission interval (TI) effectively captures the essential characteristics of packet dropout and sampling interval [8,19]. The delays are generally assumed to be smaller than the transmission intervals.

The uncertain time-varying TIs and time-varying delays provide the fundamental interactions between control and communication layers [6,19,33]. In the next section, we describe more details of mathematical modelings and assumptions.

3.1. Control Aspect

Consider a single-hop wireless network consisting with N sensor nodes and a controller, as shown in Figure 1. Note that the controller is considered as a base station of a general star network topology. We assume the slotted time with slot index $k \in \mathbb{Z}^+$. Let us denote the transmission time, $t_s, s \in \mathbb{Z}^+$, of sth successfully received packet at the controller. At each slot, the transmission scheduling policy determines which of the nodes $i \in \{1, \ldots, N\}$ can access the network. When the sensor is allowed to transmit, it measures the plant state and sends it over the wireless channel. The packet arrives after the transmission delay δ_s at the controller. Hence, the controller only updates the plant state at time $t_s + \delta_s, \forall s \in \mathbb{Z}^+$. Figure 2 illustrates the typical evolution of plant state updates at the controller.

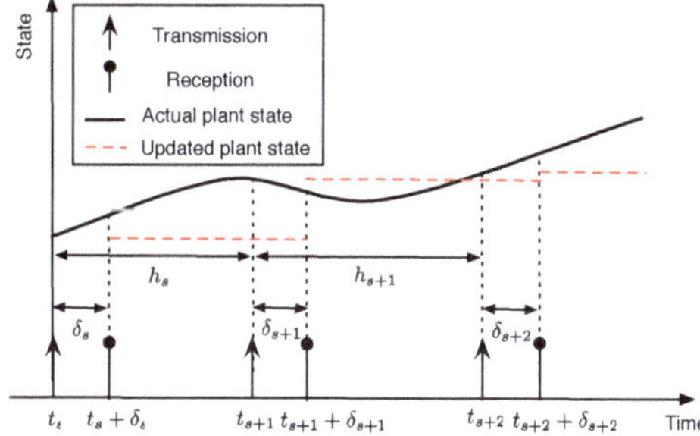

Figure 2. Illustration of a typical evolution of plant state updates at the controller.

The transmission times of successfully received packet satisfy $0 \leq t_0 < t_1 < t_2 < \ldots$ and there exists $\delta > 0$ such that the TIs $h_s = t_{s+1} - t_s$ satisfy $\delta \leq t_{s+1} - t_s \leq \bar{h}$ for $\forall s \in \mathbb{Z}^+$, where $\bar{h}$ is the maximum allowable transmission interval (MATI). Furthermore, we assume the maximum allowable delay (MAD) in the sense that $\delta_s \in [0, \bar{\delta}], \forall s \in \mathbb{Z}^+$, where $0 \leq \bar{\delta} \leq \bar{h}$. To guarantee the stability, the TI must satisfy $\delta \leq t_{s+1} - t_s \leq \bar{h}, s \in \mathbb{Z}^+$ and the delays satisfy $0 \leq \delta_s \leq \min(\bar{\delta}, t_{s+1} - t_s), \forall s \in \mathbb{Z}^+$. It implies that each transmitted packet arrives before the next sampling instance. Hence, the delay is smaller than the TI.

In the control community, many studies are conducted to analyze the stability of control systems for a given set of MATI and MAD values [6,19,33]. Hence, it is possible to derive the MATI and MAD requirements as the set of network design parameters by using the stability analysis techniques.

3.2. Communication Aspect

Denote the set of i's interfering links, $\mathcal{C}_i = \{j : j \in \mathcal{N} \setminus i\}$ where $\mathcal{N}$ is the total set of nodes for the star topology. Let $u_i(k) = 1$ if the node i transmits during slot k, whereas $u_i(k) = 0$ otherwise. When $u_i(k) = 1$, node i generates a new packet and transmits it over the wireless channel to minimize the delay. Hence, the packet delay is fixed to 1 time slot. Note that we motivate the fixed delay between traffic generation instance and transmission schedule through an illustrative example in Section 4. The scheduling constraint is

$$\sum_{i=1}^{N} u_i(k) \leq 1, \quad \forall k \in \mathbb{Z}^+ \tag{1}$$

It means that the centralized scheduling scheme selects at most one node for transmission at any given time slot k. On the other hand, each node decides its transmission for the distributed approach. We assume that the transmission scheduling algorithm is located at the controller for the centralized approach.

Let $d_i(k)$ be the random variable to indicate the successful packet transmission of node i to the controller. If node i sends a packet at slot k, $u_i(k) = 1$ and other nodes $j \in \mathcal{C}_j$ do not transmit, then $d_i(k) = 1$ with probability $p_i \in (0,1]$ and $d_i(k) = 0$ with probability $1 - p_i$. If node i does not transmit, $u_i(k) = 0$, then $d_i(k) = 0$ with probability one. We assume the heterogeneous link reliability p_i over different links. Hence, the expected delay becomes

$$\mathbb{E}[d_i(k)] = p_i \mathbb{E}[u_i(k)]. \tag{2}$$

4. Fundamental Observation

To analyze the stability of control systems, linear matrix inequality (LMI) conditions were verified on the polytopic overapproximation in [6,19,33]. The LMI conditions were verified using the YALMIP [34] and the SeDuMi solver [35]. We used the analytical technique of the control stability in [33]. This technique effectively analyzes the stability to a given linear time-invariant (LTI) plant model, a LTI controller model, and MATI and MAD bounds on the network uncertainties. In this illustrative example, we first analyzed the batch reactor system [33,36] to demonstrate how stability regions can be visualized. Then, we investigated the fundamental tradeoff between MATI and MAD for the control stability. The network is assumed to incur

- uncertain time-varying TIs $h \in [\underline{h}, \bar{h}]$; and
- uncertain time-varying network delays $d \in [\underline{d}, \min(h, \bar{d})]$.

We fixed $\underline{h} = 10$ ms and $\underline{d} = 10$ ms due to the slot duration of the typical industrial wireless standards [37].

Figure 3 shows the stability region over different MATI, $\bar{h} = 0.01, \ldots, 1$ s and MAD, $\bar{d} = 0.01, \ldots, 0.58$ s. The circle and rectangular marker present the stability and instability operating region for a given MATI and MAD value. The solid line represents the assumption of the MATI and MAD constraints, namely $\bar{d} \leq h_s \leq \bar{h}$, $\forall s \in \mathbb{Z}^+$. Clearly, as the MATI and MAD values increase, the control system becomes unstable. In other words, the lower are the MATI and MAD values, the better is the control system stability. It is not simple to approximate the boundary between stability and instability region due to the complex stability analysis techniques in Figure 3. However, in general, the MAD requirement of the stability becomes more strict as the MATI requirement is relaxed. Hence, there is a fundamental tradeoff between MATI and MAD requirements for the stability guarantee of control systems.

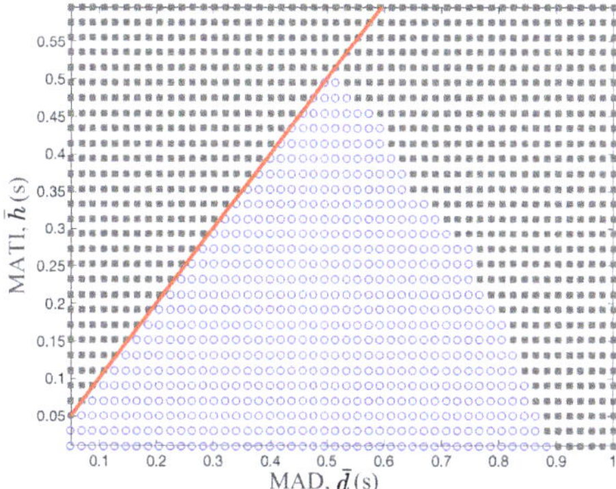

Figure 3. Stability region over different MATI and MAD values. The circle and rectangular marker present the stability and instability operating region of control systems for a given MATI and MAD value.

Since the wireless medium is shared between nodes, we analyzed the performance of typical access control schemes in the context of the control stability. In Figure 4, we provide the TI and the delay performance of two well-known schemes, namely time division multiple access (TDMA) and slotted Aloha over the stability region. The color bar shows the probability density function of the TI and the delay measurements of different access schemes. Note that the stability region in Figure 4 is equal to that in Figure 3 with different X and Y scales. The medium access scheme guarantees the control stability if it satisfies the network performance of both TI and the delay inside of the stability region. Hence, the outage probability of the control stability is a good performance metric of wireless networks in the context of the control stability.

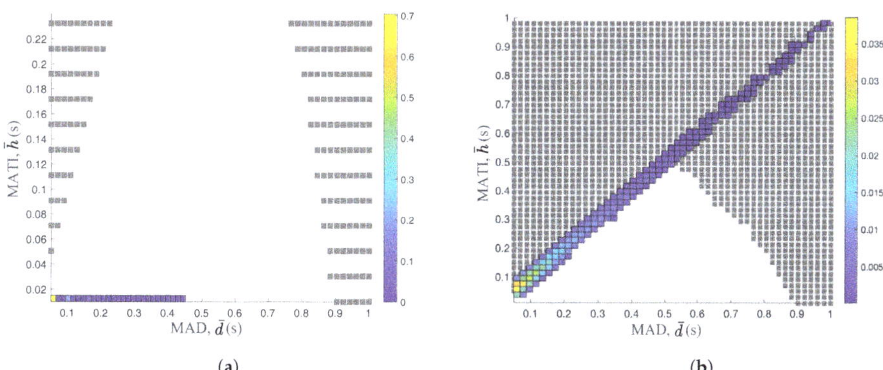

(a) (b)

Figure 4. Transmission interval and delay of both TDMA and slotted Aloha schemes over the stability region: (**a**) TDMA performance over stability region; and (**b**) slotted Aloha performance over stability region.

Both access schemes rely on the time slotted mode with $N = 5$ and $p = 0.3$. TDMA uses a simple round robin algorithm to assign the slot to each node of the network in Figure 4a. On the other hand, each node randomly decides its transmission based on the traditional slotted Aloha mechanism in Figure 4b. We set the optimal channel access probability of slotted Aloha to maximize the network throughput [38].

By comparing Figure 4a,b, we clearly observed the completely different network performances over the stability region. Since each node of TDMA transmits its corresponding values on the assigned slots, it generates the packet before the transmitting slot. Hence, the delay of the TDMA scheme is constant to 1 time slot, 10 ms, as shown in Figure 4a. On the other hand, the slotted Aloha only generates new packets after it successfully transmits the packets. Hence, the delay of the slotted Aloha is equal to TI. This is the main reason for significantly different behaviors between TDMA and slotted Aloha. In fact, the outage probability of the stability region of slotted Aloha is 0.06 while its corresponding probability is 0 for the TDMA scheme.

The MATI requirement is around 0.89 s when the delay is 10 ms for the TDMA scheme. However, as the MAD constraint is increased to 0.5 s, the MATI requirement becomes around 0.5 s, beyond which the control systems are unstable, as shown in Figure 4b. Increasing the packer delay significantly degrades the stability region. Note that the retransmission of old data to maximize the reliability increases the delay and is generally not useful for control applications [7]. As we increase the MATI requirement, the WNCS becomes more robust since it allows more number of packet losses. Hence, it is great to minimize the time delay between packet generation instance and packet transmission to maximize the MATI requirement and to simplify the protocol operation. This is the main reason of the actual packet transmission right after the packet generation in Section 3.2. Hence, the transmission scheduling policy controls both the packet generation instance and the actual packet transmission in this paper.

5. Optimization Problem Formulation

A transmission scheduling policy is an essential component to meet the MATI requirement of node i for a given network setup $(N, \bar{h}_i, p_i)$. Our objective was to design low complexity scheduling schemes to optimize the TI performance with respect to the heterogeneous MATI requirements of each node. In this section, we first introduce the performance metric called extended transmission interval (ETI) of the network. Then, we formulate a constrained optimization problem to optimize the ETI performance.

5.1. Extended Transmission Interval

Since the exact definition of TI is based on the discrete random event of packet receptions, it is not an efficient performance metric to use for the transmission scheduling policy. Hence, we introduced a continuous version of the TI metric, called the ETI metric. The extended transmission interval describes how old the information is from the controller perspective.

Figure 5 illustrates the evolution of ETI for a given sequence of packet deliveries of node i. Let $\tau_i(k)$ denote the positive integer to represent the ETI value of node i at slot k. We reset $\tau_i(k+1) = 1$ if the controller receives a packet from node i at slot k. As a reminder, the received packet was generated at the beginning of slot k. However, if the controller does not receive a packet, then we increase $\tau_i(k+1) = \tau_i(k) + 1$. Hence, the iterative update of $\tau_i(k)$ is

$$\tau_i(k+1) = \begin{cases} 1 & \text{if } d_i(k) = 1 \\ \tau_i(k) + 1 & \text{otherwise} \end{cases} \tag{3}$$

Note that the ETI linearly increases when there is no packet reception, as shown in Figure 5. The ETI constraint of each link is

$$\lim_{K \to \infty} \frac{1}{K} \sum_{k=1}^{K} \mathbb{E}[\tau_i(k)] \leq \bar{h}_i \quad \forall i \in \{1, \ldots, N\} \tag{4}$$

Since we are interested on the robust ETI performance with respect to the MATI requirement of each node, we define the ETI slack of node i as $\bar{h}_i - \tau_i(k)$ at slot k. Hence, the expected value of the ETI slack of node i is

$$\lim_{K\to\infty} \frac{1}{K}\mathbb{E}\left[\sum_{k=1}^{K} \bar{h}_i - \tau_i(k)\right] \quad \forall i \in \{1,\ldots,N\} \tag{5}$$

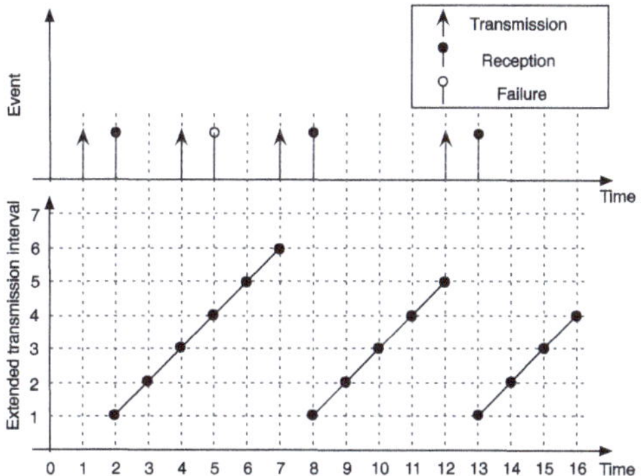

Figure 5. Illustration of a typical evolution of ETI.

5.2. Optimization Problem

By considering the ETI metric with the robustness criterion, our objective is to maximize the minimum ETI slack of the network. We formulate the constrained optimization problem

$$\min \quad \eta \tag{6a}$$

$$\text{s.t.} \quad \lim_{K\to\infty} \frac{1}{K}\mathbb{E}\left[\sum_{k=1}^{K} \tau_i(k) - \bar{h}_i\right] \leq \eta \quad \forall i \in \{1,\ldots,N\} \tag{6b}$$

$$\sum_{i=1}^{N} u_i(k) \leq 1 \; \forall k \in \mathbb{Z}^+ \tag{6c}$$

Equations (6b) and (6c) present the minimum ETI slack value constraint and the schedulability constraint, respectively. The optimal solution of the optimization problem assigns more network resources as $\bar{h}_i$ decreases, i.e., more network resources for faster control systems.

In the following section, we propose three low-complex scheduling schemes based on the optimization problem in Equation (6). Three transmission scheduling policies are centralized random access scheme, distributed random access scheme, and centralized Lyapunov-based scheduling scheme. The first two approaches are randomized methods and the third one is a deterministic approach in order to assign the slot resources to all requirement sets $\bar{h}_i, \forall i \in \{1,\ldots,N\}$.

6. Centralized Random Access Scheme

In this section, we propose the ETI optimization problem for the stationary randomized scheduling scheme in a centralized manner. The controller selects node i with probability $\alpha_i \in (0,1]$ in each time slot. Hence, the randomized scheduling scheme is the vector of scheduling probabilities $\boldsymbol{\alpha} = (\alpha_1,\ldots,\alpha_N)$. They select nodes randomly based on the fixed scheduling probabilities $\boldsymbol{\alpha}$.

Since we consider the stationary random scheduling, we first derive the expected behavior of ETI in the following proposition.

Proposition 1. *The long-term expected ETI of node i is*

$$\lim_{K\to\infty}\frac{1}{K}\sum_{k=1}^{K}\mathbb{E}[\tau_i(k)]=\frac{1}{p_i\alpha_i} \tag{7}$$

Proof. The proof of Proposition 1 is provided in Appendix A. □

By using Proposition 1 and the schedulability constraint of $\boldsymbol{\alpha}$, the optimization problem in Equation (6) is reformulated as

$$\min_{\boldsymbol{\alpha},\eta} \quad \eta \tag{8a}$$

$$\text{s.t.} \quad \frac{1}{p_i\alpha_i}-\bar{h}_i\leq\eta \quad \forall i\in\{1,\ldots,N\} \tag{8b}$$

$$\sum_{i=1}^{N}\alpha_i\leq 1 \tag{8c}$$

Note that Equation (8c) represents the schedulability constraint.

Optimal Solution

Since the proposed optimization problem in Equation (8) is a convex problem, we obtain the optimal stationary scheduling probabilities $\boldsymbol{\alpha}^*$ by analyzing the KKT conditions of the problem. The Lagrange function of the problem is

$$L(\boldsymbol{\alpha},\eta,\boldsymbol{\lambda},\gamma)=\eta+\sum_{i=1}^{N}\lambda_i\left(\frac{1}{p_i\alpha_i}-\bar{h}_i-\eta\right)+\gamma\left(\sum_{i=1}^{N}\alpha_i-1\right) \tag{9}$$

where $\boldsymbol{\lambda}=(\lambda_1,\ldots,\lambda_N),\boldsymbol{\lambda}\geq 0$ and γ are the Lagrange multipliers due to Equations (8b) and (8c), respectively.

The KKT conditions are

- Stationarity with α_i

$$\nabla_{\alpha_i}L(\eta,\alpha_i,\lambda_i,\gamma)=0 \tag{10}$$

- Stationarity with η

$$\nabla_{\eta}L(\eta,\alpha_i,\lambda_i,\gamma)=0 \tag{11}$$

- Complementary slackness of Equation (8b)

$$\lambda_i\left(\frac{1}{p_i\alpha_i}-\bar{h}_i-\eta\right)=0 \tag{12}$$

- Complementary slackness of Equation (8c)

$$\gamma\left(\sum_{i=1}^{N}\alpha_i-1\right)=0 \tag{13}$$

- Primal feasibility

$$\frac{1}{p_i \alpha_i} - \bar{h}_i - \eta \leq 0 \tag{14}$$

$$\sum_{i=1}^{N} \alpha_i \leq 1 \tag{15}$$

- Dual feasibility

$$\lambda_i \geq 0, \gamma \geq 0 \tag{16}$$

The first stationarity condition, $\nabla_{\alpha_i} L(\alpha_i, \lambda_i, \gamma) = 0$, gives

$$\frac{\lambda_i}{p_i \alpha_i^2} = \gamma \tag{17}$$

from the partial derivation. With a similar method, the second stationarity condition, $\nabla_{\eta} L(\eta, \alpha_i, \lambda_i, \gamma) = 0$, gives

$$\sum_{i=1}^{N} \lambda_i = 1\,. \tag{18}$$

We obtain that either $\gamma = 0$ or $\sum_{i=1}^{N} \alpha_i = 1$ from the complementary slackness of Equation (13). However, Equation (17) implies $\gamma > 0$ since the value of γ is zero if $\lambda_i = 0$ or $\alpha_i \to \infty$, which conflicts the conditions of Equation (18) or $\alpha_i \in (0, 1]$, respectively. Hence, we obtain

$$\sum_{i=1}^{N} \alpha_i = 1 \tag{19}$$

since $\gamma > 0$. We separate node i into two groups, namely $\lambda_i = 0$ and $\lambda_i > 0$, based on the dual feasibility $\lambda_i \geq 0$.

If node i has $\lambda_i = 0$, then $\alpha_i = 0$ from Equation (17) since $\gamma > 0$. On the other hand, if node i has $\lambda_i > 0$, then

$$\alpha_i = \frac{1}{(\bar{h}_i + \eta) p_i} \tag{20}$$

due to the complementary slackness of Equation (12). For any fixed value of $\gamma > 0$, the randomized scheduling probability of node i is given by Equation (20).

Our objective is to find the optimal values to meet the constraints of Equations (19) and (20). In Equation (20), η is only unknown variable to compute the scheduling probability. Hence, we first find the optimal value of η^* satisfying Equations (19) and (20). Now, let us derive the boundaries of η. The minimum value of η is $1/p_i - \bar{h}_i$ if $\alpha_i = 1$ from Equation (20). Hence, the feasible range of η is

$$\frac{1}{p_i} - \bar{h}_i \leq \eta \leq 0\,. \tag{21}$$

From Equation (20), a decreasing value of η, the probability α_i increases. As η converges to the lower bound, $\eta \to 1/p_i - \bar{h}_i$, Equation (19) becomes $\sum_{i=1}^{N} \alpha_i \geq 1$. On the other hand, as $\eta \to 0$, then $\sum_{i=1}^{N} \alpha_i = \sum_{i=1}^{N} \frac{1}{\bar{h}_i p_i} \ll 1$ due to Equation (20). Hence, there is a unique value of η^* to meet Equation (19) due to the monotonicity of α_i with respect to η. By gradually increasing η from $1/p_i - \bar{h}_i$ and adjusting α_i using Equation (20), we find the unique set of $\boldsymbol{\alpha}^*$ and η^* that satisfies Equations (19) and (20).

Next, we compute the optimal set of $\boldsymbol{\lambda}^*$ and γ^*. With a similar method, the boundaries of γ are

$$0 \leq \gamma \leq \bar{\gamma} = \max_{i \in \{1,\ldots,N\}} \frac{1}{p_i \alpha_i^2} \tag{22}$$

The upper bound of γ is derived from Equations (17) and (19). By gradually decreasing γ from $\bar{\gamma}$, we adjust $\lambda_i = \gamma p_i \alpha_i^2$ from Equation (17). We then obtain the optimal set of $\boldsymbol{\lambda}^*$ and γ^* for Equation (18). Note that the unique vector $(\boldsymbol{\alpha}^*, \eta^*, \boldsymbol{\lambda}^*, \gamma^*)$ fulfills the KKT conditions.

7. Distributed Random Access Scheme

In this section, we present a distributed random access where each node decides its transmission probability so that the minimum ETI slack of the network is maximized. Let us assume that each node i transmits a packet with probability β_i in each slot. As a reminder, the set of i's interfering links is C_i. Let $\boldsymbol{\beta} = (\beta_1, \ldots, \beta_N)$ be the vector of transmission probabilities of all nodes. Hence, a transmission of node $i \in \{1, \ldots, N\}$ is successful if and only if no node in C_i transmits during the transmission of node i. Hence, the successful transmission probability of node i is

$$p_i \beta_i \prod_{j \in C_i} (1 - \beta_j) \tag{23}$$

By applying Equation (23) to Proposition 1, the long-term expected ETI of node i is

$$\lim_{K \to \infty} \frac{1}{K} \sum_{k=1}^{K} \mathbb{E}[\tau_i(k)] = \frac{1}{p_i \beta_i \prod_{j \in C_i} (1 - \beta_j)} \tag{24}$$

After substituting Equation (24) into Equation (4) of the ETI constraint, we obtain

$$\frac{1}{p_i \bar{h}_i \beta_i \prod_{j \in C_i} (1 - \beta_j)} \leq 1 . \tag{25}$$

Motivating by the network utility problem [39,40], we formulate a constrained optimization problem to optimize the transmission probability of nodes. Our objective is to maximize the minimum ETI slack with respect to $\bar{h}_i$ of the network. After some manipulations of Equation (6), we propose the max-min robust optimization problem

$$\max_{\boldsymbol{\beta}, \psi} \quad \psi \tag{26a}$$

$$\text{s.t.} \quad a_i \beta_i \prod_{j \in C_i} (1 - \beta_j) \geq \psi \quad \forall i \in \{1, \ldots, N\} \tag{26b}$$

where $a_i = p_i \bar{h}_i$. Notice that $\psi > 1$ is necessary for the feasibility of ETI constraint of Equation (4). Different fairness notions corresponding to different utility functions are discussed in [39,40].

Unfortunately, the proposed max-min robust problem in Equation (26) is a non-convex problem. Next, we convert the problem in Equation (26) to a convex optimization problem in order to design the distributed random access scheme.

Proposition 2. *The max-min robust problem in Equation (26) is equivalent to the following convex programming problem*

$$\min \quad \frac{1}{2}\sum_{i=1}^{N} \nu_i^2 \tag{27a}$$

$$s.t. \quad \nu_i \leq \log a_i + \log \beta_i + \sum_{j\in\mathcal{C}_i} \log(1-\beta_j) \tag{27b}$$

$$\nu_i = \nu_j \quad \forall i \in \{1,\ldots,N\}, \forall j \in \mathcal{C}_i \tag{27c}$$

Proof. The proof of Proposition 2 is provided in Appendix B. □

Optimal Solution

A distributed algorithm can effectively obtain the optimal access probability of the previous convex optimization problem in Equation (27). In this section, we show how globally optimal access probability is obtained in a distributed manner. When we replace the equality constraints $\nu_i = \nu_j$ of Equation (27) by two inequality constraints, $\nu_i \leq \nu_j$ and $\nu_i \geq \nu_j$, the Lagrange function of the problem is

$$L(\boldsymbol{\beta},\boldsymbol{\nu},\boldsymbol{\mu},\boldsymbol{\omega}) = \frac{1}{2}\sum_{i\in N}\nu_i^2 + \sum_{i\in N}\mu_i\left(\nu_i - \log a_i - \log\beta_i - \sum_{j\in\mathcal{C}_i}\log(1-\beta_j)\right) + \sum_{i\in\{1,\ldots,N\}}\sum_{j\in\mathcal{C}_j}\omega_{i,j}(\nu_i-\nu_j) \tag{28}$$

where $\boldsymbol{\mu} = (\mu_1,\ldots,\mu_N)$ and $\boldsymbol{\omega} = (\omega_{i,j}, i \in \{1,\ldots,N\}, j \in \mathcal{C}_i)$ are the Lagrange multipliers and $\boldsymbol{\nu} = (\nu_1,\ldots,\nu_N)$.

Our basic idea is to apply the gradient project method for the dual problem $\max_{\boldsymbol{\beta},\boldsymbol{\nu}} D(\boldsymbol{\beta},\boldsymbol{\nu})$ where the dual function is $D(\boldsymbol{\beta},\boldsymbol{\nu}) = \min_{\boldsymbol{\beta},\boldsymbol{\nu}} L(\boldsymbol{\beta},\boldsymbol{\nu},\boldsymbol{\mu},\boldsymbol{\omega})$ [41]. By considering the stationarity condition of the Lagrange function, they give $\frac{\partial L}{\partial \beta_i} = 0$

$$\beta_i = \frac{\mu_i}{\mu_i + \sum_{j\in\mathcal{C}_i}\mu_j} \tag{29}$$

and $\frac{\partial L}{\partial \nu_i} = \nu_i + \mu_i + \sum_{j\in\mathcal{C}_i}\omega_{i,j} - \omega_{j,i} = 0$

$$u_i = \left[-\lambda_i - \sum_{j\in L_i}(v_{i,j} - v_{j,i})\right]^- \tag{30}$$

where $[z]^- = \min(z,0)$. Note that β_i satisfies the constraints $0 \leq \beta_i \leq 1$. The Lagrange multipliers of the gradient project method are adjusted in the direction of the gradient $\nabla D(\boldsymbol{\beta},\boldsymbol{\nu})$:

$$\mu_i(n+1) = \left[\mu_i(n) + \epsilon_n \frac{\partial D}{\partial \mu_i}\right]^+ \tag{31}$$

$$\omega_{i,j}(n+1) = \left[\omega_{i,j}(n) + \epsilon_n \frac{\partial D}{\partial \omega_{i,j}}\right]^+ \tag{32}$$

Here, $\epsilon_n > 0$ is the step size at the nth iteration, and $[z]^+ = \max(z,0)$. The gradient are $\frac{\partial D}{\partial \mu_i} = \nu_i - \log a_i - \log\beta_i - \sum_{j\in\mathcal{C}_i}\log(1-\beta_j)$ and $\frac{\partial D}{\partial \omega_{i,j}} = \nu_i - \nu_j$.

8. Centralized Lyapunov-Based Scheduling Scheme

The Lyapunov optimization theory is extensively applied to the general communication and queueing systems [42]. Using Lyapunov optimization techniques, we derive the centralized scheduling scheme for the ETI optimization problem in Equation (6). The Lyapunov-based scheduling scheme uses the feedback ETI state of nodes in order to reduce the value of the Lyapunov function. We define the Lyapunov function to give a large positive scalar when nodes have high ETI with respect to the MATI $\bar{h}_i$. Hence, the Lyapunov-based scheduling approach basically tries to minimize the growth of its function.

Next, we define the fundamental components of the Lyapunov-based scheduling scheme, namely Lyapunov function with notions of the ETI debt. Let us denote $x_i(k)$ as the ETI debt of node i at slot k. The ETI debt of node i is

$$x_i(k) = \tau_i(k) - \bar{h}_i \tag{33}$$

where $\bar{h}_i$ is the MATI value and $\tau_i(k)$ is the ETI at slot k.

When the scheduling scheme does not meet the ETI requirement $\bar{h}_i$, then the value of the ETI debt is positive. We define the positive part of the ETI debt, $x_i^+(k) = \max(x_i(k), 0)$. As increasing debt $x_i^+(k)$ indicates to the scheduling scheme that node i needs more transmission slots to meet the MATI requirement.

Let $\mathbf{S}_k = (x_1(k), \ldots, x_N(k))$ be a vector of network ETI dept states at slot k. Then, we define the Lyapunov Function by

$$V(\mathbf{S}_k) = \frac{1}{2}\sum_{i=1}^{N}\left[(x_i(k))^2 + G(x_i^+(k))^2\right] \tag{34}$$

where G is a large positive value to emphasize the ETI constraints. Remark that $V(\mathbf{S}_k)$ is large when nodes have high ETI with respect to the requirement or positive ETI debt.

Optimal Solution

To minimize the Lyapunov function, we introduce the Lyapunov drift

$$\Delta(\mathbf{S}_k) = \mathbb{E}\left[V(\mathbf{S}_{k+1}) - V(\mathbf{S}_k)|\mathbf{S}_k\right] \tag{35}$$

Note that the Lyapunov drift measures the change of the Lyapunov function over time slots. The Lyapunov-based scheduling scheme minimizes $V(\mathbf{S}_k)$ by reducing $\Delta(\mathbf{S}_k)$ in every slot k.

The high computation complexity of $V(\mathbf{S}_k)$ prevents us from deriving the exact form of $\Delta(\mathbf{S}_k)$. Hence, we derive the upper bound of $\Delta(\mathbf{S}_k)$ and apply it for the actual transmission scheduling scheme.

Proposition 3. *The upper bound of $\Delta(\mathbf{S}_k)$ is*

$$\Delta(\mathbf{S}_k) \leq -\sum_{i=1}^{N}\mathbb{E}[u_i(k)|\mathbf{S}_k]P_i(k) + Q(k) \tag{36}$$

where $P_i(k)$ and $Q(k)$ are given by

$$P_i(k) = p_i\tau_i(k)[x_i(k) + 1 + G(x_i^+(k) + 1)] \tag{37}$$

$$Q(k) = \sum_{i=1}^{N}\left[x_i(k) + \frac{\tau_i^2(k)}{2} + G\left(x_i^+(k) + \frac{\tau_i^2(k)}{2}\right)\right] \tag{38}$$

Proof. The proof of Proposition 3 is provided in Appendix C. □

We observe that both $P_i(k)$ and $Q(k)$ are functions of the network state $\mathbf{S}_k$ and setup parameters $(N, \bar{h}_i, p_i)$. However, $Q(k)$ of Equation (38) is not dependent on the scheduling decision $u_i(k)$. Hence, in each slot k, the centralized Lyapunov-based scheduling scheme selects the node with maximum value of $P_i(k)$ to minimize the upper bound of $\Delta(\mathbf{S}_k)$.

9. Performance Evaluation

In this section, we evaluate the performance of four transmission scheduling policies, namely ideal scheduling scheme, centralized random access scheme, distributed random access scheme, and centralized Lyapunov-based scheduling scheme. The ideal scheduling scheme optimizes the transmission schedule based on the entire packet loss sequences of each link. Hence, it gives the fundamental performance bounds even though it is not feasible to implement in practice. We considered a star network topology where N sensor nodes contend to send data packets to the controller. Each node has different sets of $\bar{h}_i$ and p_i. Note that each link i of the centralized scheduling approach has equal p_i for both uplink and downlink. Let us denote the maximum MATI and minimum MATI of the network as $h_{\max}$ and $h_{\min}$, respectively. With a similar method, we define the maximum link reliability and minimum link reliability of the network $p_{\max}$ and $p_{\min}$, respectively. Then, the MATI and link reliability of node i are

$$\bar{h}_i = \frac{h_{\max} - h_{\min}}{N-1}(i-1) + h_{\min} \quad \forall i \in \{1, \ldots, N\}, \tag{39}$$

$$p_i = \frac{p_{\max} - p_{\min}}{N-1}(i-1) + p_{\min} \quad \forall i \in \{1, \ldots, N\}. \tag{40}$$

The lower are the values of $\bar{h}_i$ and p_i, the more challenging are the constraints. Hence, the lower node ID has more strict MATI requirement with worse link reliability. Each simulation ran $K = N \times 10^8$ time slots.

Both centralized schemes of static random access and Lyapunov-based schedule need the resource allocation message to assign the time slot to nodes. We assumed that both centralized schemes use one additional slot to transmit the resource allocation message. The centralized solutions consume two slots for the single data transmission while the distributed random access only requires one slot for the data transmission. Since the packet delay is fixed to 1 time slot, we mainly analyzed the performance of TI of different schemes.

Figure 6 shows the cumulative density function (CDF) of slack of ideal solution, centralized random access, distributed random access, and Lyapunov-based approach with link reliability $p_{\min} = 0.9, p_{\max} = 1$, MATI requirement $h_{\min} = 70, h_{\max} = 100$, and number of nodes $N = 8$. The slack is the difference between MATI and TI measurements. As a reminder, the minimum slack of TI with respect to MATI is the objective value of the proposed optimization problem. The higher is the slack, the better is the robustness. Hence, a lower CDF is better than a higher CDF for the stability guarantee. The Lyapunov-based scheduling scheme generally gives a lower CDF than other CDFs. However, our objective was to maximize the worst slack of the network. In fact, the worst slack value of the ideal solution is greater than the one of the Lyapunov-based approach in Figure 6. Hence, the ideal solution still provides the optimal solution of our proposed optimization problem in Equation (6) using the perfect knowledge of the packet loss sequences. The Lyapunov-based approach efficiently improves the robust performance based on the feedback slack information between TI and MATI of heterogeneous multiple control systems.

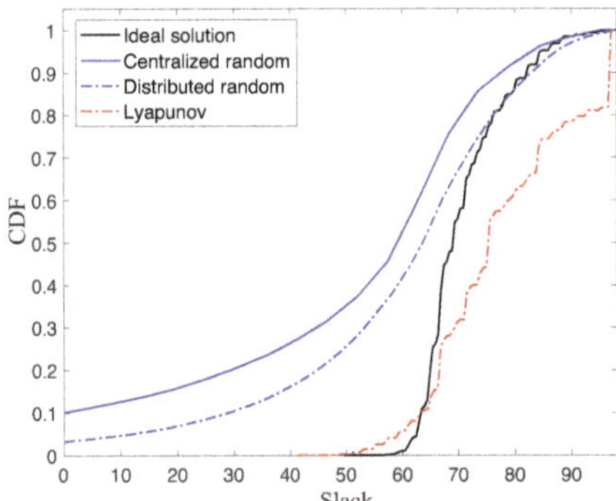

Figure 6. CDF of slack of ideal solution, centralized random access, distributed random access, and Lyapunov-based approach with $N = 8$.

On the other hand, there is a significant gap between ideal solution with two randomized scheduling approaches, namely centralized random access and distributed random access. Note that the centralized random access does not rely on any feedback information of TI. The interesting observation is that the distributed random access generally performs better than the centralized one for $N = 8$. Even though the distributed random access may incur the collisions with other transmissions, each data transmission requires only one time slot. While the centralized random access has no contention with other nodes, each data transmission consumes two time slots due to the resource allocation message. Since the collision probability is low for the small number of nodes, the distributed random access gives better robustness, as shown in Figure 6.

Another interesting point is the different percentile values between four schemes. While the four schemes have similar 50th percentile values, the percentile difference between different approaches increases as the percentage decreases. It means the worst case performance of the proposed approaches is significantly different.

Figure 7 shows the minimum slack, average TI, and outage probability of four different schemes as a function of different number of nodes $N = 3, \ldots, 30$ with link reliability $p_{\min} = 0.9, p_{\max} = 1$, and MATI requirement $h_{\min} = 50, h_{\max} = 100$. The outage probability is defined as the probability that the TI value is greater than MATI, $\bar{h}_i$.

Let us first compare the performance between ideal solution and Lyapunov-based approach. In Figure 7a,c, the ideal solution provides the lower outage probability and higher minimum slack than the one of the Lyapunov-based approach. However, the average TI of the ideal solution is slightly higher than the one of Lyapunov-based approach in Figure 7b. As a reminder, the objective of the proposed optimization problem is to maximize the minimum TI slack with respect to MATI requirements instead of minimizing average TI value of the network. The average TI is not explicitly considered in the optimization problem.

While the Lyapunov-based approach is comparable with the ideal solution, both random accesses show significantly different behaviors. In Figure 7, the centralized random access has an almost constant gap of the minimum slack, average TI, and outage probability with the Lyapunov-based approach over different number of nodes. On the other hand, the distributed random access significantly degrades these performance metrics as increasing the number of nodes. By comparing two random accesses, the distributed random approach provides better performance than the centralized random access for the small number of nodes $N \leq 9$. There is the fundamental stability limits of the distributed random access approach due to the increasing collision probability dependent on the number of nodes [38].

Figure 8 shows the MATI requirement and the average TI of each node using different transmission scheduling schemes with $N = 9$. As a reminder, the MATI requirement and link reliability become more strict as decreasing node ID of the network. We observe that the average TI of the ideal solution is decreasing as the MATI requirement becomes more strict for the lower node ID. However, the Lyapunov-based approach is quite flat over different MATI requirements with respect to the ideal solution. The node ID 1 provides the minimum slack of both ideal solution and Lyapunov-based approach of the network. This is the main reason of the greater minimum slack of the ideal solution at the cost of the higher average TI compared to the Lyapunov-based approach in Figure 7.

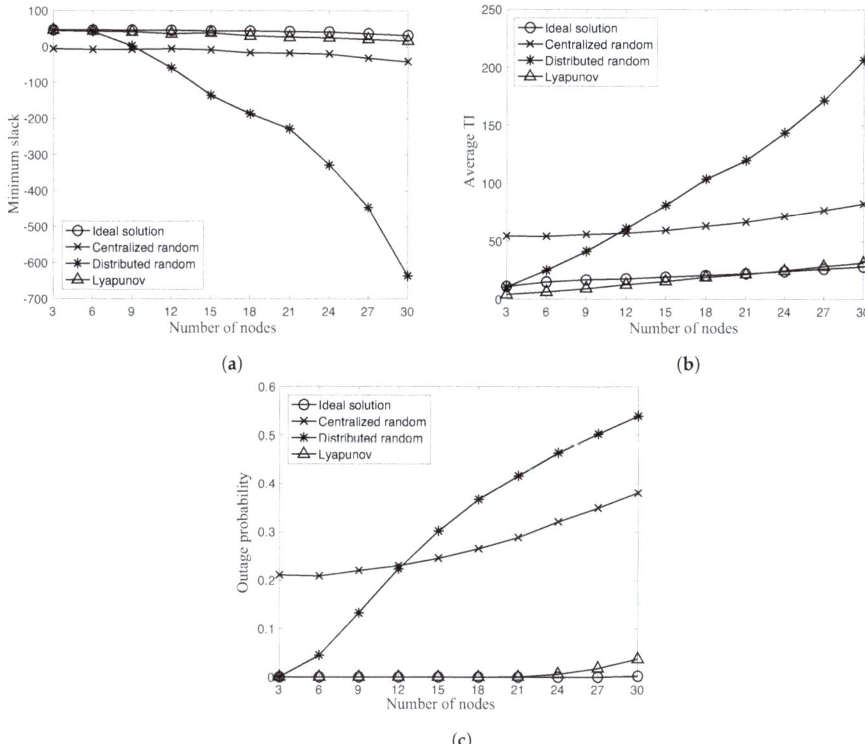

Figure 7. Minimum slack, average TI, and outage probability of ideal solution, centralized random access, distributed random access, and Lyapunov-based approach as a function of different number of nodes $N = 3, \ldots, 30$: (**a**) minimum slack vs. number of nodes; (**b**) average TI vs. number of nodes; and (**c**) outage probability vs. number of nodes.

On the other hand, both random accesses significantly increase the average TI as the MATI requirement becomes relaxed for the higher node ID. In both random accesses, the slack between MATI and average TI is decreasing as the node ID increases. In fact, the minimum slack value of both random accesses occurs at node ID 9.

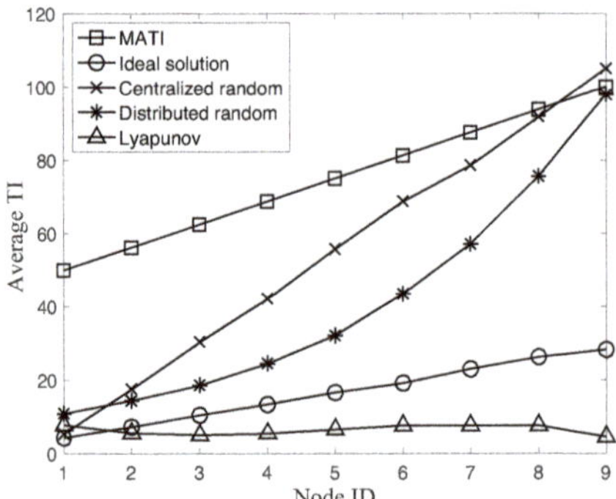

Figure 8. MATI requirement and average TI of each node using different transmission scheduling schemes with $N = 9$.

While the centralized random access supports the stationary optimal resource allocation, its operation requires the additional slots for the resource allocation message. In fact, each successful data transmission of the centralized approaches needs both successful transmissions of the resource allocation message and data packet. Hence, the performance of the random access is a complex function of number of nodes, link reliability, and operational overhead. Figure 9 compares the channel access probabilities of each node using both centralized random access and distributed random access with $N = 9$. Both solutions of random accesses assign the higher access probability to the higher priority node ID, i.e., lower node ID. While the channel access probability of distributed random access is mush smoother for different nodes, the centralized random access approach sets the very high channel access probability 0.58 for the most demanded node ID 1. In fact, both random accesses show the over-allocation of slot resources to the most demanded node as shown in Figure 8.

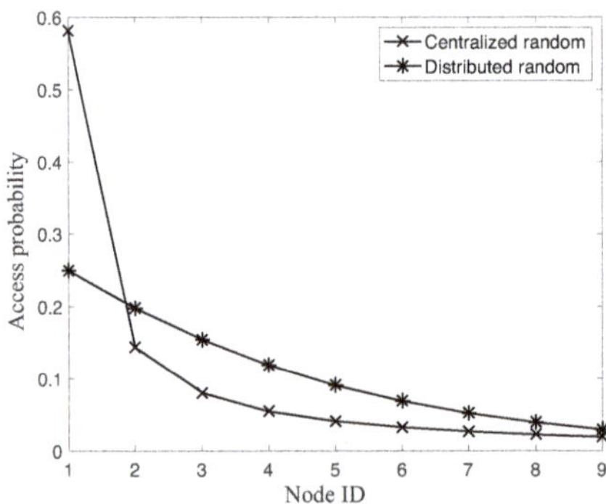

Figure 9. Access probability of each node using centralized random access and distributed random access schemes with $N = 9$.

10. Conclusions

In this paper, we consider the transmission scheduling schemes of industrial wireless sensors for heterogeneous multiple control systems over unreliable wireless channels. We first discuss the fundamental tradeoffs of the TI and the packet delay of wireless networks for the control stability. Based on the fundamental observation, we formulate the constrained optimization problem of maximizing the minimum slack of the TI with respect to the maximum allowable requirement of all network nodes. We propose three low-complex transmission scheduling schemes, namely centralized stationary random access, distributed random access, and centralized Lyapunov-based scheduling scheme, to solve the proposed optimization problem. The simulation results show that the centralized

Lyapunov-based scheduling approach provides robust performance closer that is to the ideal solution by using the feedback state information. Furthermore, the distributed random access is another good candidate of the transmission scheduling for the small number of control loops.

The practical validation of the proposed scheduling scheme is critical in the context of industrial setup [3]. Future investigations include the practical implementation of different scheduling schemes using Zolertia sensors [43] based on the specifications of the IEEE 802.15.4 standard. Furthermore, we are planning to extend the proposed framework to balance the control cost and the energy efficiency while considering the additional constraints using the energy harvesting techniques [44].

Author Contributions: B.P., I.-J.Y., and P.P. conceived the main idea and the network model; J.N., J.-Y.C., and P.P. contributed to data analysis and simulation; all authors contributed to write the paper.

Funding: The work was supported by the Basic Research Laboratory (BRL) of the National Research Foundation (NRF-2017R1A4A1015744) and the NRF-2016R1C1B1008049 funded by the Korean government.

Conflicts of Interest: The authors declare no conflict of interest.

Appendix A. Proof of Proposition 1

Proof. In each slot k, the controller successfully receives a packet if the node is scheduled with probability α_i and the link condition is good with probability p_i. Hence, the probability of successful transmission of node i is $p_i\alpha_i$. Let us denote h_i as the TI of successful received packet of node i. Then, the sum of ETI is

$$\sum_{k=1}^{K} \tau_i(k) = 1 + 2 + \ldots + h_i = \frac{h_i^2 + h_i}{2}$$

Note that h_i follows $\mathbb{P}[h_i = j] = p_i\alpha_i(1 - p_i\alpha_i)^{j-1}, \forall j \in \mathbb{Z}^+$ since it is independent and identically distributed. By using renewal theory, the expected ETI of node i becomes

$$\frac{1}{K}\sum_{k=1}^{K} \mathbb{E}[\tau_i(k)] = \frac{1}{\mathbb{E}[h_i]}\left(\frac{\mathbb{E}[h_i^2] + \mathbb{E}[h_i]}{2}\right) \quad \text{(A1)}$$

Hence, the long-term expected ETI of node i is

$$\lim_{K\to\infty} \frac{1}{K}\sum_{k=1}^{K} \mathbb{E}[\tau_i(k)] = \frac{\mathbb{E}[h_i^2]}{2\mathbb{E}[h_i]} + \frac{1}{2} = \frac{1}{p_i\alpha_i} \quad \text{(A2)}$$

due to its generalization for renewal-reward processes [45]. □

Appendix B. Proof of Proposition 2

Proof. Since the logarithmic function is strictly increasing, each ETI constraint of Equation (26b) is equivalently rewritten as $\nu \leq \log a_i + \log \beta_i + \sum_{j\in\mathcal{C}_i} \log(1 - \beta_j), \forall i \in \{1, \ldots, N\}$ where $\nu = \log \psi$. Hence, the constraint becomes

$$\nu - \log a_i - \log \beta_i - \sum_{j\in\mathcal{C}_i} \log(1 - \beta_j) \leq 0\,. \quad \text{(A3)}$$

Note that each constraint of Equation (A3) shows a convex set of $(\boldsymbol{\beta}, \nu)$ with the monotonically increasing logarithmic function. Hence, the optimization problem of Equation (26) is reformulated as the following convex problem

$$\max \quad \nu \quad \text{(A4a)}$$

$$\text{s.t.} \quad \nu \leq \log a_i + \log \beta_i + \sum_{j\in\mathcal{C}_i} \log(1 - \beta_j)\,. \quad \text{(A4b)}$$

Since we assume the fully connected graph of the single hop network, all v_i must have equal value, namely $v_i = v_j$ for all $i \in \{1,\ldots,N\}$ and $j \in \mathcal{C}_i$. By assuming $a_i\beta_i \prod_{j\in\mathcal{C}_i}(1-\beta_j) \leq 1$, we have $v \leq 0$. Hence, maximizing v is equivalent to minimize v^2 in the optimization problem. Therefore, it is possible to rewrite the non-convex optimization problem in Equation (26) to the convex problem in Equation (27). □

Appendix C. Proof of Proposition 3

Proof. In this appendix, we obtain the upper bound of the Lyapunov drift $\Delta(\mathbf{S}_k)$ of Equation (36). Let us consider the network state $\mathbf{S}_k$, the Lyapunov function $V(\mathbf{S}_k)$ in Equation (34), and the Lyapunov drift $\Delta(\mathbf{S}_k)$ in Equation (35). After plugging Equation (34) into Equation (35), we obtain

$$\Delta(\mathbf{S}_k) = \frac{1}{2}\sum_i^N \mathbb{E}[x_i^2(k+1) - x_i^2(k)|\mathbf{S}_k] + \frac{G}{2}\sum_i^N \mathbb{E}[(x_i^+(k+1))^2 - (x_i^+(k))^2|\mathbf{S}_k] \tag{A5}$$

In Equation (A5), we derive the expressions of $(x_i^+(k+1))^2 - (x_i^+(k))^2$ and $x_i^2(k+1) - x_i^2(k)$. The ETI update of Equation (3) is rewritten as

$$\tau_i(k+1) = \tau_i(k)(1-d_i(k)) + 1 \tag{A6}$$

By combining Equations (33) and (A6), the iteration of the ETI debt is

$$x_i(k+1) = x_i(k) - \tau_i(k)d_i(k) + 1, \tag{A7}$$

where $x_i(1) = 0$. Hence, $(x_i^+(k+1))^2$ becomes

$$(x_i^+(k+1))^2 = [\max(x_i(k) - \tau_i(k)d_i(k) + 1, 0)]^2 \tag{A8}$$

$$\leq [\max(x_i^+(k), 0)]^2 \tag{A9}$$

$$\leq (x_i^+(k))^2 \tag{A10}$$

Equation (A10) is rewritten as

$$(x_i^+(k+1))^2 - (x_i^+(k))^2 \leq -2\tau_i(k)(x_i^+(k)+1)d_i(k) + 2x_i^+(k) + \tau_i^2(k)\,. \tag{A11}$$

By taking the expectation of Equation (A11), we compute the upper bound

$$\mathbb{E}[(x_i^+(k+1))^2 - (x_i^+(k))^2|\mathbf{S}_k] \leq -2\tau_i(k)(x_i^+(k)+1)p_i\mathbb{E}(u_i(k)|\mathbf{S}_k) + 2x_i^+(k) + \tau_i^2(k) \tag{A12}$$

With a similar method, we get

$$\mathbb{E}[(x_i(k+1))^2 - (x_i(k))^2|\mathbf{S}_k] \leq -2\tau_i(k)(x_i(k)+1)p_i\mathbb{E}[u_i(k)|\mathbf{S}_k] + 2x_i(k) + \tau_i^2(k) \tag{A13}$$

After substituting Equations (A12) and (A13) into the Lyapunov drift of Equation (A5), we obtain the upper bound of $\Delta(\mathbf{S}_k)$ as Equations (36)–(38). □

References

1. Park, P.; Ergen, S.C.; Fischione, C.; Lu, C.; Johansson, K.H. Wireless network design for control systems: A survey. *IEEE Commun. Surv. Tutor.* **2018**, *20*, 978–1013. [CrossRef]
2. Sztipanovits, J.; Koutsoukos, X.; Karsai, G.; Kottenstette, N.; Antsaklis, P.; Gupta, V.; Goodwine, B.; Baras, J.; Wang, S. Toward a science of cyber-physical system integration. *Proc. IEEE* **2012**, *100*, 29–44. [CrossRef]
3. O'donovan, T.; Brown, J.; Büsching, F.; Cardoso, A.; Cecílio, J.; Furtado, P.; Gil, P.; Jugel, A.; Pöttner, W.-B.; Roedig, U.; et al. The GINSENG System for Wireless Monitoring and Control: Design and Deployment Experiences. *ACM Trans. Sens. Netw.* **2013**, *10*, 1–40. [CrossRef]

4. Cecílio, J.; Furtado, P. Planning for Distributed Control Systems with Wired and Wireless Sensors. In *Wireless Sensors in Industrial Time-Critical Environments*; Springer International Publishing: Cham, Switzerland, 2014.
5. Ramotsoela, D.; Abu-Mahfouz, A.; Hancke, G. A survey of anomaly detection in industrial wireless sensor networks with critical water system infrastructure as a case study. *Sensors* **2018**, *18*, 2491. [CrossRef] [PubMed]
6. Cloosterman, M.B.G.; van de Wouw, N.; Heemels, W.P.M.H.; Nijmeijer, H. Stability of networked control systems with uncertain time-varying delays. *IEEE Trans. Autom. Control* **2009**, *54*, 1575–1580. [CrossRef]
7. Hespanha, J.P.; Naghshtabrizi, P.; Xu, Y. A survey of recent results in networked control systems. *Proc. IEEE* **2007**, *95*, 138–162. [CrossRef]
8. Zhang, W.; Branicky, M.S.; Phillips, S.M. Stability of networked control systems. *IEEE Control Syst.* **2001**, *21*, 84–99.
9. Sinopoli, B.; Schenato, L.; Franceschetti, M.; Poolla, K.; Jordan, M.I.; Sastry, S.S. Kalman filtering with intermittent observations. *IEEE Trans. Autom. Control.* **2004**, *49*, 1453–1464. [CrossRef]
10. Smith, S.C.; Seiler, P. Estimation with lossy measurements: jump estimators for jump systems. *IEEE Trans. Autom. Control* **2003**, *48*, 2163–2171. [CrossRef]
11. Fujioka, H. Stability analysis for a class of networked/embedded control systems: A discrete-time approach. In Proceedings of the American Control Conference (ACC), Seattle, WA, USA, 11–13 June 2008; pp. 4997–5002.
12. Montestruque, L.A.; Antsaklis, P. Stability of model-based net-worked control systems with time-varying transmission times. *IEEE Trans. Autom. Control* **2004**, *49*, 1562–1572. [CrossRef]
13. Hetel, L.; Daafouz, J.; Iung, C. Stabilization of arbitrary switched linear systems with unknown time-varying delays. *IEEE Trans. Autom. Control* **2006**, *51*, 1668–1674. [CrossRef]
14. Gielen, R.; Olaru, S.; Lazar, M.; Heemels, W.; van de Wouw, N.; Niculescu, S.-I. On polytopic inclusions as a modeling framework for systems with time-varying delays. *Automatica* **2010**, *46*, 615–619. [CrossRef]
15. Seiler, P.; Sengupta, R. An H_∞ approach to networked control. *IEEE Trans. Autom. Control* **2005**, *50*, 356–364. [CrossRef]
16. Yue, D.; Han, Q.-L.; Peng, C. State feedback controller design of networked control systems. *IEEE Trans. Circ. Syst. II Express Brief.* **2004**, *51*, 640–644. [CrossRef]
17. Xiong, J.; Lam, J. Stabilization of linear systems over networks with bounded packet loss. *Automatica* **2007**, *43*, 80–87. [CrossRef]
18. Zhang, W.-A.; Yu, L. Modelling and control of networked control systems with both network-induced delay and packet-dropout. *Automatica* **2008**, *44*, 3206–3210. [CrossRef]
19. Heemels, W.P.M.H.; Teel, A.R.; van de Wouw, N.; Nesic, D. Networked control systems with communication constraints: Tradeoffs between transmission intervals, delays and performance. *IEEE Trans. Autom. Control* **2010**, *55*, 1781–1796. [CrossRef]
20. Baldi, M.; Giacomelli, R.; Marchetto, G. Time-driven access and forwarding for industrial wireless multihop networks. *IEEE Trans. Ind. Inform.* **2009**, *5*, 99–112. [CrossRef]
21. Antepli, M.A.; Uysal-Biyikoglu, E.; Erkan, H. Optimal packet scheduling on an energy harvesting broadcast link. *IEEE J. Sel. Areas Commun.* **2011**, *29*, 1721–1731. [CrossRef]
22. Demirel, B.; Zou, Z.; Soldati, P.; Johansson, M. Modular design of jointly optimal controllers and forwarding policies for wireless control. *IEEE Trans. Autom. Control* **2014**, *59*, 3252–3265. [CrossRef]
23. Ergen S.C.; Varaiya, P. TDMA scheduling algorithms for wireless sensor networks. *Wirel. Netw.* **2010**, *16*, 985–997. [CrossRef]
24. Park, P.; Di Marco, P.; Fischione, C.; Johansson, K.H. Modeling and Optimization of the IEEE 802.15.4 Protocol for Reliable and Timely Communications. *IEEE Trans. Parallel Distrib. Syst.* **2013**, *24*, 550–564. [CrossRef]
25. Park, P.; Di Marco, P.; Fischione, C.; Bonivento, A.; Johansson, K.H.; Sangiovanni-Vincent, A. Breath: An Adaptive Protocol for Industrial Control Applications Using Wireless Sensor Networks. *IEEE Trans. Mob. Comput.* **2011**, *10*, 821–838. [CrossRef]
26. Park, P.; Araujo, J.; Johansson, K.H. Wireless Networked Control System Co-Design. In Proceedings of the IEEE International Conference on Networking, Sensing and Control (ICNSC), Delft, The Netherland, 11–13 April 2011; pp. 486–491
27. Sadi, Y.; Ergen, S.C.; Park, P. Minimum energy data transmission for wireless networked control systems. *IEEE Trans. Wirel. Commun.* **2014**, *13*, 2163–2175. [CrossRef]

28. Sadi, Y.; Ergen, S.C. Energy and delay constrained maximum adaptive schedule for wireless networked control systems. *IEEE Trans. Wirel. Commun.* **2015**, *14*, 3738–3751. [CrossRef]
29. Park, P.; Marco, P.D.; Johansson, K.H. Cross-layer optimization for industrial control applications using wireless sensor and actuator mesh networks. *IEEE Trans. Ind. Electron.* **2017**, *64*, 3250–3259. [CrossRef]
30. Ma, Y.; Gong, B.; Sugihara, R.; Gupta, R. Energy-efficient deadline scheduling for heterogeneous systems. *J. Parall. Distrib. Comput.* **2012**, *72*, 1725–1740. [CrossRef]
31. Pang, Z.; Luvisotto, M.; Dzung, D. Wireless High-Performance Communications: The Challenges and Opportunities of a New Target. *IEEE Ind. Electron. Mag.* **2017**, *11*, 20–25. [CrossRef]
32. Tramarin, F.; Vitturi, S.; Luvisotto, M. A Dynamic Rate Selection Algorithm for IEEE 802.11 Industrial Wireless LAN. *IEEE Trans. Ind. Inform.* **2017**, *13*, 846–855. [CrossRef]
33. Donkers, M.C.F.; Heemels, W.P.M.H.; van de Wouw, N.; Hetel, L. Stability analysis of networked control systems using a switched linear systems approach. *IEEE Trans. Autom. Control* **2011**, *56*, 2101–2115. [CrossRef]
34. Lofberg, J. YALMIP: A toolbox for modeling and optimization in MATLAB. In Proceedings of the IEEE International Conference on Robotics and Automation (ICRA), New Orleans, LA, USA, 26 April–1 May 2004; pp. 284–289.
35. Sturm, J.F. Using sedumi 1.02, a matlab toolbox for optimization over symmetric cones. *Optim. Methods Softw.* **1999**, *11*, 625–653. [CrossRef]
36. Nesic, D.; Teel, A.R. Input-output stability properties of networked control systems. *IEEE Trans. Autom. Control* **2004**, *49*, 1650–1667. [CrossRef]
37. Petersen, S.; Carlsen, S. WirelessHART versus ISA100.11a: The format war hits the factory floor. *IEEE Ind. Electron. Mag.* **2011**, *5*, 23–34. [CrossRef]
38. Rom, R.; Sidi, M. Aloha Protocols. In *Multiple Access Protocols: Performance and Analysis*; Springer: Berlin, Germany, 1990.
39. Kelly, F.P.; Maulloo, A.K.; Tan, D.K.H. Rate control for communication networks: Shadow prices, proportional fairness and stability. *J. Oper. Res. Soc.* **1998**, *49*, 237–252. [CrossRef]
40. Srikant, R. *The Mathematics of Internet Congestion Control*; Boston Birkhauser: New York, NY, USA, 2004.
41. Park, P. Power controlled fair access protocol for wireless networked control systems. *Wirel. Netw.* **2015**, *21*, 1499–1516. [CrossRef]
42. Neely, M. Lyapunov Optimization. In *Stochastic Network Optimization with Application to Communication and Queueing Systems*; Morgan & Claypool: San Rafael, CA, USA, 2010.
43. Zolertia RE-Mote Revision B Internet of Things Hardware Development Platform. Available online: https://github.com/Zolertia/Resources/wiki/RE-Mote (accessed on 1 December 2018).
44. Joy, W.; Ross, Y. *Wireless Sensor Networking for the Industrial Internet of Things*; White Paper; Linear Technology: Milpitas, CA, USA, 2015.
45. Gross, D.; Harris, C.M. Advanced Markovian Queueing Models. In *Fundamentals of Queueing Theory*; Wiley: Hoboken, NJ, USA, 1998.

Article

Efficient Resource Scheduling for Multipath Retransmission over Industrial WSAN Systems

Hongchao Wang [1,*], Jian Ma [1], Dong Yang [1] and Mikael Gidlund [2]

1 School of Electronic and Information Engineering, Beijing Jiaotong University, Beijing 100044, China
2 Department of Information Systems and Technology, Mid Sweden University, Sundsvall 85170, Sweden
* Correspondence: hcwang@bjtu.edu.cn; Tel.: +86-10-51688742

Received: 16 July 2019; Accepted: 9 September 2019; Published: 12 September 2019

Abstract: With recent adoption of Wireless Sensor-Actuator Networks (WSANs) in industrial automation, wireless control systems have emerged as a frontier of industrial networks. Hence, it has been shown that existing standards and researches concentrate on the reliability and real-time performance of WSANs. The multipath retransmission scheme with multiple channels is a key approach to guarantee the deterministic wireless communication. However, the efficiency of resource scheduling is seldom considered in applications with diverse data sampling rates. In this paper, we propose an efficient resources scheduling algorithm for multipath retransmission in WSANs. The objective of our algorithm is to improve efficiency and schedulability for the use of slot and channel resources. In detail, the proposed algorithm uses the approaches of CCA (clear channel assessment)-Embedded slot and Multiple sinks with Rate Monotonic scheme (CEM-RM) to decrease the number of collisions. We have simulated and implemented our algorithm in hardware and verified its performance in a real industrial environment. The achieved results show that the proposed algorithm significantly improves the schedulability without trading off reliability and real-time performance.

Keywords: Wireless Sensor and Actuator Networks (WSANs); multipath retransmission; resource scheduling; realtime wireless communication; monitoring and control system

---

1. Introduction

The process automation industry continuously works towards minimizing the cost of production and maintenance while improving the quality of their products. A step towards that goal is to utilize wireless technologies, which offer advantages such as less maintenance, more flexibility, and easy deployment in harsh industrial environments [1,2]. As one of the key technologies, industrial Wireless Sensor-Actuator Networks (WSANs) will play a vital role in the Industry 4.0 framework and Cyber Physical System (CPS) and will also be important for future smart factories and intelligent manufacturing systems [3]. However, harsh conditions, such as complicated environments and metal interference, result in more packet loss in WSAN transmissions compared with traditional wireless sensor networks [4]. Additionally, application-driven industrial communication increases the requirements of stringent reliability as well as time efficiency [5]. Within process automation, standards such as WirelessHART, ISA100.11a, and IEEE (Institute of Electrical and Electronics Engineers) 802.15.4e have been operational for nearly a decade, and there exists several researches that address support for emerging applications, such as data aggregation [6], control, and safety [7,8]. All of the above standards employ the Times-Slotted Channel Hopping (TSCH) Medium Access Control (MAC) protocol instead of pure contention-based MAC protocol Carrier-Sense Multiple Access with Collision Avoidance (CSMA/CA). The TSCH provides the capability of allocating a specific amount of bandwidth per node in a preknown pattern [9,10]. Moreover, channel hopping addresses unreliability caused by multipath fading and narrow-band external interference. However, these issued standards do not give specific link scheduling for efficient data transmission. An obstacle faced in guaranteeing deterministic communication in industrial environments, however, is that the sensor node hardware is usually equipped with an IEEE 802.15.4 compliant half-duplex single radio transceiver [11]. The resources such as slot and channel are restricted, since the transceiver supports only up to 16 different channels [12].

Communication ranges of most sensor devices are limited to short distances due to low power of transmission. Furthermore, obstacles in the wireless link or local interferences make single-hop topology WSANs difficult to deploy. Therefore, data packets originating from a source device are relayed to a destination device on end-to-end basis [13]. Thus, relaying packets in a multi-hop wireless network should be performed by reserving timeslots to provide on-time packet delivery. The unpredictable packet loss makes the retransmission scheme inevitably considered to achieve the desired levels of reliability. Multipath routing is one of the essential retransmission schemes to improve the robustness of end-to-end transmission in error-prone wireless communication environments [14].

However, multipath retransmission needs to be pre-allocated for more communication resources. Thus, the efficiency of resource scheduling is supposed to be considered. Moreover, the industrial WSAN is a technological paradigm that enables advanced and complex services by interconnecting a possibly large number of low-complexity resource-constrained embedded devices equipped with diverse sensors and actuators [15]. The incorporation that WSANs support the coexistence of different sampling rates is imperative for a better industrial solution. At such instances, resource scheduling needs to be considered for schedulability, since more collisions may exist in the process of resource allocation.

Addressing the above problems, this paper proposes an efficient resource scheduling of CCA (clear channel assessment)-Embedded slot and Multiple sinks with Rate Monotonic scheme (CEM-RM) for multipath retransmission. The proposed CEM-RM scheme aims at periodic sensors, and we assume that multiple sampling periods exist on scheduled sensor nodes. For event-driven sensors, the CEM-RM scheme is also applicable. However, the maximum delay of packet transmission is the length of a superframe. Since the time that events occur cannot be predicted accurately, the time of waiting on communication resources increases the total end-to-end delay. The scheduling of CEM-RM aims to deal with the problem of schedulability for multipath retransmission scheduling. Therefore, first, we assume that the multipath routing graph of all the nodes can be successfully constructed. Second, the IEEE 802.15.4-based radio transceivers support 16 channels on 2.4 GHz. The scheduling of CEM-RM needs multiple channels; thus, we assume that at least eight channels can be used for scheduling without interference. Finally the last assumption is that multiple sampling periods exist on scheduled industrial flows and that the minimum period should not be less than the transmission delay of nodes, which are at the greatest distance from the gateway. The main contributions of this paper are listed as follows:

1. We analyze the resource scheduling principle for the multipath retransmission scheme based on the WirelessHART standard, and we design a link release algorithm to cater the principle of routing order in data flow.
2. We use the approaches of multiple sinks, CCA-embedded timeslot, and rate monotonic scheme to respectively decrease the negative effects of collision, path crossing, and diverse period during resources scheduling. Based on the above approaches, we propose an algorithm (CEM-RM) to improve the schedulability of resources scheduling.
3. We finally simulate our proposed algorithm to prove the improvement of schedulability. Additionally, we implement a WSAN system with our proposed scheduling in a real factory. The practical experiments are conducted to prove the reliability and real-time performance.

The rest of the paper is organized as follows. Section 2 presents the related works, and Section 3 describes the system model and the principle of multipath retransmission scheme. Moreover, we present the constraints during resources allocation. In Section 4, we present the proposed Link Release algorithm and CEM-RM algorithm in more detail. The simulation and experimental results are illustrated in Section 5, and finally, we conclude our work in Section 6.

2. Related Works

The use of multiple channels in the TSCH MAC protocol enables benefits including diversity, network scalability, and optimized scheduling. In the last couple of years, a notable trend in multichannel MAC solutions can be seen in Reference [16]. Zhao et al. [17] present multichannel, TSCH-based source-aware scheduling schemes for WSNs. The algorithm benefits from multiple channels but fails to guarantee reliability. Dobslaw et al. [5] extend the SchedEx [18] scheme to a multichannel scenario by introducing scalable integration in existing schemes. The authors also claim to cut latencies around 20% in the schedules from ShedEx. Kim et al. [19] proposed a distributed solution based on multichannel allocation that achieves max-min fairness among multiple flows. However, the fairness of channel use in such approach needs to trade off throughput. Researches, such as References [20–23], deeply exploit the use of multiple channels. The approaches proposed in these literatures focus on decreasing the interferences while using multiple channels in their specific type of WSNs. Therefore, this paper also makes full use of available channels to optimize TSCH scheduling.

Multipath routing allows multiple streams of information from source to sink, which significantly increases the reliability of data reception, especially in industrial environments. Liew, Soung-Yue et al. [24] propose a complete process of channel assignments for adaptive and energy-efficient data collection protocol, which could schedule the node connectivity based on traffic load. Van Luu et al. [25] proposed a scheduling algorithm for multipath routing structures, the objective of which is to reduce the message complexity. Mo Sha et al. [26] propose a novel channel-hopping methods for the exchange of data packages, which could prevent links, sharing the same destination from using channels with strong correlations. Xi jin et al. [27] proposed a mixed criticality scheduling algorithm, which guarantees the real-time performance and reliability requirements of data flows with different levels of criticality. Through the empirical studies, the above researches also show that graph routing leads to significant improvements over source routing in terms of worst-case reliability, where graph routing protocol is a classical multipath transmission approach in the WirelessHART standard. However, none of these researches simultaneously consider the effect of collision, path crossing, and diverse period on TSCH scheduling, which also lacks concentration on schedulability in a large WSAN.

The gateway resides at the root of all graphs in a wireless network, and it can provide multiple network access points, which are named sink nodes in this paper. The responsibility of a sink node is to maintain the connection between wireless nodes and the gateway, where most researches regard a sink node as a gateway [28]. However, the sink node is a hotspot and experiences more link conflicts than other nodes do when we schedule a multi-hop routing graph. Due to the deadline of industrial data transmission, this characteristic seriously decreases the schedulable ratio, especially in larger networks. Therefore, this paper provides multiple sink nodes to avoid the link scheduling conflicts and to improve the use of communication resources. In Reference [29], the authors apply the theory of compressive sensing on a multi-sink wireless sensor network, which could improve the capability for data gathering. However, the presented architecture of the network is not complicated enough to support routing graphs. Multiple sinks are generally used in the distributed scenario to balance the traffic load, which is responsible for packet collection as a substation [30–33]. In this paper, we realize a multi-sink gateway to achieve a highly schedulable WSAN in a practical industrial application.

3. Network Model

3.1. Network Description

Deterministic communication heavily relies on resource scheduling management. Accordingly, network management techniques adapted for industrial wireless mesh are critical. This paper adopts centralized management to form a deterministic WSAN, the structure of which is shown in Figure 1.

A network mainly consists of field nodes, multiple sinks, a gateway, and a centralized network manager. All field nodes operate over low-power radios compliant with the IEEE 802.15.4-2006 standard, which supports 16 channels in the 2.4 GHz ISM band. The TSCH MAC protocol is implemented in this paper to guarantee deterministic communication. One time slot is 10 ms long and allows for channel switching and the transmission of a single packet together with data-link acknowledgement (DL-ACK). Time slots can be either dedicated or shared, where the shared time slot is amended to CCA-Embedded Slot (CES) as proposed in Reference [34]. In a dedicated time slot (DTS), only a single transmitter–receiver pair can communicate on any given channel, thereby not allowing channel reuse. CES, in which multiple nodes are scheduled on the same channel, is mainly used for decreasing the use of slots for path crossing.

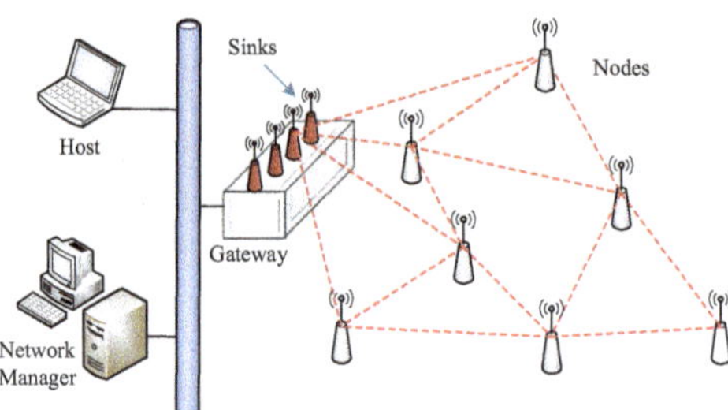

Figure 1. The structure of an industrial wireless sensor-actuator network (WSAN) system.

To guarantee the reliability of the flow transmission, we adopt the graph routing retransmission approach advised in the WirelessHART standard. As shown in Figure 2, packets can be sent by a device which has two direct neighbors at least in a graph. One is the primary routing path (PRP), and another is the alternative routing path (ARP). Each type of path has a receiving node for relaying packets. A node should preferentially send packets along the PRP. If the transmission is failed in PRP, the node will change to using ARP. In this paper, we assume that the multipath graph is already generated by the network manager. Therefore, all the possible transmissions should be involved in a slot and channel scheduling. The scheduling of slot and channel for all packet flows also has some considerations and limitations in the following based on WirelessHART.

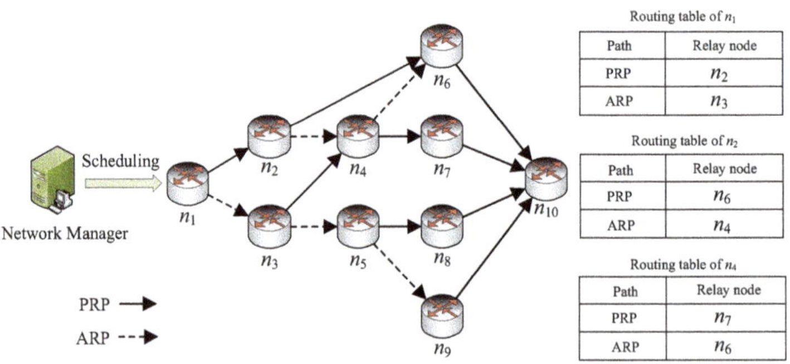

Figure 2. An example of the graph routing retransmission scheme.

- The allocation of slot and channel based on graph routing is so that a device can send a packet at most three times, including the first transmission and two retransmissions. The first transmission and retransmission are along the PRP, and the second retransmission needs to change to be along the ARP, where the changes of link includes the neighbor and channel.
- According to the above consideration, each flow consists of dedicated and abundant transmissions, all which should be allocated for the slots and channels by the network manager. Moreover, each

slot allocation should follow the routing orders, where the slot offset of next-hop transmission and retransmissions should not be allocated before the slots of the previous-hop transmission.

3.2. System Model

We describe the model of a WSAN, which consists of field devices, multiple sinks, a gateway, and a centralized network manager, as shown in Figure 1. We consider that the network can be modelled as a graph $G = (N, L)$, in which the set $N = \{n_1, n_2, \ldots, sk\}$ represents all the nodes, including field devices n_i and sinks sk. L is the link set, where the element l_{ij} in L indicates the connection state between the nodes n_i and n_j. $l_{ij} = 1$ means n_i and n_j can directly communicate with each other. $x(n_i, n_j)$ represents a transmission with the direction from node n_i to n_j. Thus, for a link l_{ij}, $x_{t,c}(l_{ij})$ represents the transmission allocated on the time slot t and channel offset c. The length of a time slot is denoted by κ. The actual channel can be calculated by $(ASN + c) \bmod |C|$, where ASN is the absolute counting of slot advance and C is an available channel set. Thus, the channel scheduling is only relevant to the assignment of channel offset. For example, $x_{t,c}(l_{ij}) = 1$ represents that node n_i transmits a packet to node n_j using channel c at slot t, otherwise, $x_{t,c}(l_{ij}) = 0$. Based on the above network model, all transmissions need to be scheduled by the TSCH MAC protocol. Thus, some basic constraints expressed by Equations (1)–(4) should be considered, which seriously influence the performance of transmission scheduling.

First, more than one transmissions cannot be occurred simultaneously in a certain slot and channel unless nodes use some contention-based mechanism. Otherwise, one transmission may generate wireless interference to interfere with the other transmissions.

$$\sum_{n_i \in N} \sum_{n_j \in N} x_{t,c}(l_{ij}) \leq 1, \forall t, \forall c \in C \tag{1}$$

Second, the DL-ACK should be sent within one slot for reliability. Thus, a node can only transmit to one neighbor at any slot.

$$\sum_{c \in C} \sum_{n_j \in N} x_{t,c}(l_{ij}) \leq 1, \forall t \tag{2}$$

Third, since a low-power node in the network only supports half-duplex wireless communication, obviously, it cannot be the transmitter and the receiver simultaneously.

$$\sum_{c \in C} \sum_{n_i \in N} x_{t,c}(l_{ij}) + x_{t,c}(l_{ji}) \leq 1, \forall t \tag{3}$$

Finally, multichannel transmission can make full use of slot and channel resources; however, the number of available channels is limited. Thus, the number of transmissions cannot be beyond the number of available channels.

$$\sum_{c \in C} \sum_{n_i \in N} \sum_{n_j \in N} x_{t,c}(l_{ij}) \leq |C|, \forall t \tag{4}$$

We consider all the end-to-end deliveries in a network as a flow set, which is denoted by $F = \{F_0, F_1, \ldots, F_\gamma\}$. The number of flows that need to be scheduled is denoted as γ. A flow can be characterized by $F_k : \langle X_k, E_k, p'_k, \phi_k \rangle$, where $X_k = \{x^k(s_1, d_1), \ldots, x^k(s_\tau, d_\tau), \ldots\}$ is the collection of all transmissions in a flow, E_k is the collection of transmission release orders related to the routing order, and ϕ_k is the routing path from the first source node to the last destination node. s_τ and d_τ are the source node and the destination node in a transmission, respectively. One of the elements $e(\alpha, \beta)$ in E_k represents that the transmission $x^k(s_\beta, d_\beta)$ is released after all the transmissions $x^k(s_\alpha, d_\alpha)$. Each F_k periodically generates a packet with the period p'_k, and the flow should be finished before its deadline d_k.

All the slots for a flow construct a superframe which is repeated periodically. The supported periods are defined as $2^a \cdot p_m$ based on WirelessHART, where a is an integer and p_m is the minimum period among all flows. Without loss of generality, we transfer the period of all flows to the regular period through Equation (5), since it can let the scheduling fit with the data period better.

$$p_k = p_m \cdot 2^{\lfloor log_2 \frac{p'_k}{p_m} \rfloor}, p'_k \geq p_m, k = 0, 1, \ldots, \gamma \tag{5}$$

For example, assuming the basic period is 20 ms, then the supported periods can be 20 ms, 40 ms, 80 ms, and so on. All the superframes constitute a hyperframe, the period of which is the maximum period among the superframes. The period is denoted as T and is expressed by Equation (6).

$$T = max\{p_k, k = 0, 1, \ldots, \gamma\} \tag{6}$$

We use $H[t][c]$ to denote which transmission is allocated on time slot t and channel c. The number of slots in a hyperframe is denoted by $|H| = T/\kappa$. As an example, if the first transmission of flow F_1 is allocated for slot 0 and channel offset 0, it can be expressed as $H[0][0] = x^1(s_1, d_1)$, otherwise, it is $H[0][0] = EMPTY$ if there is no transmission allocated to $H[0][0]$. We assume that the deadline of a flow transmission is equal to its period in this paper. Thus, all the transmissions in a flow should be scheduled within the period, where the constraint can be expressed by Equation (7).

$$t \leq \frac{p_k}{\kappa}, \forall x^k(s_\tau, n_\tau), k \in \{0, 1, \ldots, \gamma\} \tag{7}$$

4. The Proposed Scheduling

In this section, we introduce the CEM-RM algorithm for the multipath retransmission scheme based on graph routing in WirelessHART. We also present the proposed approaches to make the scheduling more efficient. The symbols and related functions used in the algorithms are listed in Table 1.

Table 1. Symbols and functions.

Symbols	Description
D_i	the number of links l_{ji} in a subgraph, where the node n_i is the destination node.
$H[t][c]$	the transmission allocated for time slot t and channel c
W	the collection of time slots, which are already allocated for transmission τ, where its element is ϖ_τ
C_t	the collection of available channels at time slot t
Ω_t	the collection of all the transmissions allocated to time slot t
Ψ_t	the collection of available sinks at time slot t
$\|C_t\|, \|\Psi_t\|$	the number of corresponding elements in the collection
Functions	**Description**
$Add(y, Y)$	the function responsible for adding an element y into the collection Y
$Del(y, Y)$	the function responsible for deleting an element y into the collection Y
$ASort(Y, z)$	the function responsible for sorting all the elements in Y in ascending order, according to feature z

4.1. Transmissions Release Algorithm

As mentioned before, the multipath scheme requires that a packet on a device is transmitted three times at most, including initial transmission and two retransmissions. Thus, we first need to generate all the transmissions and their corresponding orders for a flow before allocating slots and channels. Algorithm 1 is the function of $TransmissionRelease(F)$, which can generate all transmissions and transmission orders for a flow, according to a given graph G. Actually, the period p and the routing ϕ of the flow are already known and we mainly focus on producing the collections X and E to complete the flow. The initial transmission and the first retransmission use PRP, and the second retransmission uses ARP. A link in ϕ releases two transmissions if it is used for PRP; otherwise, it releases only one transmission for ARP. To obtain all the transmissions, we traverse all the links in ϕ from the source node to the destination node by a first in and first out ($FIFO$) pipe. Meanwhile, to generate the transmission order, we give each n_i a transmission collection $X'(n_i)$, which involves the already generated transmissions. The destination node of these transmissions is n_i. A node will be put into the $FIFO$ and obliged to release the transmission until the number of transmissions in its collection $X'(n_i)$ is equal to D_i. D_i is the in-degree of n_i, which can be calculate by Equation (8).

$$D_i = \sum l_{ji}, l_{ji} \in \phi \tag{8}$$

To illustrate the function of $TransmissionRelease(F)$, we consider the topology shown in Figure 2 as an example. The flow is generated by node n_1 and transmitted to node n_{10}. As shown in Figure 3, the generated transmissions in X are in released sequence after executing Algorithm 1. For example, the first released transmission is $x(n_1, n_2)$, and the fourth is $x(n_2, n_6)$. The transmissions in X also need sequence collection E to further explain their routing order, where $e(2,4)$ represents that the 4th transmission $x(n_2, n_6)$ must occur after the second transmission $x(n_1, n_2)$. The transmissions without orders can occur simultaneously, which means they can be allocated for the same slot with different channels. We can see that the two adjacent transmissions may not be in continuous order, such as the 6th transmission $x(n_2, n_4)$ and 7th transmission $x(n_3, n_4)$. We also notice that E has two orders, $e(5, 18)$ and $e(12, 18)$, where the transmission $x(n_6, n_{10})$ should go after the 5th and 12th transmissions. Actually, we only need to comply with the latter order, $e(12, 18)$, which should be considered in slot and channel allocation.

Algorithm 1: $TransmissionRelease(F)$

Input: $G = (N, L)$
Output: $F : \langle X, E \rangle$
Initialize a $FIFO$ and put the source node into $FIFO$;
Set each $X'(n_i) = \emptyset$ and calculate each D_i,$n_i \subset \phi$;
$\alpha \leftarrow 1$;
while $FIFO \neq \emptyset$ **do**
    Get n_i from $FIFO$;
    for *each* l_{ij},$n_j \in \phi$ **do**
        if l_{ij} *is PRP* **then**
            $s_\alpha \leftarrow n_i; d_\alpha \leftarrow n_j;$
            $s_{\alpha+1} \leftarrow n_i; d_{\alpha+1} \leftarrow n_j;$
            $Add(e(\alpha, \alpha+1), E);$
            for *each* $x(s_\beta, d_\beta) \in X'(n_i)$ **do**
                $Add(e(\beta, \alpha), E);$
            $Add(x(s_{\alpha+1}, d_{\alpha+1}), X'(n_j));$
            $\alpha \leftarrow \alpha + 2;$
            if $|X'(n_j)| = D_j$ **then**
                Put n_j into $FIFO$;
        if l_{ij} *is ARP* **then**
            $s_\alpha \leftarrow n_i; d_\alpha \leftarrow n_j;$
            $Add(e(\alpha - 1, \alpha), E);$
            $Add(x(s_\alpha, d_\alpha), X'(n_j));$
            $\alpha \leftarrow \alpha + 1;$
            if $|X'(n_j)| = D_j$ **then**
                Put n_j into $FIFO$
return X, E;

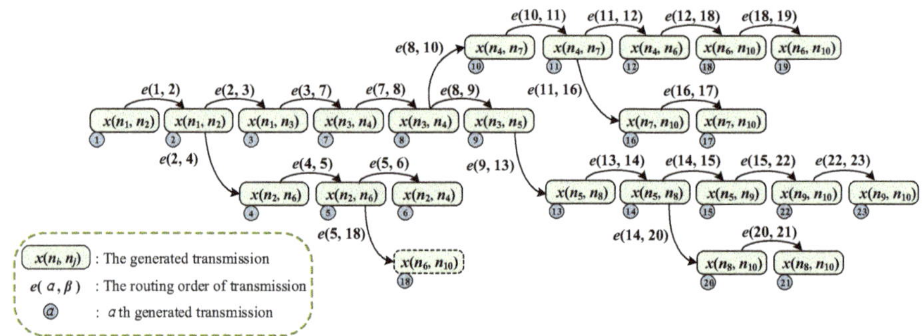

Figure 3. Generated transmissions and their transmission orders for an example flow.

4.2. CEM-RM Algorithm

The proposed scheduling algorithm CEM-RM is shown in Algorithm 2. Considering decreasing the negative effect of the diverse period, path crossing, and conflict, the algorithm adopts three approaches to improve schedulability.

Assuming that t_α and c_α respectively represent the slot offset and channel offset allocated to the transmission $x^k(s_\alpha, d_\alpha)$ within a hyperframe. For a flow F_k, if there is $e(\alpha, \beta)$, the slot allocation must satisfy $t_\alpha < t_\beta$. Meanwhile, all the slots $t_\alpha + j \cdot p_k (j \geq 0)$ are allocated to the transmission $x^k(s_\alpha, d_\alpha)$ due to periodic characteristics. Since there exists node interference in the time slot, the objective of the scheduling is to avoid these interference. However, the slot allocation for transmissions has different periods. This causes more node interference, which reduces channel utilization. The given network graph is schedulable (SCH) when the slot allocation of all the flows satisfies the above constraints expressed by Equations (1)–(4) and (7); otherwise, it is unschedulable (UNSCH). As a classical scheduling strategy, rate monotonic (RM) policy preferentially assigns communication resources to the flow which possesses the shorter period [35]. We adopt RM policy as a basic strategy, where we sort all the flows in ascending period at the beginning (line 2).

After individually tracing the RM-based scheduling, we notice that more transmissions need to be scheduled for a node that is close to the gateway, especially for the sink node. As shown in Figure 4, obviously, the sum of available communication resources is $|H| \cdot |C|$, where we regard a slot with a channel as a communication resource. On the one hand, the generated transmissions of all flows cannot exceed the number of communication resources, which is expressed as follows:

$$\sum_{k=0}^{\gamma} |X_k| \leq |H| \cdot |C|, F_k \in F \tag{9}$$

On the other hand, the number of available communication resources is $p_k \cdot |C|$ when the flow $F_k \in F$ is to be scheduled. However, some of these available resources are occupied by the previous $k-1$ allocated flows. The condition that there are enough available resources for F_k is expressd as follows:

$$\sum_{i=0}^{k-1} |X_i| \cdot \left\lfloor \frac{p_k}{p_i} \right\rfloor + |X_k| \leq \left\lfloor \frac{p_k}{\kappa} \right\rfloor \cdot |C|, \forall F_k \in F' \tag{10}$$

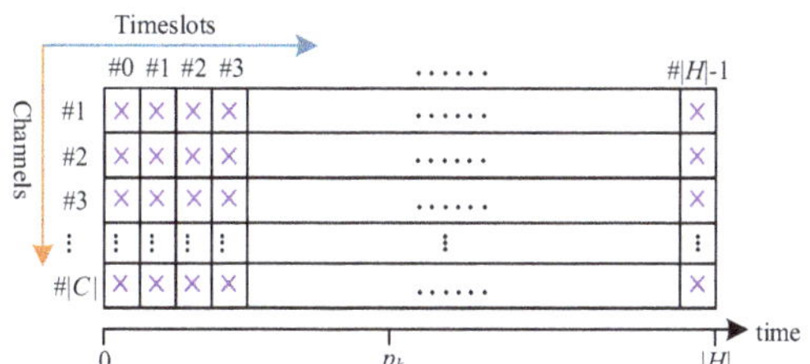

Figure 4. The communication resources for a hyperframe.

Algorithm 2: CEM-RM

Initialize $C_t \leftarrow C, \Psi_t \leftarrow \Psi, t \in [0, |H|-1]$ $F' \leftarrow ASort(F, p_k)$;
Input: G, F, Ψ
Output: $H[t][c]$, SCH or UNSCH
for *each* $F_k \in F'$ **do**
  $(X_k, E_k) \leftarrow TransmissionRelease(F_k)$;
  $t \leftarrow 0, \tau \leftarrow 1, W \leftarrow \emptyset, temp \leftarrow \tau$;
  q represents all the integers within $[0, T/p_k)$;
  while $X_k \neq \emptyset$ **do**
    if $t > p_k/\kappa$ **then**
      return UNSCH;
    if $\tau \neq temp \wedge \exists e(\mu+\tau) \in E_k$ **then**
      $temp \leftarrow \tau$;
      $\mu \leftarrow \mu' \mid max(\mu'+\tau), e(\mu', \iota) \in E_k$;
      $t \leftarrow \varpi_\mu + 1, \varpi_\mu \in W$;
    if $s_\beta \notin \{s_\tau, d_\tau\} \wedge s_\tau \notin \{s_\beta, d_\beta\} \wedge |C_t| > 0, x^{k_1}(s_\beta, d_\beta) \in \Omega_{t+q\cdot p_k}$ **then**
      if $d_\tau = d_\beta$ **then**
        if $k = k_1 \wedge d_\tau \in \Psi \wedge |\Psi_t| > 0$ **then**
          $d_\tau \leftarrow sk, sk \in \Psi_t$;
          $Del(sk, \Psi_{t+q\cdot p_k})$
        else if $k = k_1 \wedge e(\beta, \tau) \notin E_k$ **then**
          $c \leftarrow c' \mid H[t][c'] = x^k(s_\beta, d_\beta)$;
          Change $H[t][c]$ to CES;
        else
          $t \leftarrow t+1$;
          Continue;
      $H[t][c] = x^k(s_\tau, d_\tau), c \in C_t$;
      $Del(x^k(s_\tau, d_\tau), X_k)$;
      $Add(x^k(s_\tau, d_\tau), \Omega_{t+q\cdot p_k})$;
      $Del(c, C_{t+q\cdot p_k})$;
      $\varpi_\tau \leftarrow t; \tau \leftarrow t+1$;
    else
      $t \leftarrow t+1$;
return SCH,H;

The available resources are becoming more inadequate with the increase of data flow. The later scheduled flows will meet more node interferences, especially at the end hop of a flow. This is because most transmissions share the sink node. A sink node cannot be involved in two transmissions simultaneously at a slot, which results in low schedulability. Moreover, the interference caused by one sink node makes it so that only one channel can be scheduled at the end of a hyperframe. Thus, we adopt multiple sinks to reduce the interference caused by a single sink node during hyperframe scheduling. Each slot possesses multiple available sinks to receive packets on different channels. In CEM-RM, a transmission may suffer interference at one slot, which is caused by a sink node. The algorithm will use the other available sink to avoid such interferences if at least one resource exists at the slot (line 19).

Furthermore, one key adjustment for the slot employed in this paper is that we amended the common shared slot to the CES. The multipath scheme provides an alternative relay node for possible retransmissions. Actually, a packet is finally routed along one of these paths. Redundant resources need to be assigned for avoiding conflict when the different routing paths share the same relay node. Thus, merging the resources for the overlapping paths is an effective approach. However, the conflict may be caused by the DL-ACK loss problem. A packet is simultaneously transmitted along more than one path if one of the transmissions only loses a DL-ACK. The packet is still transmitted along the original path. Considering this phenomenon deteriorates the reliability and costs extra energy, we adopt an efficient approach of the CES.

At a CES, the timing structure of the transmitting node is shown in Figure 5 and the destination node still accords with the standard. The main adaptation is to add CCA units (128 µs) in the TsTxOffset, which is the 2120 ± 100 µs duration between the beginning of the slot and the start of the data frame transmission. We respectively reserve 200 µs and 340 µs at the beginning and end of TsTxOffset for some preparation of the slot and data frame. In CCA units, we adopt CCA mode 2 in IEEE 802.15.4 standard, where CCA shall report a busy medium upon the detection of a signal with the modulation and spreading characteristics of an IEEE 802.15.4 compliant signal. The maximum number of CCA units is 5 in this paper. Multiple source nodes will contend this slot by randomly selecting a CCA unit to detect the state of a channel. Once a source node detects that the channel is free, it sends a preamble signal to occupy the current slot during the remaining time TsTxOffset. Another source node will detect the preamble; then, it stops to perform CCA and to give up using the current slot. The radio chip may switch the state and channel in TsRxTx, which is less than 192 µs. We assume that there are K transmissions with different source nodes and the same destination node. The probability of successful contention is $(K/5) \cdot \sum_{j=0}^{4}(j/5)^{K-1}, K \geq 0$ and enough to guarantee the reliability of the flow transmission, since the contention at CES occurs only when the DL-ACK is lost on the previous link. The use of CES is to decrease abundant slot allocation and to increase slot use, which also results in the increase of schedulable ratios.

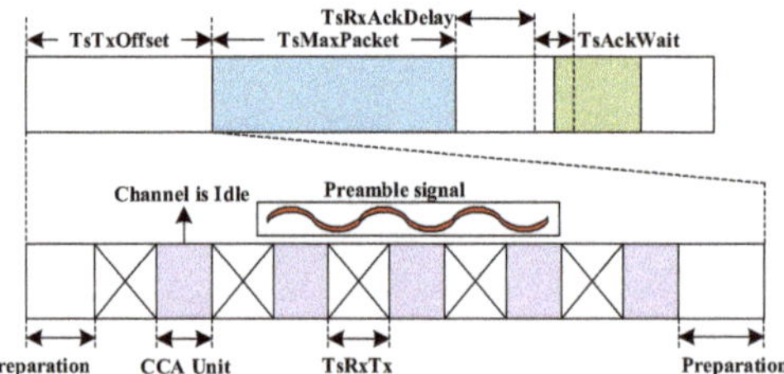

Figure 5. The structure of a CE (clear channel assessment-embedded) slot for the transmitting node.

With the use of the above three approaches, we designed the slot and channel allocation to avoid node interference. The following two situations should be solved by the above approaches when the allocation to a transmission $x^{k_1}(s_\alpha, d_\alpha)$ meets the already allocated transmission $x^{k_2}(s_\beta, d_\beta)$ in the same slot.

(1) According to constraints expressed by Equations (1)–(3), two different transmissions $(\alpha \neq \beta)$ in a flow $(k_1 = k_2)$ should be allocated for different slots if the source node of a transmission is the

same as one of the nodes in another transmission. However, if $d_\alpha = d_\beta$ and two transmissions are allocated for the same slot exactly, those transmissions can use the same channel and the slot is changed into a CES. With the assumption of $\alpha > \beta$, the condition can be expressed as follows:

$$\begin{aligned}
&\text{if} \quad s_\beta \in \{s_\alpha, d_\alpha\} \vee s_\alpha \in \{s_\beta, d_\beta\} \quad \text{then} \\
&\quad (t_\alpha \neq t_\beta) = 1 \\
&\text{else if} \quad d_\alpha \neq d_\beta \wedge e(\beta, \alpha) \notin E_{k_1} \quad \text{then} \\
&\quad ((t_\alpha = t_\beta) \wedge (c_\alpha = c_\beta) \wedge (\text{the resource is CES})) = 1 \\
&\text{else} \quad ((t_\alpha \neq t_\beta) \vee (c_\alpha \neq c_\beta)) = 1
\end{aligned}$$

(2) Accordingly, two transmissions in different flows ($k_1 \neq k_2$) also should be allocated for different slots if the source node of a transmission is the same as one of the nodes in another transmission. Also, if $d_\alpha = d_\beta \neq sk$, CES cannot be used for the transmissions in different flows. However, if $d_\alpha = d_\beta = sk$, we can change d_α to one of the other available sink nodes. In this scenario, the above two transmissions can be allocated at the same slot with different channels. With the assumption of $p_{k_1} > p_{k_2}$, q represents all integers within $[0, p_{k_1}/p_{k_2})$. Therefore, the condition can be expressed as follows:

$$\begin{aligned}
&\text{if} \quad s_\beta \in \{s_\alpha, d_\alpha\} \vee s_\alpha \in \{s_\beta, d_\beta\} \quad \text{then} \\
&\quad (t_\alpha \neq t_\beta + q \cdot p_{k_2}) = 1 \\
&\text{else if} \quad d_\alpha = d_\beta = sk \wedge |\Psi_t| > 0 \quad \text{then} \\
&\quad ((t_\alpha = t_\beta + q \cdot p_{k_2}) \wedge (c_\alpha \neq c_\beta)) = 1 \\
&\text{else} \quad ((t_\alpha \neq t_\beta + q \cdot p_{k_2}) \vee (c_\alpha \neq c_\beta)) = 1
\end{aligned}$$

The above two conditions represent the situations of node interference and scheduling interference in slot and channel scheduling. The difference between the designed algorithm and RM algorithms is that we assign the communication resources by the order of each flow, instead of the slot sequence. Thus, we traverse all the transmissions in each period-sorted flow (lines 3 and 6) and assign the slots and channels to each transmission. To avoid interference, the scheduling should be adjusted when the allocation meets these conditions. We add all assigned transmissions at time slot t into Ω_t. We define C_t and Ψ_t as the collections of available channels and sinks at time slot t, respectively. Their elements will be removed by the $Del()$ function when the corresponding channels or sinks have already assigned. Moreover, ϖ_τ is used to record the slot offset allocated to the transmission τ in every flow, and it is responsible for finding the earliest available slot which does not break the routing order (lines 11–13). There could be several transmissions at a CES. Thus, changing a normal slot to a CES means that the slot expands for multiple transmissions. With the above considerations, we can finally obtain the scheduling table H when all transmissions are scheduled within their period.

In the algorithm, we define $|N_F|$ and $|X_F|$ as the maximum number of nodes and transmissions, respectively, and they are all related to the length of the flow. For example, for a four-hop flow in a routing tree topology, $|N_F|$ is less than 24. All the nodes have a PRP except the sink node and have an ARP except the first-hop nodes and the sink node. Thus, $|X_F|$ is less than $2 \times (2^4 - 1) + 1 \times (2^3 - 1) = 37$. The number of iterations of the "**for**" loop in line 3 and the "**while**" loop in line 6 are $O(\gamma)$ and $O(|H|)$, respectively. Moreover, the time complexities in line 2, line 4, line 11, and line 14 are $O(\gamma log(\gamma))$, $O(|N_F|^3)$, $O(|X_F|)$, and $O(T/p_m)$, respectively. Therefore, the time complexity in a worst case is $O(\gamma log(\gamma) + \gamma(|N_F|^3 + |H||X_F| + |H| \cdot T/p_m))$. In practicality, the periods of different applications could not have much difference; thus, $|H| \cdot T/p_m$ could be ignored.

5. Simulations and Experiments

In this section, we evaluate the performance of the proposed algorithm by comparing with two widely used real-time scheduling policies, RM and Least Laxity First (LLF) [36]. RM schedules a transmission based on the packet's absolute deadline, while LLF schedules a transmission based on the packet's laxity, which is defined as the transmission's deadline minus remaining transmission time. The schedules of individual RM and LLF are not suitable with our applications due to the quantity of nodes and period. Therefore, we also add multiple sinks into each compared scheduling policy for fairness. Thus, we respectively denote the two schedules as M-RM and M-LLF in the following simulations and experiments.

Different topology structures need different quantities of communication resources. Thus, four kinds of topology structures are generated for comparison, which are denoted as Tp1, Tp2, Tp3, and Tp4. We assume that the furthest leaf node is four hops away from the gateway in all topology structures. In industrial wireless networks, since there is the requirement of reliable and real-time performance, on the one hand, the big hop size can increase the rate of packet loss and, on the other hand, the large network cannot cater the demand of transmission deadlines. In our experience, four hops can overlap the general factory. Meanwhile, four hops can exactly overlap the experimental factory in this paper. The difference between each topology structure is that the number of nodes on each hop is generated according to different probabilities. For example, to generate one of the Tp1 topologies, we first traverse all the nodes, where the total number of nodes is determined previously. The traverse approach stipulates that a node has a 0.3 probability of being on the first hop, a 0.3 probability of being on the second, a 0.3 probability of being on the third, and a 0.1 probability of being on the fourth. Then, each node on the lower hop randomly selects two higher-level nodes as their parents to be the major path and redundant path. Thus, all the topologies of the Tp1 structure may be different from each other. Without loss of generality, we will stochastically generate enough topologies of every kind of structure. With the same method, the other kinds of topologies are generated according to the corresponding probabilities, which are listed in Table 2.

Table 2. The proportions of different types of topology.

Type of Topology	Probability			
	1st Hop	2nd Hop	3rd Hop	4th Hop
Tp1	0.5	0.3	0.1	0.1
Tp2	0.5	0.2	0.2	0.1
Tp3	0.4	0.3	0.2	0.1
Tp4	0.3	0.3	0.3	0.1

5.1. Simulations

Our simulations are presented to demonstrate the schedulability and effectiveness of our scheduling. The algorithms have been written in C language, and the simulations have been performed on a Windows machine with a 3.1-GHz Inter Core 3 processor. The following metrics are used for performance analysis: (a) Schedulable ratio is measured as the percentage of test cases for which a scheduling policy is able to schedule all transmissions. (b) Scheduling time is the total time required to successfully make a complete hyperframe for all transmissions. (c) Average normalized bandwidth is the ratio between the number of allocated resources and the total resources within a hyperframe after scheduling. In simulations, we assume that every node generates a packet periodically and sends it to the gateway along the routing graph. A packet is regarded as a flow without aggregation. The period is randomly selected in the set $\{p_m \cdot 2^a \mid \forall a \in [0, b]\}$, where b is the maximum exponent. A packet belonging to a fourth-hop node needs at most 22 transmissions using the scheduling of multipath scheme; thus, three values of $p_m(0.25\ \text{s}, 0.5\ \text{s}, 1\ \text{s})$ are tested in the simulations.

Figure 6 shows the schedulable ratio of the three scheduling policies with different parameter configurations. The number of nodes ranges from 50 to 100. We randomly generate 8000 cases for each node quantity and use the three algorithms to schedule the generated topologies separately. The number of available channels is set to be the maximum, 16, for a better comparison, since several cases using less channels are hardly scheduled with the existence of diverse periods. In Figure 6a–d, the three algorithms, which execute on each type of topology with $p_m = 1$ s and $b = 0$, can successfully schedule the majority of graphs. Obviously, the schedulable ratio drops in pace with the increase of nodes; however, we can see that our proposed scheduling is almost unaffected with these configurations. With the period shortened and diversified, the number of schedulable graphs decreases dramatically. Figure 6e–h and Figure 6i–l show the results under the period configurations of $p_m = 0.5$ s, $b = 1$ and $p_m = 0.25$ s, $b = 2$, respectively. They clearly indicate that the schedulable ratio of CEM-RM is higher than that of M-LLF and M-RM. Especially, increasing the number of nodes makes the gaps between CEM-RM and the others even bigger.

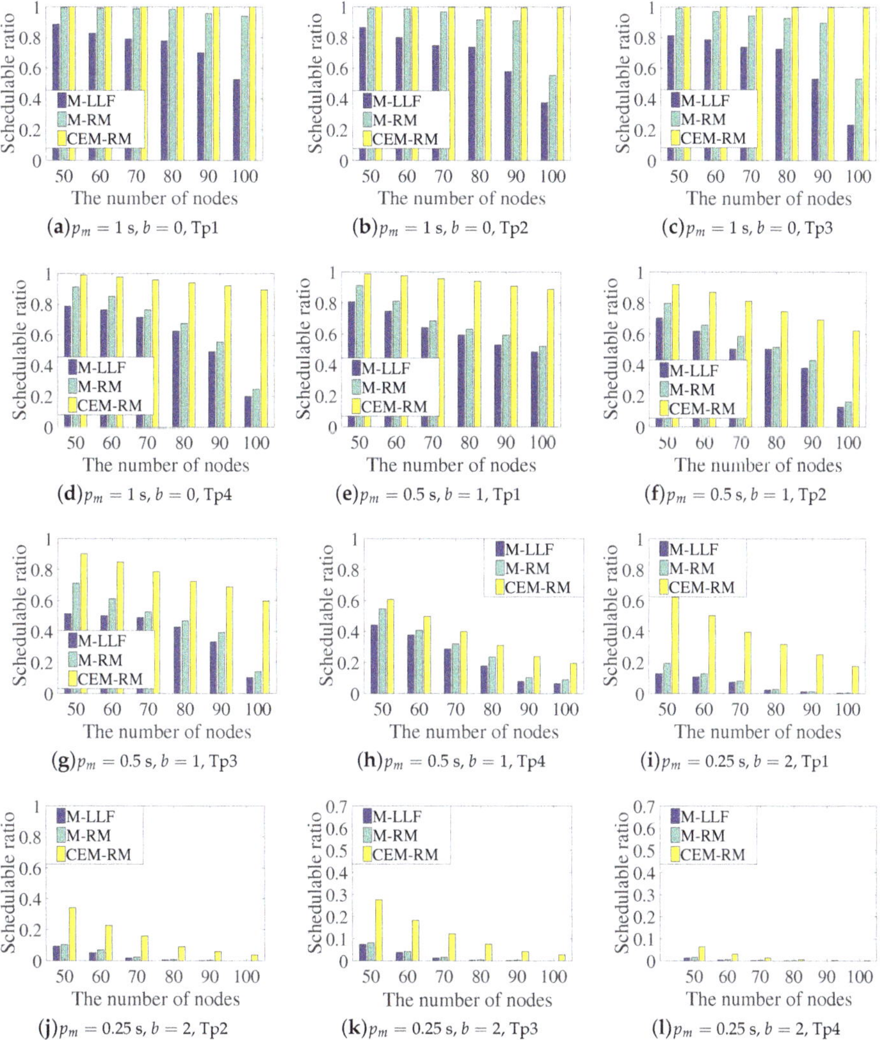

Figure 6. A comparison of the schedulable ratios of the three algorithms with different periods and topologies.

The execution time of a scheduling can affect the performance of network formation and rescheduling; thus, we have measured the average run time of the three algorithms. Each test with a different number of nodes and algorithms includes 100 successfully scheduled graphs. To guarantee that the tested graphs are more schedulable, we set $p_m = 0.5$ s and $b = 1$ in all tests. As displayed in Figure 7, the markers represent the average time of different algorithms with different numbers of nodes and the lines show the trend of scheduling time as the number of nodes is increased. We can obtain that the average execution times of M-LLF and M-RM are lower than 1264 ms, while the time of our proposed scheduling is lower than 371 ms. The results indicate that CEM-RM is better for the performance improvement of network formation or rescheduling.

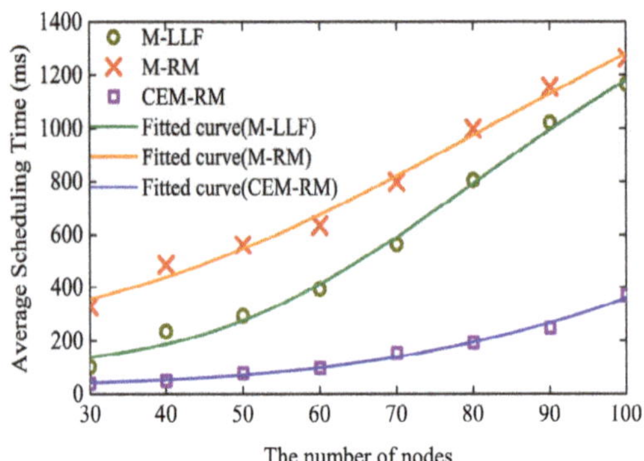

Figure 7. Comparison of scheduling time of the three algorithms.

Moreover, Figure 8 shows the average requirement of the wireless bandwidth for our scheduling and for the others. The normalized bandwidth is calculated by the ratio between the number of allocated resources and total resources in a hyperframe. We also select the successfully scheduled hyperframe with 100 nodes and the configuration of $p_m = 0.5$ s, $b = 1$. As can be seen, the average normalized bandwidths of Tp4 of M-LLF and M-RM are 0.918 and 0.906, respectively, which are similar to each other, while CEM-RM only requires 0.627. The same conclusion also applies to other types of topologies. This is because the number of transmissions is determined by a graph in M-LLR and M-RM. However, CEM-RM adopts CES to solve the problem of overlapping paths and to reduce the phenomenon of DL-ACK loss, which can decrease the number of dedicated slots additionally. The remaining unscheduled resources can be allocated for network management to improve network stability.

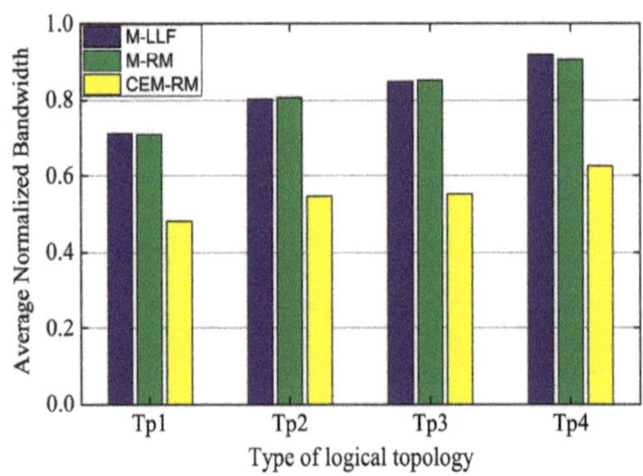

Figure 8. Comparison of average normalized bandwidth.

5.2. Experiments

We also test the network performance of the scheduling policies in practice, where the performance includes the reliability and end-to-end packet delay. Since the scheduling of M-LLF and M-RM follows the basic policy (the number of retransmissions) in WirelessHART, their reliability and end-to-end packet delay are only affected by the channel condition. Thus, we make comparisons only between our scheduling and M-RM. The algorithms are executed on a developed WSAN system, which is compatible with WirelessHART and also supports our applications. As shown in Figure 9, the experiments are conducted in a 50×30 m^2 factory in which the packet error rate is about 8% mainly caused by metal and electromagnetic interference.

Figure 9. Experimental environment and deployment of field devices.

Our wireless nodes used to collect corresponding information are implemented on a LPC1769 mote with an AT86rf231 radio transceiver. In reality, we have three kinds of information to be collected, which are temperature and humidity (HT) with a 1-s period, electrical information (EI) with a 0.5-s period, and machine data (MD) with a 0.25-s period. Each kind of information is collected by different versions of wireless node, respectively shown in Figure 10a–c. A total of 60 nodes consist of 30 HT nodes (HTN), 15 EI nodes (EIN), and 15 MD nodes (MDN). The information of MD is collected through a CAN (Controller Area Network) bus, and the EI is transmitted by RS485. The radio transmission power is 3 dBm, and we use 10 of the available channels with lower packet error rates. The multi-sink-based gateway with 8 sinks, a hard disk, and a wireless adaptor is shown in Figure 10d. The communication between multiple sinks and the gateway is constructed by USB port. The metal case is used to avoid dust and water to increase the lifetime. The generated topology is shown in Figure 11, and the remaining experimental parameters are listed in Table 3.

Table 3. Experiment parameters.

Parameters	Description
The number of nodes	60
The number of nodes in each hop	1-hop:26, 2-hop:18, 3-hop:10, 4-hop:6
Transmission rate	250 kb/s
Transmission rate	3 dBm
The number of CCA units at a CES	5
Maximum frame retries	3
Packet length	80 bytes

(a) HTN (b) MDN (c) EIN (d) Gateway

Figure 10. The hardware of wireless nodes and gateway.

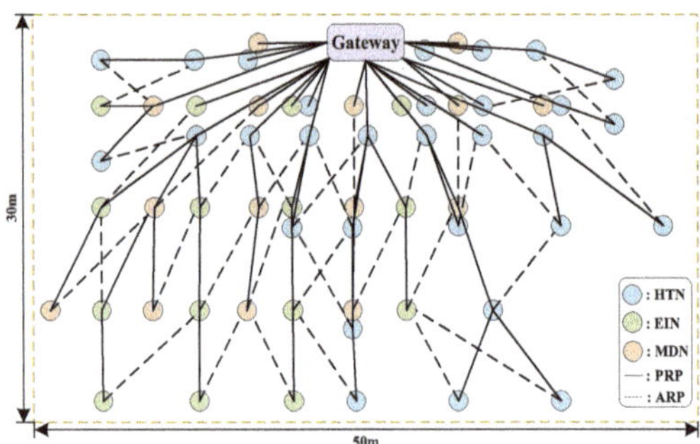

Figure 11. Experiment topology.

On the one hand, we first measure the packet delivery ratio (PDR) of M-RM and CEM-RM, as shown in Figure 12. The PDR represents the ratio between the number of packets that succeed in arriving to the gateway on time and the total number of generated packets. Since the period of packet generation is known previously, we can get the total number of packets of each node. Then, we gather the statistic of the number of packets that successfully reach the gateway within their own deadline. An experiment remains 20 min, and we assess the PDR every minute. The average PDRs of M-RM and CEM-RM for 20 min are 0.971 and 0.973, respectively. We can observe that their PDR for every minute vibrates over their average values because the condition of channels may change with time. The jitter of our CEM-RM is a little higher than that of M-RM, where the minute-PDR standard variations of M-RM and CEM-RM are 0.0042 and 0.0055, respectively. The time-varying channel has an effect on the adopted CES to a certain extent. However, the average PDR is mainly the same, which means that the schedulability of our proposed scheduling is improved but does not sacrifice reliability.

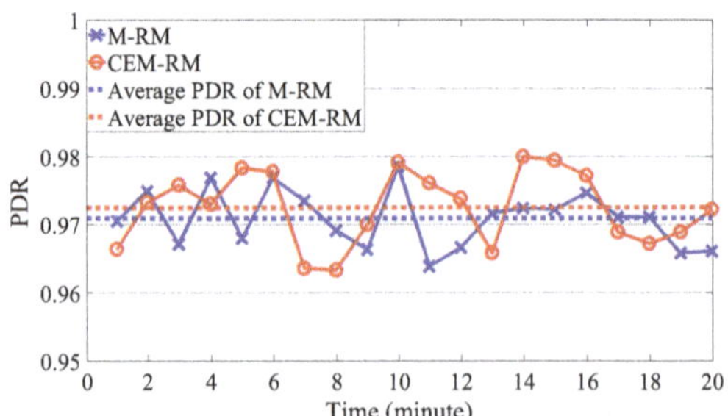

Figure 12. The packet delivery ratios of CEM-RM and M-RM.

On the other hand, we measured the delay of end-to-end delivery between each hop node and the gateway. The starting time of a packet is recorded at the time of its first transmission, while the ending time is recorded at the packet reception in the gateway. Figure 13 shows the comparison of average end-to-end delays between M-RM and CEM-RM. Obviously, the delay of CEM-RM is generally lower than that of M-RM. Moreover, as the routing hop increased, the average delay of our proposed scheduling was much lower than that of M-RM. For the packets of four-hop nodes, the average delay of our scheduling is about 32 ms lower than that of M-RM. The use of CES and multiple sinks effectively decreases the end-to-end delay, and CES can also reduce the impact of the DL-ACK loss phenomenon. The experimental results roughly indicate that our proposed scheduling not only improves the efficiency of scheduling but also decreases the end-to-end delay.

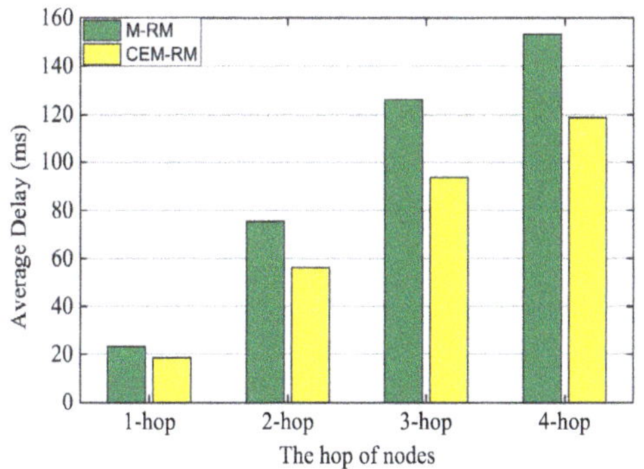

Figure 13. Comparison of average end-to-end delay between CEM-RM and M-RM.

6. Conclusions

In this paper, we study the problem of how to improve the schedulability of communication resources in industrial WSANs. Existing centralized approaches for multipath retransmission scheme do not consider schedulability in the coexistence of diverse industrial applications. In this paper, we propose a resources-scheduling algorithm based on the multipath retransmission scheme in the WirelessHART standard. We use the approaches of RM policy, CES, and multiple sinks to reduce the negative impact of diverse periods and to avoid node interference as far as possible. By simulation and implementation in hardware, we compare our proposed CEM-RM algorithm with M-LLF and M-RM scheduling policies. The achieved results show that our proposed solution outperforms the aforementioned scheduling policies in terms of less use of communication resources, end-to-end delay, and increase in schedulable ratio without jeopardizing network reliability and timeliness of data packets.

Author Contributions: Conceptualization, H.W. and D.Y.; methodology, H.W. and J.M.; software, H.W. and J.M.; validation, H.W. and D.Y.; formal analysis, J.M.; investigation, J.M.; resources, H.W. and D.Y.; data curation, J.M.; writing—original draft preparation, J.M.; writing—review and editing, H.W., D.Y., and M.G.; visualization, H.W. and J.M.; supervision, H.W.; project administration, H.W.; funding acquisition, H.W.

Funding: This research was partially funded by The National Key Research and Development Program of China grant number 2018YFB1702001 and by the National Natural Science Foundation of China grant numbers 61701018 and 61771040.

Conflicts of Interest: The authors declare no conflict of interest.

References

1. Lu, C.; Saifullah, A.; Li, B.; Sha, M. Real-Time Wireless Sensor-Actuator Networks for Industrial Cyber-Physical Systems. *Proc. IEEE* **2016**, *104*, 1013–1024. [CrossRef]
2. Queiroz, D.V.; Alencar, M.S.; Gomes, R.D.; Fonseca, I.E.; Benavente-Peces, C. Survey and Systematic Mapping of Industrial Wireless Sensor Networks. *J. Netw. Comput. Appl.* **2017**, *97*. [CrossRef]
3. Li, X.; Li, D.; Wan, J.; Vasilakos, A.V.; Lai, C.F.; Wang, S. A review of industrial wireless networks in the context of Industry 4.0. *Wirel. Netw.* **2015**, *23*, 23–41. [CrossRef]

4. Raza, M.; Aslam, N.; Le-Minh, H.; Hussain, S.; Cao, Y.; Khan, N.M. A Critical Analysis of Research Potential, Challenges, and Future Directives in Industrial Wireless Sensor Networks. *IEEE Commun. Surv. Tutor.* **2017**, *20*, 39–95. [CrossRef]
5. Dobslaw, F.; Zhang, T.; Gidlund, M. End-to-end reliability-aware scheduling for wireless sensor networks. *IEEE Trans. Ind. Inf.* **2016**, *12*, 758–767. [CrossRef]
6. Nguyen, N.T.; Liu, B.H.; Pham, V.T.; Liou, T.Y. An Efficient Minimum-Latency Collision-Free Scheduling Algorithm for Data Aggregation in Wireless Sensor Networks. *IEEE Syst. J.* **2017**, *12*, 2214–2225. [CrossRef]
7. Supramaniam, T.R.; Ibrahim, R.; Chung, T.D.; Hassan, S.M. Towards the development of WirelessHART adaptor for process control applications. In Proceedings of the International Conference on Intelligent and Advanced Systems, Kuala Lumpur, Malaysia, 15–17 August 2016; pp. 1–6.
8. Yang, D.; Ma, J.; Xu, Y.; Gidlund, M. Safe-WirelessHART: A Novel Framework Enabling Safety-Critical Applications over Industrial WSNs. *IEEE Trans. Ind. Inf.* **2018**, *14*, 3513–3523. [CrossRef]
9. Ouanteur, C.; Aïssani, D.; Bouallouche-Medjkoune, L.; Yazid, M.; Castel-Taleb, H. Modeling and performance evaluation of the IEEE 802.15. 4e LLDN mechanism designed for industrial applications in WSNs. *Wirel. Netw.* **2017**, *23*, 1343–1358. [CrossRef]
10. Hermeto, R.T.; Gallais, A.; Theoleyre, F. Scheduling for IEEE802. 15.4-TSCH and slow channel hopping MAC in low power industrial wireless networks: A survey. *Comput. Commun.* **2017**, *114*, 84–105. [CrossRef]
11. Wang, H.; Chen, Y.; Dong, S. Research on efficient-efficient routing protocol for WSNs based on improved artificial bee colony algorithm. *IET Wirel. Sens. Syst.* **2016**, *7*, 15–20. [CrossRef]
12. Hashimoto, M.; Wakamiya, N.; Murata, M.; Kawamoto, Y.; Fukui, K. End-to-end reliability-and delay-aware scheduling with slot sharing for wireless sensor networks. In Proceedings of the 2016 8th International Conference on Communication Systems and Networks, Bangalore, India, 5–10 January 2016; pp. 1–8.
13. Saifullah, A.; Xu, Y.; Lu, C.; Chen, Y. End-to-end communication delay analysis in industrial wireless networks. *IEEE Trans. Comput.* **2015**, *64*, 1361–1374. [CrossRef]
14. Tian, X.; Zhu, Y.H.; Chi, K.; Liu, J.; Zhang, D. Reliable and Energy-Efficient Data Forwarding in Industrial Wireless Sensor Networks. *IEEE Syst. J.* **2017**, *11*, 1424–1434. [CrossRef]
15. Xia, C.; Jin, X.; Kong, L.; Zeng, P. Bounding the demand of mixed-criticality industrial wireless sensor networks. *IEEE Access* **2017**, *5*, 7505–7516. [CrossRef]
16. Soua, R.; Minet, P. Multichannel assignment protocols in wireless sensor networks: A comprehensive survey. *Pervasive Mob. Comput.* **2015**, *16*, 2–21. [CrossRef]
17. Zhao, J.; Qin, Y.; Yang, D.; Rao, Y. A source aware scheduling algorithm for time-optimal convergecast. *Int. J. Distrib. Sens. Netw.* **2014**, *10*, 251218. [CrossRef]
18. Raza, M.; Ahmed, G.; Khan, N.M. Experimental evaluation of transmission power control strategies in wireless sensor networks. In Proceedings of the 2012 International Conference on Emerging Technologies (ICET), Islamabad, Pakistan, 8–9 October 2012; pp. 1–4.
19. Kim, W.; Yoon, W. Multi-constrained max–min fair resource allocation in multi-channel wireless sensor networks. *Wirel. Pers. Commun.* **2017**, *97*, 5747–5765. [CrossRef]
20. Phung, K.H.; Lemmens, B.; Goossens, M.; Nowe, A.; Tran, L.; Steenhaut, K. Schedule-based multi-channel communication in wireless sensor networks: A complete design and performance evaluation. *Ad Hoc Netw.* **2015**, *26*, 88–102. [CrossRef]
21. Wu, Y.; Cardei, M. Multi-channel and cognitive radio approaches for wireless sensor networks. *Comput. Commun.* **2016**, *94*, 30–45. [CrossRef]
22. Wu, Y.; Liu, K.S.; Stankovic, J.A.; He, T.; Lin, S. Efficient multichannel communications in wireless sensor networks. *ACM Trans. Sens. Netw.* **2016**, *12*, 3. [CrossRef]
23. Fernández de Gorostiza, E.; Berzosa, J.; Mabe, J.; Cortiñas, R. A Method for Dynamically Selecting the Best Frequency Hopping Technique in Industrial Wireless Sensor Network Applications. *Sensors* **2018**, *18*, 657. [CrossRef]
24. Liew, S.Y.; Tan, C.K.; Gan, M.L.; Goh, H.G. A Fast, Adaptive, and Energy-Efficient Data Collection Protocol in Multi-Channel-Multi-Path Wireless Sensor Networks. *IEEE Comput. Intell. Mag.* **2018**, *13*, 30–40. [CrossRef]
25. Van Luu, H.; Tang, X. An efficient algorithm for scheduling sensor data collection through multi-path routing structures. *J. Netw. Comput. Appl.* **2014**, *38*, 150–162. [CrossRef]
26. Sha, M.; Gunatilaka, D.; Wu, C.; Lu, C. Empirical Study and Enhancements of Industrial Wireless Sensor–Actuator Network Protocols. *IEEE Internet Things J.* **2017**, *4*, 696–704. [CrossRef]

27. Jin, X.; Xia, C.; Xu, H.; Wang, J.; Zeng, P. Mixed Criticality Scheduling for Industrial Wireless Sensor Networks. *Sensors* **2016**, *16*, 1376. [CrossRef] [PubMed]
28. Jain, T.K.; Saini, D.S.; Bhooshan, S.V. Lifetime optimization of a multiple sink wireless sensor network through energy balancing. *J. Sens.* **2015**, *2015*, 921250. [CrossRef]
29. Zheng, H.; Xiao, S.; Wang, X.; Tian, X.; Guizani, M. Capacity and delay analysis for data gathering with compressive sensing in wireless sensor networks. *IEEE Trans. Wirel. Commun.* **2013**, *12*, 917–927. [CrossRef]
30. Safa, H.; Moussa, M.; Artail, H. An energy efficient Genetic Algorithm based approach for sensor-to-sink binding in multi-sink wireless sensor networks. *Wirel. Netw.* **2014**, *20*, 177–196. [CrossRef]
31. Tan, D.D.; Kim, D.S. Dynamic traffic-aware routing algorithm for multi-sink wireless sensor networks. *Wirel. Netw.* **2014**, *20*, 1239–1250. [CrossRef]
32. Li, G.; Wang, Y.; Wang, C.; Liu, Y. Unbalanced threshold based distributed data collection scheme in multisink wireless sensor networks. *Int. J. Distrib. Sens. Netw.* **2016**, *12*, 8527312. [CrossRef]
33. Dobslaw, F.; Zhang, T.; Gidlund, M. QoS-aware cross-layer configuration for industrial wireless sensor networks. *IEEE Trans. Ind. Inf.* **2016**, *12*, 1679–1691. [CrossRef]
34. Yang, D.; Gidlund, M.; Shen, W.; Xu, Y.; Zhang, T.; Zhang, H. CCA-Embedded TDMA enabling acyclic traffic in industrial wireless sensor networks. *Ad Hoc Netw.* **2013**, *11*, 1105–1121. [CrossRef]
35. I.E.Commission. *Industrial Communication Networks-Wireless Communication Network and Communication Profiles-WirelessHART*; IEC Standard: Geneve, Switzerland, 2010.
36. Saifullah, A.; Xu, Y.; Lu, C.; Chen, Y. Real-Time Scheduling for WirelessHART Networks. In Proceedings of the Real-Time Systems Symposium, San Diego, CA, USA, 30 November–3 December 2010; pp. 150–159.

Article

A New Method of Priority Assignment for Real-Time Flows in the WirelessHART Network by the TDMA Protocol

Yulong Wu [1,†], Weizhe Zhang [1,2,*,†], Hui He [1,†] and Yawei Liu [1,†]

1 School of Computer Science and Technology, Harbin Institute of Technology, Harbin 150001, China; yulongwu@hit.edu.cn (Y.W.); hehui@hit.edu.cn (H.H.); liuyawei@hit.edu.cn (Y.L.)
2 Cyberspace Security Research Center, Pengcheng Laboratory, Shenzhen 518055, China
* Correspondence: wzzhang@hit.edu.cn; Tel.: +86-0451-86418272
† Current address: Harbin Institute of Technology, Harbin 150001, Heilongjiang, China.

Received: 17 October 2018; Accepted: 29 November 2018; Published: 3 December 2018

Abstract: WirelessHART is a wireless sensor network that is widely used in real-time demand analyses. A key challenge faced by WirelessHART is to ensure the character of real-time data transmission in the network. Identifying a priority assignment strategy that reduces the delay in flow transmission is crucial in ensuring real-time network performance and the schedulability of real-time network flows. We study the priority assignment of real-time flows in WirelessHART on the basis of the multi-channel time division multiple access (TDMA) protocol to reduce the delay and improve the ratio of scheduled. We provide three kinds of methods: (1) worst fit, (2) best fit, and (3) first fit and choose the most suitable one, namely the worst-fit method for allocating flows to each channel. More importantly, we propose two heuristic algorithms—a priority assignment algorithm based on the greedy strategy for C (WF-C) and a priority assignment algorithm based on the greedy strategy for U(WF-U)—for assigning priorities to the flows in each channel, whose time complexity is $O(max(N * m * log(m), (N - m)^2))$. We then build a new simulation model to simulate the transmission of real-time flows in WirelessHART. Finally, we compare our two algorithms with WF-D and HLS algorithms in terms of the average value of the total end-to-end delay of flow sets, the ratio of schedulable flow sets, and the calculation time of the schedulability analysis. The optimal algorithm WF-C reduces the delay by up to 44.18% and increases the schedulability ratio by up to 70.7%, and it reduces the calculation time compared with the HLS algorithm.

Keywords: WirelessHART network; delay analysis; real-time systems; multi-channel processing; simulation modeling

1. Introduction

With the introduction of the Industry 4.0 concept, IoT devices face many challenges such as sustainability and security, among which one of the most important is real-time performance [1,2]. In this paper, we propose two heuristic algorithms to assign priority to real-time flows in WirelessHART networks, and focus on the real-time performance of these two algorithms. The WirelessHART network is widely used in many fields, such as industrial manufacturing, automation control, and information collection. WirelessHART includes HART, EDDL, and IEEE 802.15.4 protocols. The TDMA protocol is widely used in wireless sensor networks, and it consists of the following devices [3].

1. *Field devices* are nodes in the WirelessHART network that can be regarded as source, transmission, or destination nodes. These nodes can collect, transmit, or receive information if they function as source, route, or destination nodes, respectively.

2. *Gateway* limits the range of the WirelessHART network and allocates an IP address to every node to ensure that every node's IP is unique. Hence, the source node that sends the information in the network can be determined. Backtracking the information through the gateway is convenient and accurate. The WirelessHART network has only one gateway.
3. The *network manager* is a global administrative unit for the WirelessHART network. It contains all of the topological structure information of the network. When a new node wishes to join the network, it has to obtain its own IP address from the gateway through the network manager. The information of this new node, such as its neighbor nodes, is saved at the host by the network manager.

Among the many protocols of the WirelessHART network, TDMA is to the most suitable for the real-time feature, the veracity of dates, and the power requirement of the WirelessHART network [4]. In general, we consider the flows to be periodic; i.e., all flows send packets again after a certain interval. Similarly, we can think of an aperiodic real-time flow as a real-time flow with an infinite length period. Hence, the TDMA protocol is also suitable for acyclic real-time flows. The TDMA protocol allows the time to be divided into several periodic frames, and each frame is divided into an equal numbers of slots. To guarantee the real-time characteristic, a device is only allowed to send and receive data in a specified slot. The TDMA protocol allows a source node to send many of the same packages to the destination node through different routes. This guarantees the accuracy of data transmission by increasing redundancy. Hence, we think of these same packets in different routes as separate flows in this study. In other words, if a source node sends the same packets through k routes, then we think of them as k-independent flows with the same period and deadline. The TDMA protocol also allows nodes that do not send or receive data to enter sleep mode and meet the power requirement of the WirelessHART network.

Real-time flow: The information collected by the source node in the WirelessHART network is transmitted as a packet. We define a real-time flow (referred to as flow) as a packet sent from the source node to the destination node. The most obvious difference between real-time flow and regular flow is that real-time flow has strong real time. The real-time flow must be finished before its deadline. Otherwise, there will be serious consequences, such as a system crash.

Priority assignment of flow: We consider a set of N flows, $F = \{F_1, F_2, ..., F_N\}$, and a priority assignment strategy $f = \{1, 2, ..., N\}$. The number of priorities is equal to the number of flows, and the priorities are in a strict decreasing order, i.e., $f_1 > f_2 > ... > f_N$. Hence, no two flows have the same priority. Specifically, the priority of flow F_j is higher than that of flow F_i if and only if $j < i$. We consider flow F_i with priority j. If the condition $\forall F_i \in F, \exists j \in f, s.t. f : F_i = j$ is met. Furthermore, we say a priority assignment strategy f is satiable if the flows in the set all meet their deadline.

Optimal priority assignment: We consider a schedulability test S and a priority assignment algorithm A. If is a priority assignment policy is available for a flow set that is satisfied in schedulability test S, A can also provide a satisfactory priority assignment. We then call A an optimal priority assignment. In other words, if a flow set cannot be satisfied by using A assignment priority in the schedulability test S, then no assignment algorithm can satisfy the flow set in schedulability test S.

The contributions of our work are as follows: (1) We transform the multi-channel flow model of the WirelessHART network into a multi-processor real-time model of the CPU. We allocate the flows to channels by using the worst fit algorithm. (2) We show that the problem of priority assignment in a single channel makes the set of flows acceptable, which is an NP-complete (NPC) problem. Then we propose two heuristic algorithms, namely, a priority assignment algorithm based on the greedy strategy for C (WF-C) and a priority assignment algorithm based on the greedy strategy for U(WF-U). (3) We build a new WirlessHART network simulation model according to the actual environment. (4) We compare our two algorithms with WF-D and HLS algorithms in terms of the average value of the total end-to-end delay of flow sets, ratio of schedulable flow sets, and calculation time of the schedulability analysis.

2. Background

We consider the WirelessHART network model with m channels based on the time division multiple access (TDMA) protocol. Each channel is divided into several periodical frames, and each frame is divided into slots with the same number. The length of every slot is 10 ms [5]. All information in the WirelessHART network is transmitted as packets. We call the packet that is sent from the source node to the destination node a flow.

Each node in the WirelessHART network can be viewed as a source, middle, or destination node [6]. Every node in a slot can receive or send a packet. That is, flow F_i sent from node a to node b costs one slot, and node b receiving it and sending it to node c also costs one slot. Each node can only perform one type of function (either sending or receiving a packet) in one slot.

To ensure transmission accuracy, we allow a packet to be sent from the source node to the destination node in different routes. The advantage of this policy is that can avoid blocking by the disabled node, bad routes, or blacklisting. For example, we assume that a packet will be set from node C to node D in Figure 1a. We send the same packet by two routes: (1) $C \rightarrow H \rightarrow I \rightarrow D$ and (2) $C \rightarrow D$. This packet cannot be transferred through Route (1) if the node I belongs to the blacklisting. However, this packet can still be transferred through Route (2). Hence, we guarantee that this packet can be received by node D by adding a route. In other words, this policy guarantees the accuracy of data transmission by increasing redundancy.

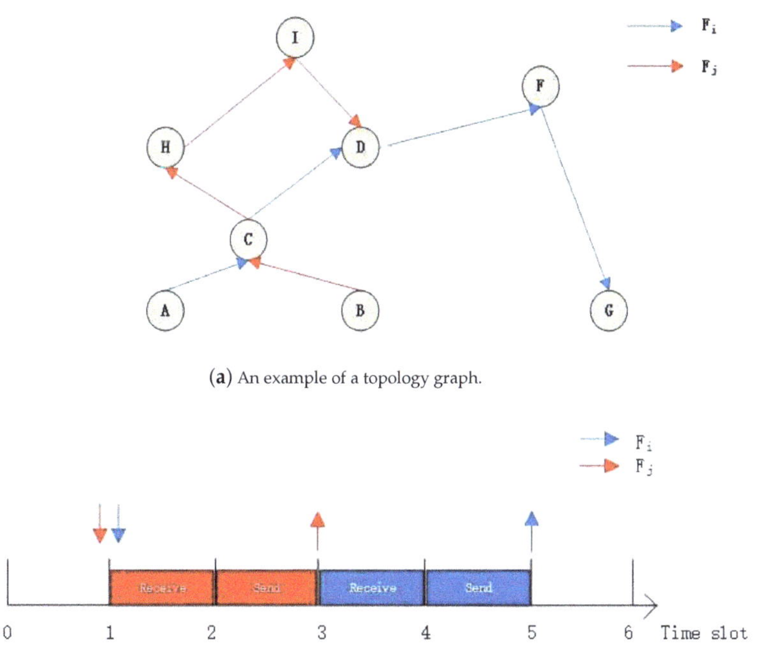

(**a**) An example of a topology graph.

(**b**) The transmission operation of node C.

Figure 1. *Cont.*

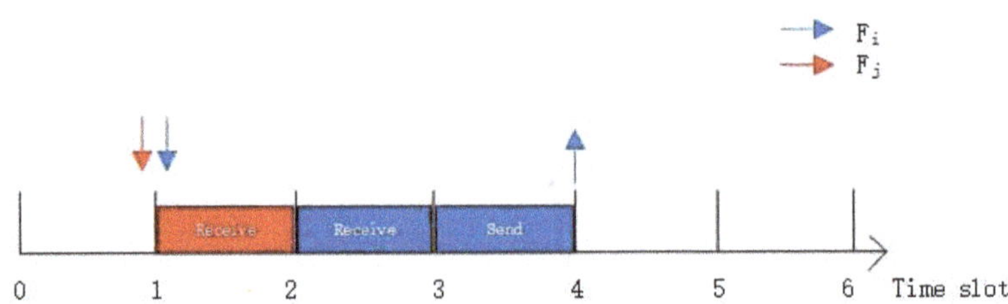

(c) The transmission operation of node D.

Figure 1. An example of transmission conflicts.

3. Real-Time Flow Model

We consider the set F with N flows mentioned above. A flow $F_i, F_i \in F$ consists of three parameters $\langle C_i, T_i, D_i \rangle$. That is, for a flow F_i, C_i is its execution time that contains the transmission time, T_i is its minimum interval time between two consecutive jobs, and D_i is its relative deadline. We also consider that the real-time flow model is deadline-constrained, i.e., $D \leq T$.

The utilization of flow F_i is denoted by $U_i = C_i / T_i$. We let r_i and a_i be the release and arrival time of F_i, respectively. The r_i is the instant that flow F_i sends its packet from the source node. a_i is the instant that the packet arrives at the destination node. Hence, the real transmission time of F_i is $t_i = a_i - r_i + 1$. We denote the end-to-end delay as $delay = C_i + t_i$ shown in Figure 2. The worst-case end-to-end delay (WETED) of F_i as R_i.

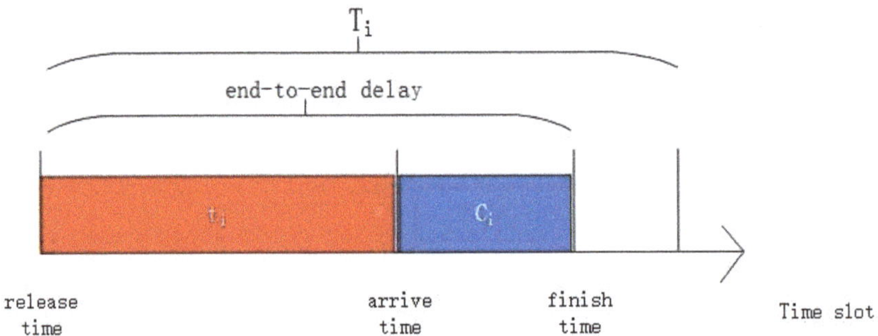

Figure 2. The diagram of the relationship between parameters.

The multi-channel real-time flow model of the WirelessHART network can be transformed into a multi-processor real-time model of a CPU [7,8]. On the basis of the model described above, m channels can be transformed into m processors in the multi-processor model. n flows can be transformed into n real-time tasks, and each task of the multi-processor model also has three parameters $\langle C_i, T_i, D_i \rangle$. WETED can be transformed into the worst-case response time (WCRT) of the multi-processor model.

4. The End-to-End Delay Analyze

In the WirelessHART network, the end-to-end delay is the time when a job of flow F_i is sent from the source node to the destination node to receive and execute the carried packet. In this paper, the slot is the smallest unit of time, which means that high-priority flows can no longer preempt low-priority flows in one slot. A flow with high priority must wait until flows with low priority finish this slot.

Flow F_i is blocked in two ways, namely, (1) channel contention and (2) transmission conflicts, by flows with higher priorities.

4.1. Channel Contention

We consider a real-time flow model with m channels. When a free channel is present, the newly released job of a flow is sent to the channel for execution. When no channel is idle, the newly released high-priority job of a flow preempts the channel that transmits the low-priority job of the flow in the beginning of the next slot. We use $\Omega_i(t)$ to represent the total channel competition caused by F_i for all the jobs of flows with a priority that is higher than that of flow F_i in time t.

For flow F_i, $\Omega_i(t)$ cannot be accurately calculated if we use a global priority assignment strategy in the multi-channel model. Considering that we do not know when the level-k busy period [9] starts, we use the $\Omega_i(t)$ upper bound [10–12] to calculate the upper bound of the WETED and set it to R_i.

4.2. Transmission Conflicts

In the WirelessHART network, when the topology is determined, the transmission conflict between two flows with overlapping transmission routes is a constant [7]. For example, two flows are shown in Figure 1a, F_i and F_j. We assume that F_j has higher priority than F_i, i.e., $j < i$. Flow F_i sends a packet from node A, through nodes C, D, and F, to node G. Flow F_i sends a packet from node B, through nodes C, H, and I, to node D. The common nodes of two flows are C and D. There are two cases of transmission conflicts: (1) full transmission conflicts and (2) half transmission conflicts. The two cases are going to cost two slots and one slot, respectively.

The example of the first case is shown in Figure 1a with node C. We assume that F_i and F_j arrive at node C at the same time. The transmission operation of node C is shown in Figure 1b. We assume that F_i and F_j arrive at Time Slot 1. Node C first costs two slots to receive the packet of flow F_j from node A and sends it from node C to node D, respectively. F_i must wait for two slots because the priority of F_i is lower than that of F_j. The packet of flow F_j then starts to be processed. Hence, the first case of transmission conflicts of F_i is blocked for two slots by F_j at node C.

The example of the second case is shown in Figure 1a with node D. Like the first case, we also assume that F_i and F_j arrive at node D at the same time. The transmission operation of node D is shown in Figure 1c. We assume that F_i and F_j arrive at Time Slot 1. Different form the first case, F_i only needs to wait for one slot because flow F_j ends at node D, i.e., node D do not need cost one slot to send the packet of F_j to other nodes. Hence, the second case of transmission conflicts of F_i is blocked for one slot by F_j at node D.

Therefore, F_i is blocked for three slots by F_j in this topology structure in the worst case. We define $\Delta(i,j)$ as the transmission conflicts caused by F_j to F_i, where F_j has a higher priority than F_i.

4.3. Schedulability Test

If we know the priority assignment for the flow set, we can determine whether the flow set is schedulable or not by calculating the end-to-end delay in the worst case [8]. We can obtain the end-to-end delay in the worst case by calculating the minimum value of y in Equation (1), where t^* is the minimum t in Equation (2).

If one of a flow's t or y is greater than its deadline, we call it unschedulable, or schedulable at that priority. If all the flows in the set are schedulable, we say that this priority assignment policy is satiable.

$$y = t^* + \sum_{j \in hp(i)} \left\lceil \frac{y}{T_j} \right\rceil \Delta(i,j) \tag{1}$$

$$t = \left\lfloor \frac{\Omega_i(t)}{m} \right\rfloor + C_i. \tag{2}$$

5. The Priority Assignment of Real-Time Flows Based on the Worst-Fit

In this section, we first introduce several algorithms (worst-fit, best-fit, and first-fit) to allocate the flows and explain why we use the worst-fit. We then introduce several priority assignment algorithms based on the worst-fit.

We think of the problem of allocating the flows to channels as a bin packing problem, since each channel is equivalent to a processor and the utilization upper bound of the processor is 1. Hence, we assume the channel as a bin with a capacity of 1. Each flow is viewed as cargo whose volume equals its utilization. There are three state-of-the-art existing allocation methods: (1) worst-fit, (2) best-fit, and (3) first-fit. The descriptions and characters of them are as follows:

1. Worst-fit is to allocate the flow to the channel with the maximum remaining capacity of all channels each time. This method can make the use part of each channel more even.
2. Best-fit is to allocate the flow to the channel with the minimum remaining capacity of all channels each time. This method can reduce the remaining resources of each channel as much as possible.
3. First-fit is to allocate the flow to the first channel, and the remaining utilization of this channel is not less than the utilization of this flow. This method can reduce search time.

To make efficient use of each channel, our aim is to distribute the flow to each channel as evenly as possible, while ensuring that the capacity of each channel is within the allowed range. Hence, the worst-fit method is the best choice.

The multi-channel fixed priority transmission scheduling problem for WirelessHART networks can be mapped to the CPU scheduling on a global multiprocessor platform [8] by Saifullah et al. Similarly, we regard each channel for multiple channels WirelessHART networks as a homogeneous and independent processor for CPU scheduling; i.e., each channel is mapped to a processor and there is no interference between any two processors. We then distribute the flows to the channels, i.e., the flow can be transmitted in a channel if and only if it has been distributed in this channel. At last we assign priority to the set of flows in each channel.

This method avoids the analysis of channel contention and uses a new approach to analyze the WETED accurately in Equation (3), where $hp(i)$ is a set of flows that have a higher priority than F_i. R_i can be calculated with C_i as the starting value.

$$R_i = C_i + \sum_{j \in hp(i)} \left\lceil \frac{y}{T_j} \right\rceil (C_j + \Delta(i,j)). \tag{3}$$

5.1. Method of Allocating Flows to Channels based on the Worst Fit

We assume that a set of N flows $F = \{F_1, F_2, ..., F_N\}$ exists in the WirelessHART network with m channels. We denote U^k_{sum} as the total of the utilization of the k-th channel, and it can be calculated with Equation (4). Similarly, we denote U_{sum} as the total utilization of the set, which can be calculated with Equation (5).

$$U^k_{sum} = \sum_{F_i \in channel\ k} U_i. \tag{4}$$

$$U_{sum} = \sum_{k=1}^{m} U^k_{sum}. \tag{5}$$

A necessary condition for flow set schedulability on a multi-processor platform is $U_{sum} \leq m$, where m is the number of processors. This condition means that the set can never be scheduled if $U_{sum} > m$. We also determine that the set can never be scheduled for a single channel if $U^k_{sum} > 1$. We define several symbols to describe the worst-fit method as follows:

- Re^k : the current remaining utilization of the k-th channel;
- Ch^k : the set of flows that belong to Channel k;
- UCh^k : the set of flows that have not been assigned priority in the k-th channel;
- ACh^k : the set of flows that have already been assigned priority in the k-th channel;
- ReF : the set of flows that have not been allocated a channel;
- C_i^j : the execution time of flows F_j at priority i;
- R_i^j : the WETED of flows F_j at priority i.

Before the worst-fit algorithm begins, no flows are allocated to the channel. All flows are stored in the set ReF. The flow in the set ReF is continuously allocated to the channel with the implementation of the worst-fit algorithm. It terminates until all flows in ReF are allocated or until no channel can accept the current allocated flow.

The steps to allocate flows to the channels are as follows:

1. Initialize $Re^k = 1, Ch^k = \emptyset$, where $k \in m$, and then proceed to Step 2.
2. If set ReF is empty, then the priority assignment ends. Return *True* and the algorithm stops. Otherwise, jump to Step 3.
3. Sort the remaining utilization in a descending order for each channel, and then jump to Step 4.
4. Allocate one flow F_i from set ReF into Channel 1. If $Re^1 - U_i < 0$, then we denote the flow set as unschedulable and return *False*. Otherwise, $Re^1 = Re^1 - U_i$. Delete F_i from ReF and jump to Step 2.

Specifically, no channel can accommodate flow F_i if Channel 1 cannot accommodate it because the channels have already been sorted in Step (2). We present a pseudo-code of the worst-fit method as Algorithm 1.

Algorithm 1 The worst-fit algorithm.

```
Input: The set of flows F = {F_1, F_2, ..., F_N}
Output: bool WF
function WORST-FIT
    Re^k ← 1, where k = {1, 2, ..., m}
    Ch^k ← ∅, where k = {1, 2, ..., m}
    for i = 1 to n do
        sort Re in descending order
        if Re^1 − U_i < 0 then
            return False
        else
            Ch^1 = Ch^1 ∪ F_i
            ReF = ReF − F_i
        end if
    end for
    return True
end function
```

We can assign the priority of flows independently in the single channel if the worst-fit method returns *True* in Algorithm 1.

We then consider the priority assignment in a single channel. As shown in Theorem 1, the problem of priority assignment in a single channel (PAS) makes the set of flows acceptable, which is an NPC problem. Hence, we propose two heuristic algorithms: a priority assignment algorithm based on the greedy strategy for C (WF-C) and a priority assignment algorithm based on the greedy strategy for U(WF-U). Both are based on an analysis of Equation (3).

Theorem 1. *The PAS problem is an NPC problem.*

Proof of Theorem 1. First, we prove that the PAS problem is an NP problem. If we assign a priority assignment strategy for a set of flows, we can easily determine the delay of each flow in polynomial time according to Equation (3). By summing the delay of each flow, we can verify whether the total delay can be satisfied or not under the priority assignment strategy. Thus, the PAS problem is an NP problem.

Second, we prove that PAS is an NPC problem. During the maximum continuous working time [13], if all flows have only one job instance, i.e., the inequality in Equation (6) is satisfied, where T_i is the period of $F_i, F_i \in F$, then Equation (7) can be used to calculate the delay of flow F_i. Therefore, solving the PAS problem can be transformed into solving the minimum delay of the lowest priority (DLP) problem.

$$T_i \leq \sum_{j \in hp(i)} [C_j + \Delta(i,j)] . \tag{6}$$

$$R_i = \sum_{j \in hp(i)} (C_j + \Delta(i,j)) . \tag{7}$$

We know that the traveling salesman problem (TSP) is an NPC problem [14]. We stipulate that n cities in the TSP problem correspond to n flows in the DLP problem. We define $\Delta(i,j) + C_j$ as the cost from city i to city j. We define the path from the starting city in the TSP problem as the sequence from low priority to high priority in the DLP problem. Assuming that the starting city of the TSP problem is s, the last arriving city is l, and the solution is ST, then the solution of the DLP problem is $ST - \Delta(l,s)$. In summary, the TSP problem is reduced to a PAS problem. Thus, the PAS problem is an NPC problem. □

5.2. WF-C Algorithm and WF-U Algorithm

Through an analysis of Equation (3), we find that one of the most important factors that influence WETED is the execution time of flow F_j with a higher priority than flow F_i. Thus, we start by assigning the lowest priority to the flow set. According to the greedy strategy, the current priority is assigned to the flow in UCh^k that has the maximum execution time in the schedulable flows. This algorithm can avoid high priority flows with a long execution time causing too many blocks to low priority flows and missing their deadline. The steps of the WF-C algorithm are as follows:

1. Initialize $ACh^k = \emptyset$ and $UCh^k = Ch^k$ and then jump to Step 2.
2. The priority is assigned from the lowest priority, and if set UCh^k is empty, then the priority assignment policy is returned and the algorithm ends. Otherwise, jump to Step 3.
3. Calculate the WETED of all flows in UCh^k at the current priority. If no flow meets its deadline, then the algorithm returns *False* and the algorithm ends. Otherwise, jump to Step 4.
4. Assign the current priority to the flow in Step 2 that meets its deadline and has the longest execution time. Then raise the current priority and jump to Step 2.

The pseudo-code of the WF-C method is shown in Algorithm 2.

Algorithm 2 The WF-C algorithm.

Input: Ch^k
Output: the scheme of priority f
1: **function** WF-C
2: $ACH^k \leftarrow \varnothing$
3: $UCH^k \leftarrow ACH^k$
4: **for** $i = length(CH_k)$ *down to* 1 **do**
5: Calculate all $R_i^j, where\ j \in UCh^k$
6: **if** all $R_i^j > D_j, where\ j \in UCh^k$ **then**
7: **return** *False*
8: **else**
9: $F_j \leftarrow i$ // where $j = max\left\{C_i^j\right\}$
10: **end if**
11: **end for**
12: **return** f
13: **end function**

The WF-U algorithm is similar to the WF-C algorithm. The difference between these two algorithms is the condition for determining priority of flows. The first three steps of the WF-U algorithm are the same as those of the WF-C algorithm in Section 5.2, and the longest execution time is replaced with the maximum utilization in Step 4.

The pseudo-code of the WF-U method is shown in Algorithm 3.

Algorithm 3 The WF-U algorithm.

Input: Ch^k
Output: the scheme of priority f
1: **function** WF-U
2: $ACH^k \leftarrow \varnothing$
3: $UCH^k \leftarrow ACH^k$
4: **for** $i = length(CH_k)$ *down to* 1 **do**
5: Calculate all $R_i^j, where\ j \in UCh^k$
6: **if** all $R_i^j > D_j, where\ j \in UCh^k$ **then**
7: **return** *False*
8: **else**
9: $F_j \leftarrow i$ // where $j = max\left\{U_i^j\right\}$
10: **end if**
11: **end for**
12: **return** f
13: **end function**

5.3. Time Complexity

In this section, we discuss the time complexity of the WF-C and WF-U algorithms. According to Algorithms 2 and 3, we can find that the calculation steps of WF-C and WF-U are basically the same, except that the condition for determining priorities are different. Hence, we just discuss the time complexity of WF-C.

We assume there are m channels and a flow set with N flows. According to Algorithm 2, we know that WF-C contains two parts, which are to allocate flows to channels and to assign priority to flows. The time complexity of the first part, $O(N * m * \log(m))$, is equivalent to the time complexity of ordering N times for m processors (as is known, the time complexity of the quicksort of m elements is $m * \log(m)$).

According to Equation (3), we know that the number of calculations for a certain flow F_i depends on the number of elements in $hp(i)$. For example, we assume that $|hp(i)| = k$. We can then calculate that the number of calculations for F_i is $3k$ (the 3 indicates three basic operations, the plus at the end of Equation (3), the division, and the ceil). There are at most $N - m + 1$ flows in a channel by using the worst-fit algorithm. We can derive that the time complexity of the second part is $O((N - m)^2)$ from Algorithm 1.

From what has been discussed above, the time complexity of WF-C is $O(X)$, where X is $\max(N * m * \log(m), (N - m)^2)$. Moreover, the WF-U has the same time complexity as WF-C.

6. Simulation Experiment

In this section, we propose a new simulation model based on the reality, and conduct many comparative experiments on the four algorithms, namely, a heuristic search (HLS) algorithm (which adds two discriminant conditions to find a feasible solution based on changing priorities of two tasks locally) [7], the worst case fit strategy for the deadline monotonic (WF-D) algorithm (which sends flows into channels based on Algorithm 1, and then assigns priority by DM strategy), the WF-C algorithm, and the WF-U algorithm.

We then compare the performance of these algorithms in terms of the average value of the total end-to-end delay of flow sets, the ratio of schedulable flow sets, and the calculation time of the schedulability analysis according to the experimental results.

On the basis of the actual environment [15–17], we define several symbols to describe our simulation model in Table 1. We assume that the number of flows in each flow set is represented by N. The number of channels in the WirelessHART network is $m = 12$ [7]. The total utilization of the flow sets is generated from 0.05 to 0.9, with 0.05 as the step length. The range of the period is from 2^6 to 2^9 for every flow in the set [18]. Each flow's execution time can be calculated by $C_i = U_i \times T_i$. Through Equation (3), we find that the transmission conflicts caused by F_j to F_i can be converted into the execution time of flow F_j by proportion β, where $\beta = \Delta(i,j)/C_j$. Hence, we assume the upper bound of the transmission conflicts $T^{up} = \beta \times C$. We define α as the cross ratio of a flow to other flows in the set, that is, the number of flows in the collection that have common nodes with that flow in the WirelessHART network.

Table 1. The symbols of simulation model.

Symbol	Description
N	The number of flows in the set.
m	The number of channels.
α	The cross ratio of a flow.
β	The factor for upper bound of transmission conflicts.
δ	The ratio of D to T.
T^{up}	The upper bound of the transmission conflicts.

Under each utilization, we generate 1000 sets of flows and use the average value to represent the characteristics of the flow sets. The experimental parameters are $N = 100, \alpha = 0.1, \beta = 1, \delta = 1$.

6.1. Average Value of the Total End-to-End Delay of Flow Sets

The average value of the total end-to-end delay of the flow sets are shown in Figure 3. The delay of the four algorithms increases with the increase in the total utilization of flow sets. We set the average value of total end-to-end delay of sets to zero, when none of the 1000 sets of tasks we generate are available for scheduling. The average value of total end-to-end delay of flow sets becomes zero, when the total utilization exceeds 0.65, by using HLS to assign priority because no set can be scheduled. This means that there is not one set of the 1000 sets we generated that can be scheduled when the utilization exceeds 0.65. In the same way, there is no set that can be scheduled by using WF-D,

WF-C, or WF-U when the total utilization exceeds 0.85. At each utilization rate, the delay of the four algorithms is sorted from high to low in the order of HLS, WF-D, WF-U, and WF-C. Among the four algorithms, the optimal algorithm WF-C can reduce the delay by up to 44.18% compared with the worst algorithm HLS when the utilization is 0.6.

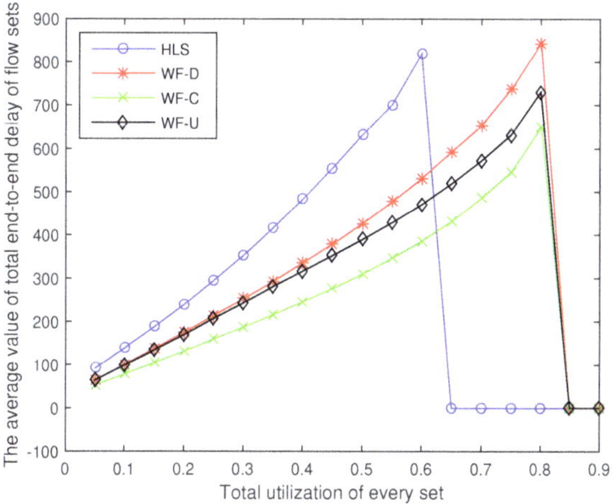

Figure 3. The average value of total end-to-end delay of flow sets.

6.2. Ratio of Schedulable Flow Sets

The ratio of schedulable flow sets refers to the ratio of the scheduled flow sets in the 1000 experimental sets. As shown in Figure 4, the ratio of schedulable flow sets by using the four algorithms decreases from large to small as the total utilization of flow sets increases. When the total utilization of the flow sets is 0.2, the HLS algorithm exhibits unschedulable flow sets, and when the total utilization exceeds 0.6, all flow sets are unschedulable. Similarly, unschedulable flow sets appear when the utilization of WF-D, WF-U, and WF-C algorithms is 0.25, 0.35, and 0.35, respectively. When the utilization rate exceeds 0.85, none of the three algorithms has a schedulable flow set. The WF-C algorithm increases the schedulability ratio by up to 70.7% compared with the HLS algorithm when the utilization is 0.5.

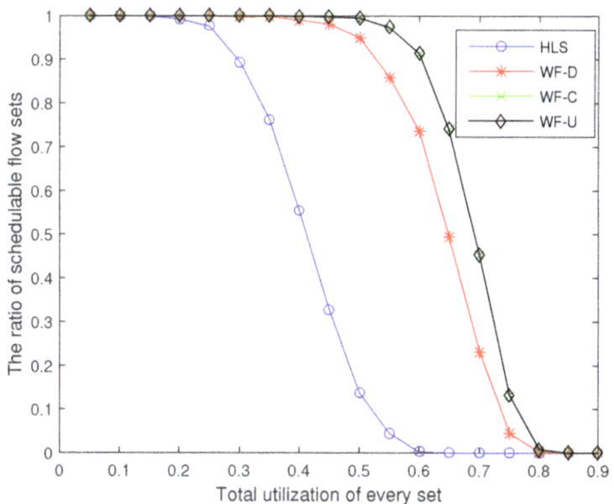

Figure 4. The ratio of scheduled of sets.

6.3. Calculation Time of the Schedulability Analysis

The calculation time of the schedulability analysis represents the time consumed to verify that a flow set is schedulable. The calculation time is shown in Figure 5. We review the WF-D algorithm above. The WF-D only assigns priority according to the deadline of flows in each channel. Different from WF-D, each time WF-C and WF-U assign a priority, they will calculate whether the flow is schedulable under the current priority and choose the highest C and U from all the schedulable flows to assign this priority. In other words, the WF-D algorithm has a pre-allocated priority according to the deadline of the flows and does not need to spend time assigning the priority of the flows. That is why the WF-D algorithm consumes the least amount of time among the four algorithms. As the utilization rate increases, the feasible tree generated by the HLS algorithm increases, and searching for a feasible solution consumes much time. When the utilization exceeds a certain value, due to the high utilization, the number of initial feasible solutions generated by the HLS algorithm becomes smaller, as does the solution space, so the calculation time becomes greatly reduced. Hence, the calculation time of the HLS sharply drops when the utilization exceeds 0.8. The calculation times of the WF-C and WF-U are much shorter than that of the HLS when the utilization is larger than 0.3.

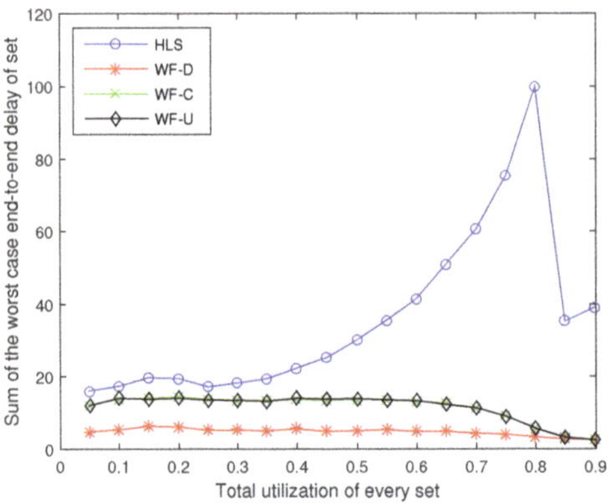

Figure 5. The calculation time of the schedulability analysis.

6.4. Effect of Other Parameters on the Model

To check the stability of our simulation model and algorithms (WF-C and WF-U), we performed several experiments with changing input parameters. Figure 6a–c show the average value of the total end-to-end delay of the flow sets, the ratio of schedulable flow sets, and the calculation time of the schedulability analysis with $N = 100, \alpha = 0.05, \beta = 1$, and $\delta = 1$, respectively. The experiments with $N = 50, \alpha = 0.05, \beta = 2$, and $\delta = 0.8$ are shown in Figure 7.

By observing and comparing the experimental results, we find that the overall trend of the four algorithms remains unchanged even when the input is changed. In other words, the two algorithms we proposed are superior to WF-D and HLS.

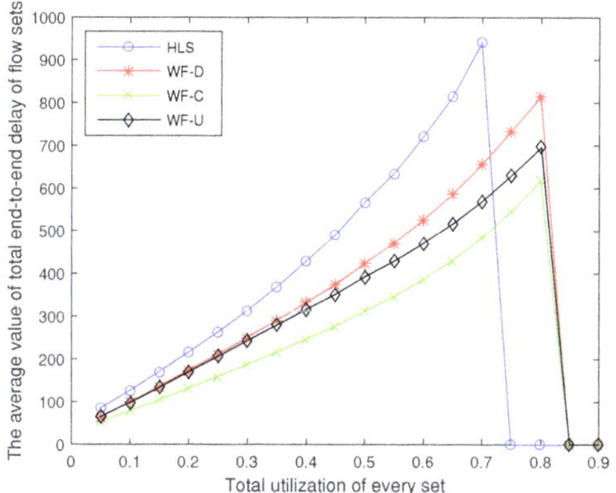

(**a**) The average value of worst-case end-to-end delay of sets.

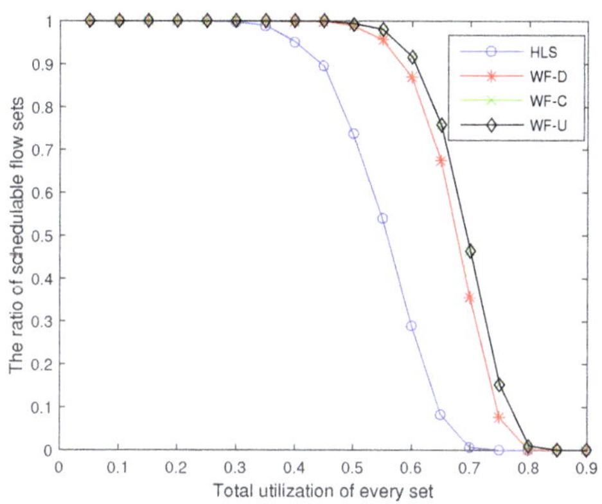

(**b**) The ratio of scheduled of sets.

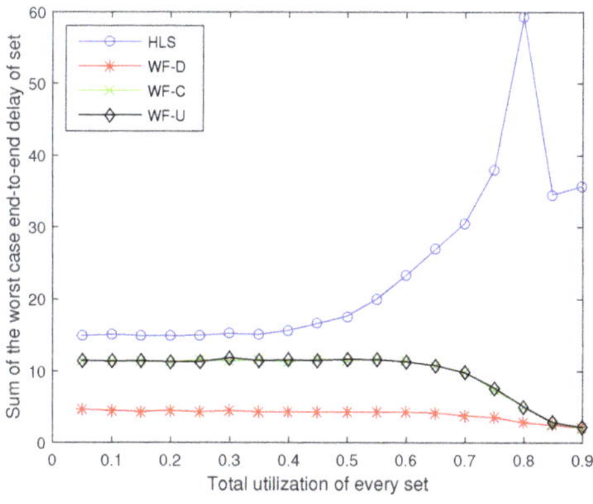

(**c**) The calculation time of the schedulability analysis.

Figure 6. Experiments with $N = 100, \alpha = 0.05, \beta = 1$, and $\delta = 1$.

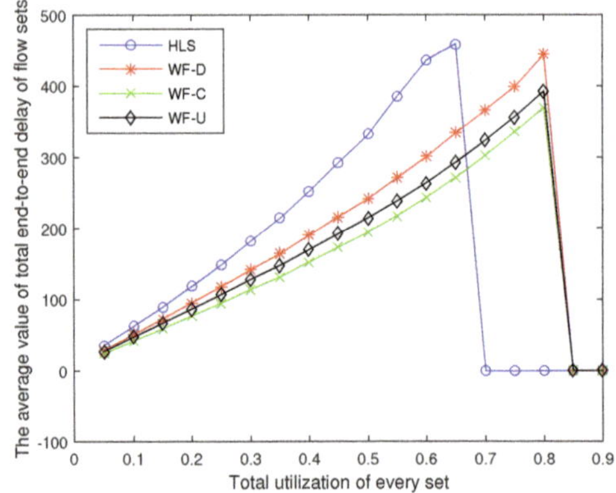

(a) The average value of the worst-case end-to-end delay of sets.

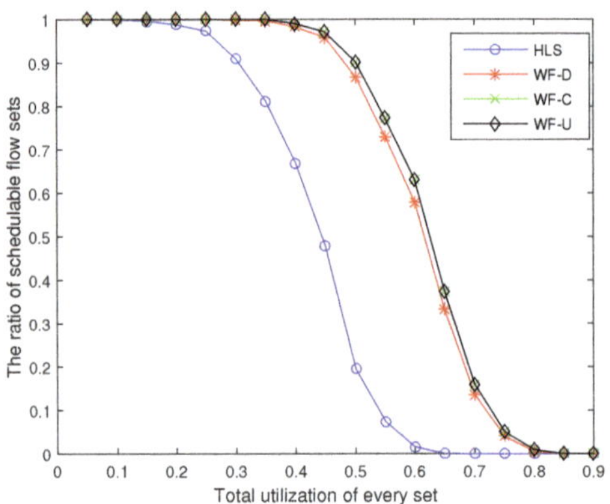

(b) The ratio of scheduled of sets.

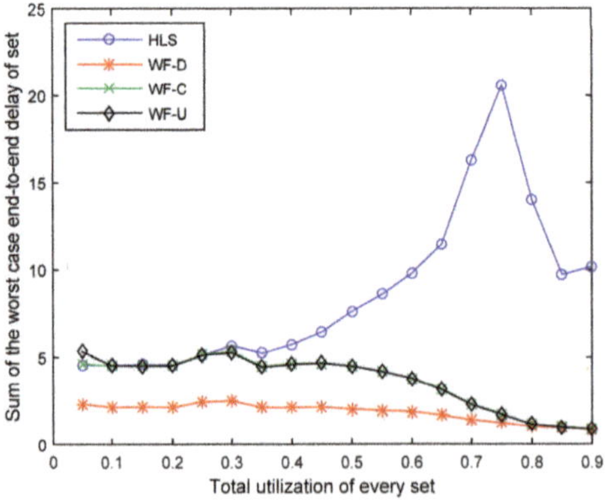

(c) The calculation time of the schedulability analysis.

Figure 7. Experiments with $N = 50, \alpha = 0.05, \beta = 2$, and $\delta = 0.8$.

Comparative experiments indicate that the four algorithms are similar, even with different input parameters. However, WF-C and WF-U are better than the HLS algorithm.

7. Related Work

Priority assignment and schedulability analysis of multi-channel real-time flows in the WirelessHART network are usually treated as problems for multi-processor real-time tasks in a CPU [8]. As a result, we do not know when the level-*k* busy period [9] begins. Therefore, the upper bound of multiprocessor interference is usually calculated instead of the exact solution [19]. Tao et al. proposed a mechanism which can allocate the requirements to user channels based on the different priority levels and ensure that the user with the highest priority will immediately gain channel access [20]. Hossam et al. proposed a new protocol called SS-MAC which can reduce nearly 30% in the worst-case delay [21]. Both Tao and Hossam assume that the deadline of every flow is fixed at 250 ms. Compared with their model, our method, which randomly generates deadlines based on utilization, can be adapted to the system, whose flow deadlines change dynamically. Wei Shen et al. proposed a new scheduling policy called SAS-TDMA to improve the quality of service for the network and reduce the delay for real-time communication [22]. This method improves the reliability of the system in the heavy-noise environment. It is different from our method, which improves reliability by adding redundancy. Furthermore, we use multiple channels to improve the efficiency of the system compared to their method.

The fixed priority model has two global priority assignment strategies, namely, LS and HLS [7]. The LS and HLS algorithms are based on creating a feasible tree of priority assignment strategies and cutting in accordance with the upper and lower bounds of WETED. Several priority assignment strategies are available for multi-processors in the arbitrary deadline model [23–25].

Considering the reality, the energy of IoT devices (including wireless sensors) are powered by batteries (IEEE 802.15.4). Therefore, reducing power has a very important value [26]. Weizhe et al. proposed a trusted real-time scheduling model and a successful meta-heuristic method called shuffled frog leaping algorithm (SFLA) [27] to reduce energy consumption. The node residual energy after data flow transmission in linear wireless sensor networks from source nodes and relay nodes was evaluated by Wang et al. [28].

8. Conclusions and Future Work

In this paper, we transform the multi-channel TDMA real-time flow model in the WirelessHART network into a real-time task scheduling model in the CPU. On the basis of this transformation, we use the worst-fit method to allocate flows to each channel and provide the calculation formula of the WETED of flows in a single channel. Afterward, we present two heuristic algorithms, WF-C and WF-U, to assign priority. Using the simulation model we built, we find that the two proposed algorithms are more efficient than WF-D and HLS algorithms, and the WF-C algorithm is the most efficient. Compared with HLS, the average value of the total end-to-end delay of WF-C can be reduced by up to 44.18%, and the ratio of schedulable flow sets can be increased by up to 70.7%.

In this paper, we assume that the real-time flows are preemption fixed priority flows. However, there are many kinds of real-time flows such as non-preemption, preemption latency fixed priority flows, and many kinds of dynamic priority real-time flows. The scheduling policy of these real-time flows and the schedulability analysis of these real-time flows can be improved in the future. We intend to propose a new schedulability analysis in the future, so as to reduce the calculation time and reduce the time complexity of schedulability analysis, thus improving the computing efficiency of the system. We also intend to propose a new method of priority assignment for other kinds of real-time flows in the future, so as to improve the ratio of scheduled of sets and thus make the system more stable.

Author Contributions: Formal analysis, Y.W.; Investigation, Y.W.; Methodology, Y.W.; Project administration, W.Z.; Resources, W.Z., H.H. and Y.L.; Validation, Y.W.; Writing—original draft, Y.W.; Writing—review and editing, Y.W., W.Z. and H.H.

Funding: This work was supported by the National Key Research and Development Plan under grant No. 2016YFB0800801, the National Science Foundation of China (NSFC) under grant No.61672186, 61472108.

Conflicts of Interest: The authors declare no conflict of interest.

References

1. Sisinni, E.; Saifullah, A.; Han, S.; Jennehag, U.; Gidlund, M. Industrial internet of things: Challenges, opportunities, and directions. *IEEE Trans. Ind. Inform.* **2018**, *14*, 4724–4734. [CrossRef]
2. Wang, H.; He, H.; Zhang, W. Demadroid: Object Reference Graph-Based Malware Detection in Android. *Secur. Commun. Netw.* **2018**, *2018*, 7064131. [CrossRef]
3. Dang, K.; Shen, J.Z.; Dong, L.D.; Xia, Y.X. A graph route-based superframe scheduling scheme in WirelessHART mesh networks for high robustness. *Wirel. Pers. Commun.* **2013**, *71*, 2431–2444. [CrossRef]
4. Khader, O.; Willig, A.; Wolisz, A. *A Simulation Model for the Performance Evaluation of Wirelesshart Tdma Protocol*; Technical Report, Telecommunication Networks Group, Technical University Berlin, TKN Technical Report Series TKN-11-001; Technical University Berlin: Berlin, Germany, 2011.
5. Han, S.; Zhu, X.; Mok, A.K.; Chen, D.; Nixon, M. Reliable and real-time communication in industrial wireless mesh networks. In Proceedings of the 2011 17th IEEE Real-Time and Embedded Technology and Applications Symposium (RTAS), Chicago, IL, USA, 11–14 April 2011; pp. 3–12.
6. Qu, F.; Zhang, J.; Shao, Z.; Qi, S. Research on resource allocation strategy of industrial wireless heterogeneous network based on IEEE 802.11 and IEEE 802.15. 4 protocol. In Proceedings of the 2017 3rd IEEE International Conference on Computer and Communications (ICCC), Chengdu, China, 11–13 December 2017; pp. 589–594.
7. Saifullah, A.; Xu, Y.; Lu, C.; Chen, Y. Priority assignment for real-time flows in WirelessHART networks. In Proceedings of the 2011 IEEE 23rd Euromicro Conference on Real-Time Systems, Porto, Portugal, 5–8 July 2011; pp. 35–44.
8. Saifullah, A.; Xu, Y.; Lu, C.; Chen, Y. End-to-end delay analysis for fixed priority scheduling in WirelessHART networks. In Proceedings of the 2011 17th IEEE Real-Time and Embedded Technology and Applications Symposium (RTAS), Chicago, IL, USA, 11–14 April 2011; pp. 13–22.
9. Breaban, G.; Koedam, M.; Stuijk, S.; Goossens, K. Time synchronization for an emulated CAN device on a Multi-Processor System on Chip. *Microprocess. Microsyst.* **2017**, *52*, 523–533. [CrossRef]
10. Chipara, O.; Lu, C.; Roman, G.C. Real-time query scheduling for wireless sensor networks. In Proceedings of the 28th IEEE International Real-Time Systems Symposium RTSS 2007, Tucson, AZ, USA, 3–6 December 2007; pp. 389–399.
11. Chipara, O.; Lu, C.; Roman, G.C. Real-time query scheduling for wireless sensor networks. *IEEE Trans. Comput.* **2013**, *62*, 1850–1865. [CrossRef]
12. Bertogna, M.; Cirinei, M. Response-time analysis for globally scheduled symmetric multiprocessor platforms. In Proceedings of the 28th IEEE International Real-Time Systems Symposium RTSS 2007, Tucson, AZ, USA, 3–6 December 2007; pp. 149–160.
13. Guan, N.; Ekberg, P.; Stigge, M.; Yi, W. Resource sharing protocols for real-time task graph systems. In Proceedings of the 2011 23rd IEEE Euromicro Conference on Real-Time Systems (ECRTS), Porto, Portugal, 5–8 July 2011; pp. 272–281.
14. Papadimitriou, C.H. The Euclidean travelling salesman problem is NP-complete. *Theor. Comput. Sci.* **1977**, *4*, 237–244. [CrossRef]
15. Künzel, G.; Cainelli, G.P.; Pereira, C.E. A Weighted Broadcast Routing Algorithm for WirelessHART Networks. In Proceedings of the 2017 VII Brazilian Symposium on IEEE Computing Systems Engineering (SBESC), Curitiba, Brazil, 6–10 November 2017; pp. 187–192.
16. Hassan, S.M.; Ibrahim, R.; Saad, N.; Asirvadam, V.S.; Chung, T.D. Implementation of real-time WirelessHART network for control application. In Proceedings of the 2016 6th International Conference on IEEE Intelligent and Advanced Systems (ICIAS), Kuala Lumpur, Malaysia, 15–17 August 2016; pp. 1–6.
17. Nobre, M.; Silva, I.; Guedes, L.A. Performance evaluation of WirelessHART networks using a new network simulator 3 module. *Comput. Electr. Eng.* **2015**, *41*, 325–341. [CrossRef]
18. Pathan, R.M.; Jonsson, J. Interference-aware fixed-priority schedulability analysis on multiprocessors. *Real-Time Syst.* **2014**, *50*, 411–455. [CrossRef]

19. Baruah, S.; Bertogna, M.; Buttazzo, G. *Multiprocessor Scheduling for Real-Time Systems*; Springer: Berlin, Germany, 2015.
20. Zheng, T.; Gidlund, M.; Åkerberg, J. WirArb: A new MAC protocol for time critical industrial wireless sensor network applications. *IEEE Sens. J.* **2016**, *16*, 2127–2139. [CrossRef]
21. Farag, H.; Gidlund, M.; Österberg, P. A Delay-Bounded MAC Protocol for Mission-and Time-Critical Applications in Industrial Wireless Sensor Networks. *IEEE Sens. J.* **2018**, *18*, 2607–2616. [CrossRef]
22. Shen, W.; Zhang, T.; Gidlund, M.; Dobslaw, F. SAS-TDMA: A source aware scheduling algorithm for real-time communication in industrial wireless sensor networks. *Wirel. Netw.* **2013**, *19*, 1155–1170. [CrossRef]
23. Zhang, F.; Burns, A.; Baruah, S. Sensitivity analysis for edf scheduled arbitrary deadline real-time systems. In Proceedings of the 2010 IEEE 16th International Conference on Embedded and Real-Time Computing Systems and Applications (RTCSA), Macau, China, 23–25 August 2010; pp. 61–70.
24. Lee, J.; Chwa, H.S.; Lee, J.; Shin, I. Thread-level priority assignment in global multiprocessor scheduling for DAG tasks. *J. Syst. Softw.* **2016**, *113*, 246–256. [CrossRef]
25. de Oliveira, R.S.; Carminati, A.; Starke, R.A. On using adversary simulators to evaluate global fixed-priority and FPZL scheduling of multiprocessors. *J. Syst. Softw.* **2013**, *86*, 403–411. [CrossRef]
26. Sarwesh, P.; Shet, N.S.V.; Chandrasekaran, K. ETRT–cross layer model for optimizing transmission range of nodes in low power wireless networks—An internet of things perspective. *Phys. Commun.* **2018**, *29*, 307–318.
27. Zhang, W.; Bai, E.; He, H.; Cheng, A.M. Solving energy-aware real-time tasks scheduling problem with shuffled frog leaping algorithm on heterogeneous platforms. *Sensors* **2015**, *15*, 13778–13804. [CrossRef] [PubMed]
28. Wang, N.; Fu, Y.; Zhao, J.; Chen, L. Node importance measure in linear wireless sensor networks. *Adv. Mech. Eng.* **2016**, *8*. [CrossRef]

Article

CA-CWA: Channel-Aware Contention Window Adaption in IEEE 802.11ah for Soft Real-Time Industrial Applications

Yujun Cheng, Huachun Zhou * and Dong Yang

School of Electronic and Information Engineering, Beijing Jiaotong University, Beijing 100044, China
* Correspondence: hchzhou@bjtu.edu.cn

Received: 12 June 2019; Accepted: 5 July 2019; Published: 8 July 2019

Abstract: In 2016, the IEEE task group ah (TGah) released a new standard called IEEE 802.11ah, and industrial Internet of Things (IoT) is one of its typical use cases. The restricted access window (RAW) is one of the core MAC mechanisms of IEEE 802.11ah, which aims to address the collision problem in the dense wireless networks. However, in each RAW period, stations still need to contend for the channel by Distributed Coordination Function and Enhanced Distributed Channel Access (DCF/EDCA), which cannot meet the real-time requirements of most industrial applications. In this paper, we propose a channel-aware contention window adaption (CA-CWA) algorithm. The algorithm dynamically adapts the contention window based on the channel status with an external interference discrimination ability, and improves the real-time performance of the IEEE 802.11ah. To validate the real-time performance of CA-CWA, we compared CA-CWA with two other backoff algorithms with an NS-3 simulator. The results illustrate that CA-CWA has better performance than the other two algorithms in terms of packet loss rate and average delay. Compared with the other two algorithms, CA-CWA is able to support industrial applications with higher deadline constraints under the same channel conditions in IEEE 802.11ah.

Keywords: wireless local area network; IEEE 802.11ah; industrial IoT; medium access control; timeliness

1. Introduction

In recent years, wireless communication has been widely adopted in the field of industrial communication systems [1,2]. Compared with traditional wired industrial communication systems [3] (e.g., Fieldbus and Industrial Ethernet), wireless communication does not require the deployment of expensive communication cables, and therefore they are cost-effective and easy to maintain. Thus, it is also very attractive for industrial soft real-time applications, such as soft real-time control systems [4] and multimedia embedded systems [5]. In soft real-time industrial systems, slight deadline misses are tolerable, as long as their impact is below some functional threshold, although this may affect quality of service and system accuracy to some extent. The tolerance degree depends on the different requirements of underlying industrial applications. Thus, when designing soft real-time systems, it is important to consider the deadline constraint and keep it below the threshold.

As one of the most widely deployed wireless technologies, IEEE 802.11 Wireless Local Area Network (WLAN) becomes a good candidate for various industrial wireless applications with different requirements [5,6]. However, WLAN was originally designed for high throughput applications. When it is adopted in the industrial context, a few issues are still to be resolved, such as energy efficiency, transmission range, interference and real-time performance. To provide a better support for IoT communications, the IEEE task group ah (TGah) released a new standard, called IEEE 802.11ah (marketed as Wi-Fi HaLow), and industrial automation is one of its typical use cases [7]. IEEE 802.11ah operates in the frequency band below 1 GHz, and supports up to 8192 nodes (sensors) in a WLAN with the transmission range up to 1 km. To address the collision problem for such a dense wireless network, the standard introduces a novel access mechanism called Restricted Access Window (RAW). The core idea of RAW is to limit the number of stations accessing the channel by a grouping-based medium access control (MAC) protocol. The stations are partitioned into groups, and the channel is split into slots to decrease collision probability in networks with thousands of stations.

Timeliness is usually dealt with at the MAC layer. In 802.11ah MAC, RAW mechanism can significantly reduce collisions and improve real-time performance. However, in each RAW slot, stations still need to contend for the channel by Distributed Coordination Function and Enhanced Distributed Channel Access (DCF/EDCA). While these MAC layer channel access schemes provide good real-time performance under light traffic, they have severe problems under congested network conditions when applied to real-time applications [8–10]. The original design intention of DCF/EDCA does not consider the deadline requirements, leading to unpredictable real-time performance of the industrial systems. Moreover, high external interference exists in the real industrial scenario, which brings high bit error rates in device communication [11]. The interference can seriously affect the network performance, which makes it harder to meet the real-time requirements of various industrial applications.

In this paper, the authors intend to improve the performance of the IEEE 802.11ah-based soft real-time networks in industrial scenario. The main contributions of this paper are summarized as follows:

- We propose a channel-aware contention window adaption (CA-CWA) algorithm, which dynamically increases and decreases the CW according to the channel status in order to improve the real-time performance of the RAW mechanism in IEEE 802.11ah.
- To eliminate the influence of the interference in real wireless environment, the CW adaption process is integrated with an external interference discrimination method. This method can improve the performance of CA-CWA algorithm effectively in the wireless environment with interference.

The remainder of the paper is organized as follows. Section 2 reviews the background and related work on this research. Section 3 presents the proposed CA-CWA algorithm in detail. Section 4 shows simulation results of the proposed algorithm in detail. Finally, conclusions are given in Section 5.

2. Background

2.1. IEEE 802.11 DCF and EDCA

As a fundamental MAC mechanism of the IEEE 802.11, the Distributed Coordination Function (DCF) is a simple and flexible scheme to share the medium among multiple stations. As shown in Figure 1, in DCF, stations contend for the chance of channel access by Carrier Sense Multiple Access mechanism with Collision Avoidance (CSMA/CA). When collisions happen, DCF adopts a Binary Exponential Backoff (BEB) algorithm to alleviate the congestion [12]. Two separate and distinct carrier-sensing functions are defined in IEEE 802.11 standard: Clear Channel Assessment (CCA) and the Network Allocation Vector (NAV). CCA is physical carrier sense, which determines whether the medium is idle or not, based on energy thresholds from the radio interface. NAV is virtual carrier sense, which is an indicator for the station to avoid potential conflicts by overhearing stations.

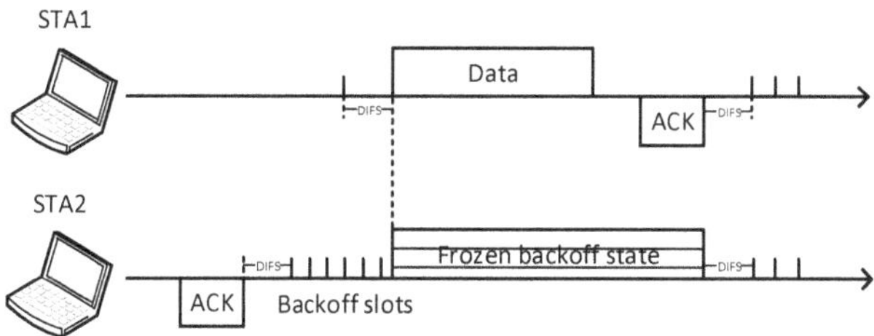

Figure 1. Example of the DCF mechanism.

Before starting a new transmission, each station must sense the status of the channel. The station is permitted to initiate its transmission only if it finds the channel is idle in an additional random backoff period plus a DCF Interframe Space (DIFS) duration. Otherwise, the transmission must be frozen until the medium is idle again. The backoff duration is composed of a multiple of time slots. Each active station generates a uniformly random backoff value from $[0, CW-1]$, where CW is the contention window size. The backoff value is the number of time slots that a station has to wait before transmission. In the first transmission, CW is set to minimal value CW_{min}, which is defined in the standard. When the transmission fails, the CW is doubled until it reaches maximum CW value CW_{max}. Once the CW reaches CW_{max}, the contention window is maintained at CW_{max} even if the next transmissions are still unsuccessful. The CW is set back to CW_{min} after a successful data transmission or when the retransmission counter exceeds the retry limit.

To provide priority-based QoS for real-time applications, IEEE 802.11e task group enhances the DCF through a new channel access mechanism: Enhanced Distributed Channel Access (EDCA). As illustrated in Figure 2, four Access Categories (ACs) are defined in EDCA, namely voice (AC_VO), video (AC_VI), best-effort (AC_BE), and background (AC_BK) traffic. In EDCA, higher priority traffic uses shorter arbitration inter-frame space (AIFS). When an internal traffic collision happens, the higher priority access category obtains the data transmission chance, while the other ACs should restart the backoff procedures. With EDCA, high-priority traffic has a higher transmission chance than low-priority traffic by differentiating the backoff parameters for different ACs (Table 1).

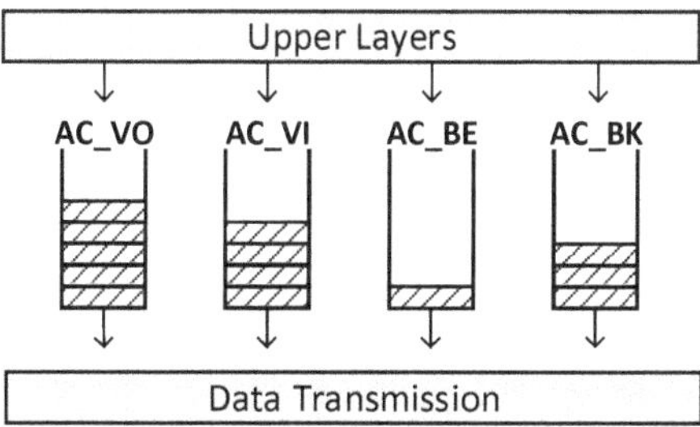

Figure 2. EDCA traffic priorities mapping.

Table 1. IEEE 802.11E EDCA parameter set.

AC	CW_{min}	CW_{max}	$AIFSN$	T_{AIFS}
AC_VO	3	7	2	28 µs
AC_VI	7	15	2	28 µs
AC_BE	15	1023	3	37 µs
AC_BK	15	1023	7	73 µs

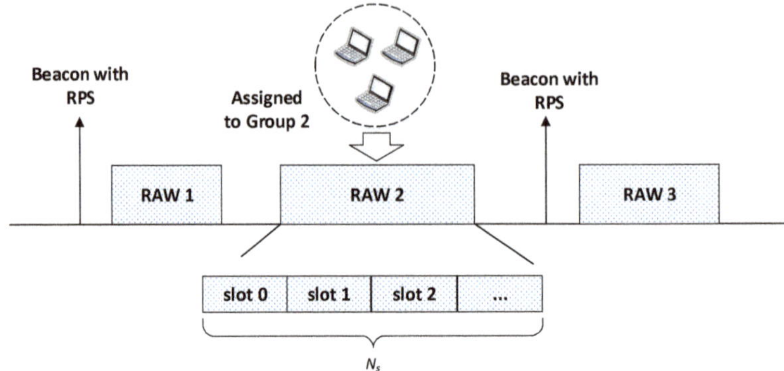

Figure 3. IEEE 802.11ah RAW structure.

2.2. IEEE 802.11ah RAW

Although IEEE 802.11ah inherits most of the basic IEEE 802.11 MAC features, several innovative MAC mechanisms are proposed to support the general requirements of the IoT applications. One of these novel MAC features is Restricted Access Window (RAW).

The RAW mechanism aims to mitigate collisions in dense wireless networks, where a large number of stations are contending for channel access simultaneously. Specifically, the channel time is split into several intervals, namely the RAW periods and the shared channel airtime. As shown in Figure 3, only a portion of stations, namely the RAW stations, from a specific group are allowed to contend the channel in a particular RAW period. By contrast, all stations can compete for the channel in the shared channel airtime. The AP is responsible for assigning each RAW period to a group of stations by a beacon frame carrying a RAW Parameter Set (RPS), which is an information element that specifies the RAW related information, including the stations belonging to the group and the group start time. Besides, the RPS also contains the slot format, the number of RAW slots (N_s) and slot duration count sub-fields, which jointly determine the RAW slot duration as follows:

$$D = 500\ \mu s + C \times 120\ \mu s, \tag{1}$$

where C is slot duration count sub-field, and D is the RAW slot duration. The number of RAW slots N_s and C are determined by the slot format sub-field. If the slot format sub-field is set to 1, each RAW period consists of at most eight RAW slots and the maximum value of C is 2047. Otherwise, each RAW period consists of at most 64 RAW slots and the maximum value of C is 255.

To make EDCA compatible with the RAW mechanism, each station adopts two backoff states of EDCA to manage data transmission inside and outside its assigned RAW slot, respectively [13]. The first backoff state is adopted outside RAW slots, in which all stations are permitted to compete for the channel. For the first backoff state, the station freezes its backoff timer at the start of each RAW period, and resumes the backoff timer at the end of the RAW period. The second backoff state is adopted inside RAW slots, where only the designated group of stations is permitted to contend for channel access. For the second backoff state, stations start backoff procedure at the start of their own RAW slot, and terminate their backoff procedure at the end of their RAW slot.

2.3. Contention Window Adaption

In IEEE 802.11 networks, it is extensively accepted that the backoff algorithm plays a significant role in achieving a high throughput and less medium access delay [14,15]. The IEEE 802.11 adopts a binary exponential backoff algorithm by default. As described in Section 2.1, when collisions happen, the BEB scheme simply exponentially doubles CW value to avoid repeated collisions, while it always resets the CW value to CW_{min} after a successful transmission, assuming that the network is no longer congested. The fundamental problem is that BEB has no perception of the channel state, thus the algorithm does not know how to obtain an appropriate CW value to provide a better network performance.

Thus, considerable effort was devoted to improve the efficiency of the IEEE 802.11 backoff protocol. Several feedback-based schemes [14–19] have been proposed for adapting the station backoff to the present network conditions. In [16], an additional control is introduced on frame transmission for adaptation of CW according to the present network congestion. Further, the authors of [17] tuned CW by runtime estimation of the congestion condition and network status. In [14], the authors proposed a solution CCCW, which dynamically adjusts the CW size in both of saturated and unsaturated traffic conditions. The CW adaptation process in CCCW aims to achieve the optimal throughput. Recently, the authors of [15] proposed a delay-aware CW scheme adaption scheme called D2D, which tunes CW by the present delay level and channel congestion status of the network. However, these CW adaption schemes do not consider the influence of error-prone channel on the CW adjustment. On the other hand, several schemes [20,21] proposed the optimal configuration of the CW parameters in EDCA according to a predefined set of performance criteria. However, the theoretically derived optimal values rely on the actual measurement of network parameters such as number of contending stations, which is difficult to measure precisely in the real dynamic wireless environment.

3. The Proposed CA-CWA Algorithm

To fulfill the requirements of the industrial soft real-time applications, it is a good solution to improve the real-time performance of the IEEE 802.11ah networks by a better backoff algorithm. As discussed in Section 2.3, the most critical issue in designing backoff algorithm is to make it fully aware of the channel and network status. In industrial scenario, high external interference exists, which brings high bit error rates in device communication. The interference will significantly degrade the performance of the backoff algorithm in wireless channels [18,22]. Moreover, the designed CW adaption algorithm also needs to be carefully optimized for the two distinct backoff states of the RAW mechanism in IEEE 802.11ah, due to their different characteristics. To address these challenges, in this section, we first introduce a congestion status estimation scheme with interference discrimination ability by several observation measures. Then, a contention window adaption algorithm called CA-CWA is designed based on the congestion status estimation. Finally, the proposed CW adaption algorithm is integrated with the IEEE 802.11ah networks to provide better real-time performance.

3.1. Congestion Status Estimation

To provide a backoff algorithm in the 802.11 WLAN based on the congestion status, we need to first consider how to estimate the current network congestion level based on the available observation measures. In this work, we choose the parameter called channel busyness ratio (ρ) for capturing the channel status, which refers to the literature [14,19]. The channel busyness ratio is defined as the rate that a station finds the channel is busy during a certain time interval. Let T_i be the slot length of the ith slot, in a given time interval which has n time slots, and the ratio ρ can be calculated as follows:

$$\rho = \frac{\sum_{i=1}^{n} \alpha_i T_i}{\sum_{i=1}^{n} T_i}. \tag{2}$$

where α_i is the indicator function expressed as:

$$\alpha_i = \begin{cases} 1, & \text{if } i\text{th slot is busy} \\ 0, & \text{if } i\text{th slot is idle} \end{cases}. \tag{3}$$

In the IEEE 802.11 standard, every station has the ability of carrier sensing. Thus, α_i is easy to obtain without any additional hardware modification. However, in a real wireless scenario, both external interferences and transmission collisions can cause busy channels, which was not taken into account in the mentioned previous work. To obtain a more accurate estimation of α_i, a method should be designed to discriminate between the external interferences and transmission collisions. Let N be the number of transmissions in a given period of time for an arbitrary station, and S of them are transmitted successfully (ACK is received). On the other hand, suppose there are R slots in which the station does not transmit, and I of them are idle, we can conduct the estimation of the collision probability p_c and the channel error probability p_e by maximum likelihood estimation method [23]:

$$p_c = \frac{R-I}{R}, \tag{4}$$

$$p_e = 1 - \frac{1-(N-S)/N}{1-p_c}. \tag{5}$$

With the collision probability p_c and the channel error probability p_e, the modified channel busyness ratio (ρ') can be calculated as follows:

$$\rho' = \frac{p_c}{p_c + p_e} \frac{\sum_{i=1}^{n} \alpha_i T_i}{\sum_{i=1}^{n} T_i}. \tag{6}$$

3.2. The Contention Window Adaption Scheme

In this section, we present the contention window adaption scheme based on the channel status estimation. In a general backoff scheme, a station should randomly select a backoff value from the interval [0, CW] before each transmission in order to avoid collision. Let W_k be the CW size of the kth transmission attempt. The value of W_k in BEB algorithm can be calculated as follows. If the CW limit CL ($CL = [log_\omega \frac{CW_{max}}{CW_{min}}]$) is greater than the retry limit RL, then,

$$W_k = \omega^k CW_{min}. \tag{7}$$

otherwise,

$$W_k = \begin{cases} \omega^k CW_{min}, & \text{for } k \in [0, CL] \\ CW_{max}, & \text{for } k \in [CL, RL] \end{cases}, \tag{8}$$

where CW_{max}, CW_{min}, and ω are the maximum CW size, the minimum CW size, and the backoff stage factor, respectively. In CA-CWA, only the CW update procedure is different from BEB after a success transmission and a failure transmission. In a fixed interval T_ρ, each station should observe the channel, and calculate the modified channel busyness ratio ρ' by Equation (6). Besides, to minimize the estimation bias introduced by burst traffic or interference, CA-CWA adopts an Exponentially Weighted Moving Average (EWMA) estimator to smoothen the estimated ρ'. In an arbitrary interval T_ρ^j, the value of ρ'_j is updated according to the following rules:

$${\rho_{avg}^{j}}' = (1-\pi) \times \rho'_j + \pi {\rho_{avg}^{j-1}}', \tag{9}$$

where ρ'_j is the estimated ρ' in the interval T_ρ^j, ${\rho_{avg}^{j}}'$ is the smoothed ρ' value for CW adaption, and π is the smoothing factor of the EWMA estimator, which determines the preserved number of historical values in the smoothing process. The ${\rho_{avg}}'$ is updated continuously in each estimating interval T_ρ. The value of T_ρ should be set appropriately to reflect the recent channel state better.

As mentioned in Section 2.1, EDCA defines four traffic types with different priorities. To ensure the priority mechanism still works properly, CA-CWA defines a decrement factor θ with different values for each type of traffic. Based on θ, the station is able to adjust its CW dynamically according to ${\rho_{avg}^{j}}'$. The decrement factor of nth priority traffic (θ_n) is defined as:

$$\theta_n = \min\{\omega^n \rho_{avg}^{j\,\prime}, \theta_{max}\}, \quad n \in [1,4] \tag{10}$$

The traffic priority decreases gradually from $n = 1$ to $n = 4$, which ensures the higher priority class is able to adjust the CW parameter with a smaller θ. θ_{max} is a parameter that keeps the value of θ not be too large. An excessive value of θ might cause the reset CW value to be greater than the previous CW value. For each class n, CW_n is updated after each successful transmission according to Equation (11):

$$CW_{new}^n = \max\{CW_{min}^n, \theta_n CW_{old}^n\}, \quad n \in [1,4] \tag{11}$$

where CW_{old}^n is the CW value before an arbitrary successful transmission for class n, CW_{new}^n is the updated CW value after the successful transmission. After each unsuccessful transmission, the CW of each class is doubled as long as the value of CW is less than CW_{max}, which is as same as the mechanism in BEB. Based on the channel status estimation, the aforementioned design of backoff adaption in CA-CWA follows a few principles which provide improved network performance in terms of timeliness. In a highly congested channel, a station should avoid blind resetting its CW to CW_{min} after a successful transmission. In CA-CWA, a station will select an appropriate (relatively high) CW value according to the high value of θ in this situation. On the other hand, if the channel congestion reduces, a station will also reduce the value of θ and ensure a lower CW to minimize the access delay. In addition, to integrate CA-CWA into the IEEE 802.11ah protocol, CA-CWA defines a multiplier factor λ for the first backoff state (free contention period) in IEEE 802.11ah. For each station, the minimal CW is initialized to the product of the default CW_{min} and λ because the channel is more likely to be congested. The CA-CWA algorithm is summarized in Algorithm 1, and a flowchart is illustrated in Figure 4 to make the algorithm process more clear. The two functions defined in Algorithm 1 are the two most upper arrows that appear in Figure 4. Besides, the specific values of the parameters in the algorithm will be introduced in the simulation part.

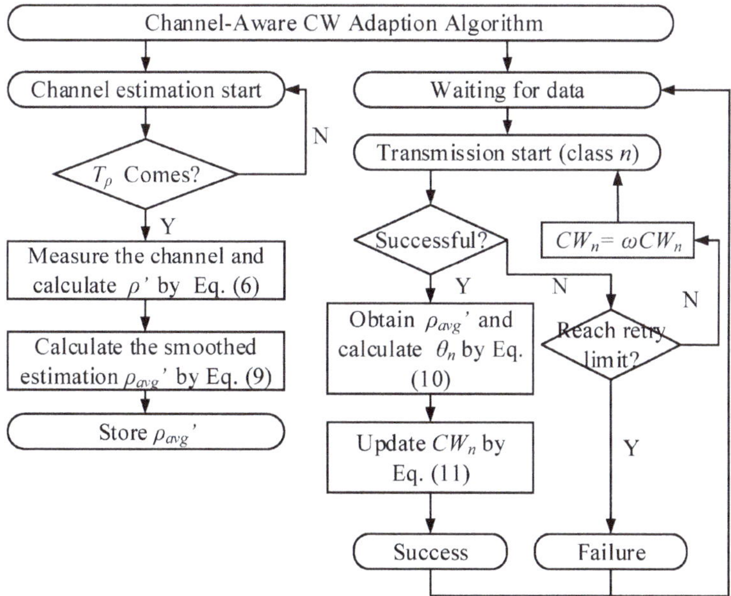

Figure 4. CA-CWA algorithm.

Algorithm 1 Channel-aware CW adaption algorithm.

function ESTMATION
  for each fixed interval T_ρ **do**
    calculate ρ' by Equation (6)
    $\rho_{avg}' \leftarrow$ calculate the smoothed ρ' by Equation (9)
    update ρ_{avg}'
  end for
end function

function CA-CWA
  /*initialization process*/
  if second backoff state **then**
    $CW_{min} \leftarrow CW_{min}$
  else
    $CW_{min} \leftarrow \lambda CW_{min}$
  end if
  initialize other parameters

  while (1) **do**
    ...
    /*after a transmission triggered*/
    if successful **then**
      for $n \in [1, 4]$ **do**
        obtain ρ_{avg}'
        calculate θ_n by Equation (10)
        update CW_n by Equation (11)
        transmission end (successful)
      end for
    else
      if reach retry limit **then**
        transmission end (failed)
      else
        $CW_n \leftarrow \omega CW_n$
      end if
    end if
  end while
end function

4. Performance Evaluation

In this section, we present our simulation results and analysis to demonstrate the real-time performance of the CA-CWA algorithm in IEEE 802.11ah networks.

4.1. Simulation Environment

The CA-CWA algorithm was implemented in the NS-3 simulator with IEEE 802.11ah modules, which is proposed in [24]. The simulation process was based on a general WLAN scenario, where one AP was located in the center, and other stations were randomly distributed around it within its communication range. Each station was installed with a UDP application that generated traffic with the interval of 0.05 s, and the packet size was set to 100 bytes due to the characteristics of small packet size in industrial scenario [25]. Only one transmission queue (AC_BE) was retained to focus on the competition among stations under the RAW mechanism, and CW_{min} and CW_{max} were set to the default values. The values of other network parameters were set by default according to Tian et al. [24]. On the other hand, the algorithm parameters were mainly determined by tests. For example, the smoothing factor π adopted in EWMA is usually recommended to be in the interval between 0.75 and 0.95 according to Lucas and Saccucci [26]. To obtain the optimal value of π, the relationship between the smoothing factor π and the network average delay was obtained through a test simulation. In the simulation, we adopted 10 non-RAW stations and 20 RAW stations, and the number of RAW slots N_s was fixed to 1. The results are shown in Figure 5, and we finally set the value of π to 0.9 for the good performance. The main network parameters in the simulations are listed in Table 2.

Table 2. Basic parameters in simulation.

Basic Parameters	
Reception energy threshold	−116.0 dbm
CCA threshold	−119.0 dbm
Noise figure	3 db
Channel bandwidth	1 MHz
Path loss model	Log-distance
Path loss exponent	3.67
Data rate	2.4 Mbps
Maximal distance between AP and stations	250 m
CW_{min}	15
CW_{max}	1023
UDP traffic interval	0.05 s
Packet payload size	100 bytes
RAW Parameters	
RAW slot format	0
C	100
D	12.5 ms
Number of group	1
Algorithm Parameters	
T_p	5 ms
λ	0.2
ω	2
π	0.9
θ_{max}	0.82

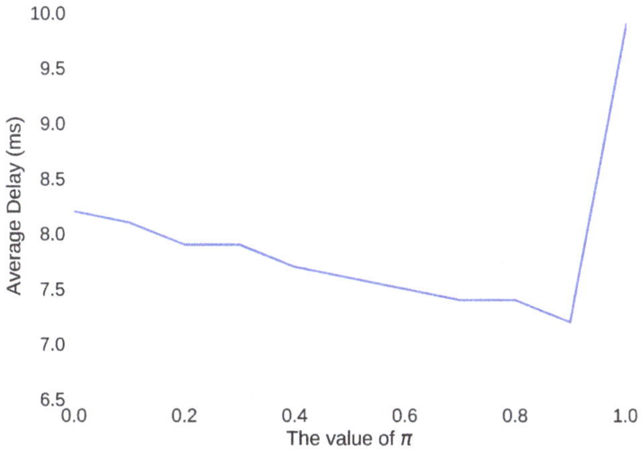

Figure 5. Smoothing factor effect on average delay.

4.2. Simulation Results

Because the proposed CA-CWA algorithm is mainly intended for industrial soft real-time applications, the simulation environment was adjusted to make it similar to the wireless conditions in industrial scenarios. According to the authors of [11,27], high interference exists in unstable and harsh industrial environments, which causes high bit error rates (BER = 10^{-2}–10^{-6}) in industrial wireless communication. Thus, we adopted a loss model to introduce a packet loss ratio at approximately 2.5%. Besides, CA-CWA was compared with two other backoff algorithms, namely BEB and CCCW [14]. BEB is the default backoff algorithm in IEEE 802.11 protocol, and CCCW is another CW adaption algorithm based on the channel congestion status, which is described in Section 2.3.

In the simulation, we mainly focused on the real-time performance improvement that CA-CWA algorithm can bring for IEEE 802.11ah in industrial scenario. Thus, we chose average delay, packet loss ratio, and the delay distribution, which is the most important metric for the industrial soft real-time systems, to test the network performance. These metrics can reflect the real-time performance of the network very well. To eliminate random drift of the simulation results, each simulation was conducted five times, and the results are the average of the five.

To validate the characteristics of the algorithms in IEEE 802.11ah more precisely, the simulation was conducted in two simulation scenarios for the two backoff states.In the first simulation scenario, the number of non-RAW stations N_{NRAW} (the stations which not support RAW) was fixed to 10. The number of RAW stations N_{RAW} varied dynamically to obtain the real-time performance of the algorithms in the second backoff state (the backoff used in the RAW period). Besides, as one of the key parameters in the RAW mechanism, the number of RAW slots N_s was also set to distinct values (N_s = 1, 2 and 4) to show the influence of different RAW parameters on the network real-time performance. In fact, the influence of N_s on network real-time performance has been discussed in the literature [28,29]. In simple terms, when N_s increases and the other RAW parameters remain unchanged, the network delay will increase slightly under a lower load, but decrease slightly under a heavier load. The reason is elaborated in in detail below along with the simulation results.

Table 3 illustrates the change of the values of the core parameters in CA-CWA when varying the number of RAW stations, and N_s was set to 1. To show the channel state estimation process of the CA-CWA in the simulation, we recorded the smoothed modified channel busyness $\rho_{avg}^{j}{}'$. In each interval T_ρ, $\rho_{avg}^{j}{}'$ was updated by Equation (9). Here, we use $\rho_{avg}^{100}{}'$ to show the channel state because $\rho_{avg}{}'$ tended to be stable after one hundred updates. Besides, the backoff decrement factor θ_1, which was calculated by $\rho_{avg}^{100}{}'$, is also listed in the table. We observed that the value of $\rho_{avg}^{100}{}'$ and θ_1 increased as N_{RAW} increased, which is consistent with our intuition. The values of the $\rho_{avg}^{100}{}'$ and θ_1 were quite similar when N_s was set to 2 and 4 in our simulation, and, therefore, the table only shows the change of the values when N_s was set to 1. Moreover, it is easy to find the network real-time performance (average delay and packet loss ratio) under each value of $\rho_{avg}^{100}{}'$ and θ_1 by analyzing Figures 6 and 7. For example, when N_{RAW} equaled to 20, $\rho_{avg}^{100}{}'$ and θ_1 were 89% and 0.82, respectively, and the corresponding delay and packet loss ratio were 7.2 ms and 0, respectively.

Table 3. The calculated core algorithm value in the first scenario (N_s = 1).

N_{RAW}	1	5	10	15	20	25	30	35	40	45	50
$\rho_{avg}^{100}{}'$	2%	14%	35%	53%	89%	92%	94%	95%	98%	98%	98%
θ_1	0.02	0.14	0.35	0.53	0.82	0.82	0.82	0.82	0.82	0.82	0.82

Figures 6 and 7 show the average delay and packet loss ratio of the different backoff algorithms under different RAW parameter settings. We observed that the objective values associated with each algorithm increased as the number of RAW stations increased. This is intuitively expected because the network is more congested if there are more stations transmitting their data. Comparing Figure 6a–c, we observed that the number of the RAW slots (N_s) can influence the delay performance slightly. For example, when the network was not congested ($N_{RAW} < 15$), the average delay increased if the number of the RAW slots N_s increased. When N_{RAW} = 15, the average delay shown in Figure 6a–c was 1.2 ms, 2.24 ms and 7 ms, respectively. The reason is that the RAW mechanism only allows stations to transmit its data in its assigned slot. A station will wait for a longer time to transmit if there are more slots in a RAW. On the other hand, when the network is heavily loaded, adopting RAW can slightly alleviate network congestion, and obtain a lower average delay (although it is not obvious). The similar analysis result can be obtained for packet loss ratio when comparing Figure 7a–c.

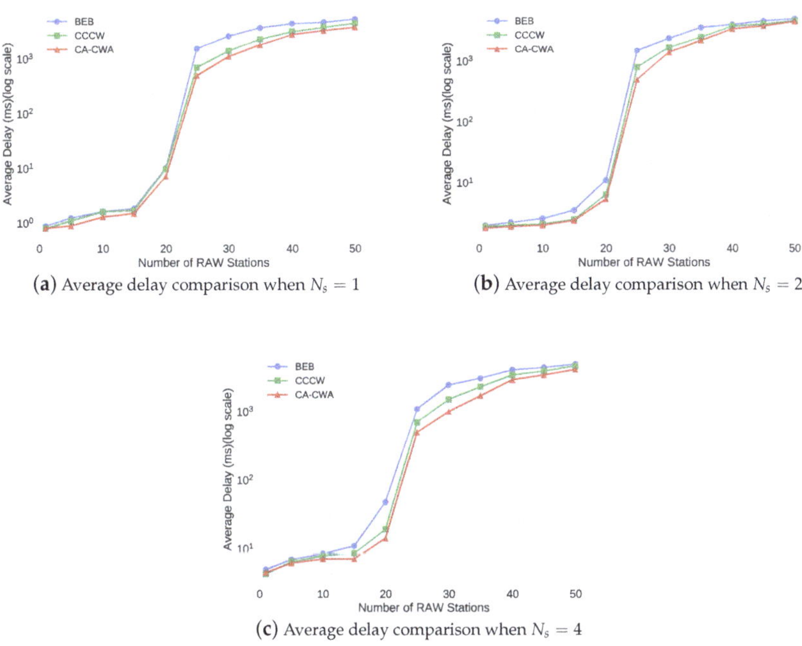

(a) Average delay comparison when $N_s = 1$

(b) Average delay comparison when $N_s = 2$

(c) Average delay comparison when $N_s = 4$

Figure 6. Performance of average delay of each algorithm.

Figures 6 and 7 further show that the CA-CWA algorithm has better performance on average delay and packet loss ratio than the other two algorithms, regardless of the different number of RAW slots. We take the results in Figure 6a as an example; when N_{RAW} was less than 15, all the three backoff algorithms performed well due to the non-congested channel condition. However, when N_{RAW} was more than 20, the average delay of the network rose rapidly. In BEB, each station blindly reset its contention window to CW_{min} after a successful transmission, which further aggravates the degree of channel congestion. Thus, BEB has the worst performance on average delay among the three algorithms. CCCW algorithm aims to optimize the network throughput by adaptive usage of contention window size. It significantly reduces collisions of the stations, and the average delay was up to 40% lower than the performance in BEB. However, CCCW does not consider the collisions caused by interference, which results in the adjusted CW value not matching the real channel condition. Thus, the average delay in CCCW was at most 30% higher comparing with the average delay in CA-CWA. On the other hand, Figure 7a shows the packet loss performance of the three backoff algorithms. The packet loss ratio of BEB increased dramatically from 5% to 23% when N_{RAW} was 40. Compared with BEB, the packet loss ratio decreased up to 37.5% and 62.3% in CCCW and CA-CWA, respectively, when the network was congested ($N_{RAW} >= 35$). A similar conclusion about average delay and packet loss ratio can also be drawn from the results in Figures 6b,c and 7b,c.

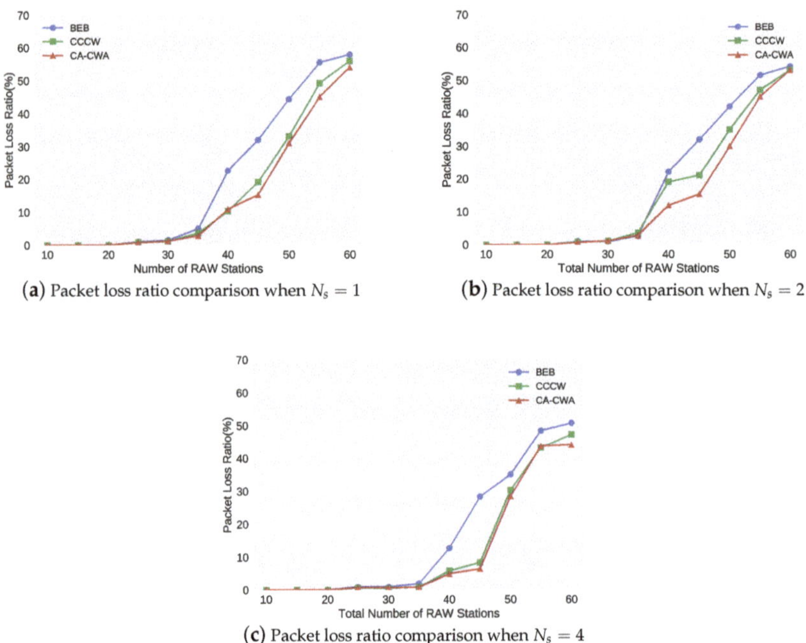

(**a**) Packet loss ratio comparison when $N_s = 1$

(**b**) Packet loss ratio comparison when $N_s = 2$

(**c**) Packet loss ratio comparison when $N_s = 4$

Figure 7. Performance of packet loss ratio of each algorithm.

To indicate the real-time performance of the three algorithms more intuitively, we also plot histograms (Figure 8) to show the discrete distribution of the transmission delay in 10 RAW stations (Figure 8a), 22 RAW stations (Figure 8b) and 40 RAW stations (Figure 8c) scenarios, respectively. In Figure 8a, the end-to-end delay of 95% packets was below 10 ms in BEB, CCCW and CA-CWA, and only 5% packet delay was distributed between 10 and 100 ms. This is because the channel was not crowded in this scenario, and all the three backoff algorithms performed well. In the scenario shown in Figure 8b, the network was slightly congested. The end-to-end delay in BEB was mainly distributed between 100 and 1000 ms, which had the poorest performance among the three algorithms. By contrast, the end-to-end delay of CCCW and CA-CWA was mostly less than 10 ms due to their feature of CW adaption. Compared with BEB and CCCW, CA-CWA had 578% and 23.4% performance boost, respectively, when considering the percentage of delay distributed below 10 ms. In the scenario shown in Figure 8c, the end-to-end delay of the three algorithms was mainly distributed in the interval of more than 1000 ms, due to the heavily congested channel condition. However, compared with the other two algorithms, CA-CWA still had the largest proportion (17.8% for < 10 ms, and 21.3% for 10–100 ms) when considering the end-to-end delay distribution between 0 and 100 ms. In summary, CA-CWA is able to support industrial applications with higher deadline constraints under the same channel conditions in IEEE 802.11ah.

The second simulation scenario was to verify the real-time performance of the first backoff state (the backoff used in the free contention period). Table 4 shows change of the values of the core parameters in CA-CWA when varying the number of non-RAW stations. In this scenario, the number of RAW stations N_{RAW} was fixed to 20, and N_{NRAW} varied from 0 to 40. As shown in Figure 9a,b, we observed that the average delay and packet loss ratio rose rapidly when N_{NRAW} only equaled 5 and 20, respectively. This is because all stations could compete for channels in the first backoff state, which led to more congestion of the channel. The performance comparison of the three algorithms was similar to the discussion in the first scenario, which is disccussed in the previous paragraph. Figure 9a,b shows that the real-time performance of CA-CWA was superior to the other two algorithms too.

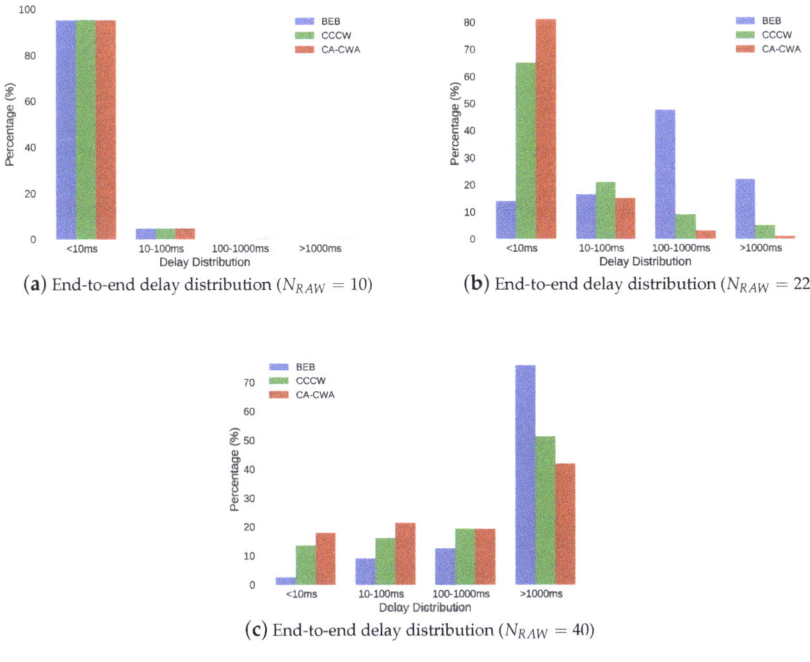

(a) End-to-end delay distribution ($N_{RAW} = 10$)

(b) End-to-end delay distribution ($N_{RAW} = 22$)

(c) End-to-end delay distribution ($N_{RAW} = 40$)

Figure 8. End-to-end delay distribution when $N_s = 2$.

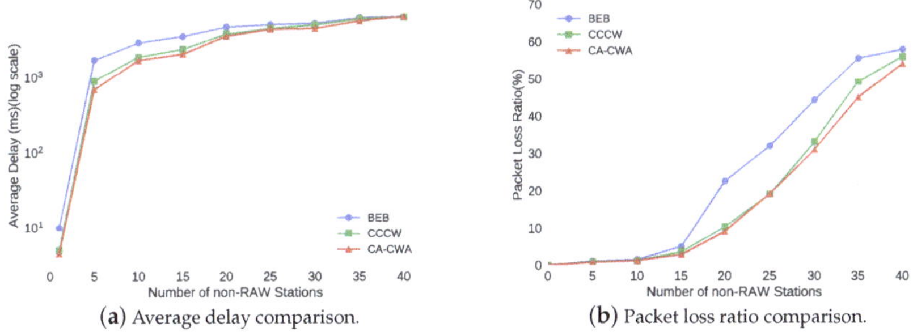

(a) Average delay comparison.

(b) Packet loss ratio comparison.

Figure 9. The real-time performance in the first backoff state.

Table 4. The calculated core algorithm value in the second scenario.

N_{RAW}	1	5	10	15	20	25	30
$\rho_{avg}^{100'}$	80%	92%	94%	95%	97%	98%	98%
θ_1	0.80	0.82	0.82	0.82	0.82	0.82	0.82

5. Conclusions

In this paper, we propose a channel aware contention window adaption (CA-CWA) algorithm for the real-time performance improvement of the IEEE 802.11ah-based industrial applications. The CA-CWA scheme adapts the CW according to a measurement based parameter called channel busyness ratio. Moreover, to eliminate the influence of the interference in real wireless environment on the algorithm, the channel busyness ratio is then modified with an interference discrimination method. To validate the performance of the proposed algorithm, we compared the real-time performance of CA-CWA with the other two algorithms, namely BEB and CCCW, in NS-3 simulator with IEEE 802.11ah modules. The results illustrate that CA-CWA has better performance than the other two algorithms in terms of packet loss rate and average delay. Moreover, CA-CWA has a lower delay distribution in the congested wireless condition. Thus, compared with the other two algorithms, CA-CWA is able to support industrial applications with higher deadline constraints under the same channel conditions in IEEE 802.11ah. As for the future research work, we plan to introduce a model considering both backoff schemes and RAW mechanism to provide a theoretical analysis of the network real-time performance, as well as to develop a new channel state estimation method based on several novel mechanisms, such as machine/deep learning techniques, to provide more accurate channel information.

Author Contributions: Conceptualization, Y.C., H.Z.; validation, Y.C.; writing–original draft preparation, Y.C.; writing–review and editing, Y.C., D.Y., H.Z.

Funding: This paper was partly supported by NSAF under Grant No. U1530118, partly by National High Technology of China ("863 program") under Grant No. 2015AA015702, partly by the National Key Research and Development Program of China under Grant No.2018YFB1702001 and partly by National Natural Science Foundation of China under Grant No. 61771040.

Conflicts of Interest: The authors declare no conflict of interest.

References

1. Magno, M.; Boyle, D.; Brunelli, D.; O'Flynn, B.; Popovici, E.; Benini, L. Extended wireless monitoring through intelligent hybrid energy supply. *IEEE Trans. Ind. Electron.* **2014**, *61*, 1871–1881. [CrossRef]
2. Cheng, Y.; Yang, D.; Zhou, H. Det-LB: A Load Balancing Approach in 802.11 Wireless Networks for Industrial Soft Real-Time Applications. *IEEE Access* **2018**, *6*, 32054–32063. [CrossRef]
3. Wang, T.; Qiu, J.; Yin, S.; Gao, H.; Fan, J.; Chai, T. Performance-based adaptive fuzzy tracking control for networked industrial processes. *IEEE Trans. Cybern.* **2016**, *46*, 1760–1770. [CrossRef] [PubMed]
4. Silvestre-Blanes, J.; Almeida, L.; Marau, R.; Pedreiras, P. Online QoS management for multimedia real-time transmission in industrial networks. *IEEE Trans. Ind. Electron.* **2010**, *58*, 1061–1071. [CrossRef]
5. Seno, L.; Cena, G.; Valenzano, A.; Zunino, C. Bandwidth management for soft real-time control applications in industrial wireless networks. *IEEE Trans. Ind. Inform.* **2017**, *13*, 2484–2495. [CrossRef]
6. Seno, L.; Cena, G.; Scanzio, S.; Valenzano, A.; Zunino, C. Enhancing communication determinism in Wi-Fi networks for soft real-time industrial applications. *IEEE Trans. Ind. Inform.* **2017**, *13*, 866–876. [CrossRef]
7. Park, M. IEEE 802.11 ah: Sub-1-GHz license-exempt operation for the internet of things. *IEEE Commun. Mag.* **2015**, *53*, 145–151. [CrossRef]
8. Sojka, M.; Molnár, M.; Hanzálek, Z. Experiments for real-time communication contracts in IEEE 802.11e EDCA networks. In Proceedings of the IEEE International Workshop on Factory Communication Systems, Dresden, Germany, 21–23 May 2008; pp. 89–92.
9. Moraes, R.; Portugal, P.; Vasques, F.; Fonseca, J.A. Limitations of the IEEE 802.11e EDCA protocol when supporting real-time communication. In Proceedings of the IEEE International Workshop on Factory Communication Systems, Dresden, Germany, 21–23 May 2008; pp. 119–128.
10. Cena, G.; Seno, L.; Valenzano, A.; Zunino, C. On the performance of IEEE 802.11e wireless infrastructures for soft-real-time industrial applications. *IEEE Trans. Ind. Inform.* **2010**, *6*, 425–437. [CrossRef]
11. Gungor, V.C.; Hancke, G.P. Industrial wireless sensor networks: Challenges, design principles, and technical approaches. *IEEE Trans. Ind. Electron.* **2009**, *56*, 4258–4265. [CrossRef]
12. Ali, R.; Kim, S.W.; Kim, B.S.; Park, Y. Design of MAC layer resource allocation schemes for IEEE 802.11 ax: Future directions. *IETE Tech. Rev.* **2018**, *35*, 28–52. [CrossRef]
13. Shahin, N.; Ali, R.; Kim, Y.T. Hybrid Slotted-CSMA/CA-TDMA for Efficient Massive Registration of IoT Devices. *IEEE Access* **2018**, *6*, 18366–18382. [CrossRef]
14. Hong, K.; Lee, S.; Kim, K.; Kim, Y. Channel condition based contention window adaptation in IEEE 802.11 WLANs. *IEEE Trans. Commun.* **2012**, *60*, 469–478. [CrossRef]

15. Khatua, M.; Misra, S. D2D: Delay-aware distributed dynamic adaptation of contention window in wireless networks. *IEEE Trans. Mob. Comput.* **2016**, *15*, 322–335. [CrossRef]
16. Bononi, L.; Conti, M.; Donatiello, L. Design and performance evaluation of a distributed contention control (DCC) mechanism for IEEE 802.11 wireless local area networks. *J. Parallel Distrib. Comput.* **2000**, *60*, 407–430. [CrossRef]
17. Bononi, L.; Conti, M.; Gregori, E. Runtime optimization of IEEE 802.11 wireless LANs performance. *IEEE Trans. Parallel Distrib. Syst.* **2004**, *15*, 66–80. [CrossRef]
18. Deng, D.J.; Ke, C.H.; Chen, H.H.; Huang, Y.M. Contention window optimization for IEEE 802.11 DCF access control. *IEEE Trans. Wirel. Commun.* **2008**, *7*, 5129–5135. [CrossRef]
19. Heusse, M.; Rousseau, F.; Guillier, R.; Duda, A. Idle sense: An optimal access method for high throughput and fairness in rate diverse wireless LANs. In Proceedings of the ACM SIGCOMM Computer Communication Review, Philadelphia, PA, USA, 22–26 August 2005; Volume 35, pp. 121–132.
20. Serrano, P.; Banchs, A.; Patras, P.; Azcorra, A. Optimal configuration of 802.11e EDCA for real-time and data traffic. *IEEE Trans. Veh. Technol.* **2010**, *59*, 2511–2528. [CrossRef]
21. Patras, P.; Banchs, A.; Serrano, P.; Azcorra, A. A control-theoretic approach to distributed optimal configuration of 802.11 WLANs. *IEEE Trans. Mob. Comput.* **2011**, *10*, 897–910. [CrossRef]
22. Ma, H.; Roy, S. Contention window and transmission opportunity adaptation for dense IEEE 802.11 WLAN based on loss differentiation. In Proceedings of the IEEE International Conference on Communications, Beijing, China , 19–23 May 2008; pp. 2556–2560.
23. Malone, D.; Clifford, P.; Leith, D.J. MAC layer channel quality measurement in 802.11. *IEEE Commun. Lett.* **2007**, *11*, 143–145. [CrossRef]
24. Tian, L.; Deronne, S.; Latré, S.; Famaey, J. Implementation and Validation of an IEEE 802.11 ah Module for ns-3. In Proceedings of the Workshop on ns-3, Seattle, WA, USA, 15–16 June 2016; pp. 49–56.
25. Moyne, J.R.; Tilbury, D.M. The emergence of industrial control networks for manufacturing control, diagnostics, and safety data. *Proc. IEEE* **2007**, *95*, 29–47. [CrossRef]
26. Lucas, J.M.; Saccucci, M.S. Exponentially Weighted Moving Average Control Schemes: Properties and Enhancements. *Technometrics* **1990**, *32*, 1–12, doi:10.1080/00401706.1990.10484583. [CrossRef]
27. Willig, A.; Kubisch, M.; Hoene, C.; Wolisz, A. Measurements of a wireless link in an industrial environment using an IEEE 802.11-compliant physical layer. *IEEE Trans. Ind. Electron.* **2002**, *49*, 1265–1282. [CrossRef]
28. Raeesi, O.; Pirskanen, J.; Hazmi, A.; Levanen, T.; Valkama, M. Performance evaluation of IEEE 802.11 ah and its restricted access window mechanism. In Proceedings of the IEEE International Conference on Communications Workshops (ICC), Sydney, Australia, 11–14 June 2014; pp. 460–466.
29. Badihi, B.; Del Carpio, L.F.; Amin, P.; Larmo, A.; Lopez, M.; Denteneer, D. Performance evaluation of IEEE 802.11 ah actuators. In Proceedings of the IEEE 83rd Vehicular Technology Conference (VTC Spring), Nanjing, China, 15–18 May 2016; pp. 1–5.

Article

Virtualization of Industrial Real-Time Networks for Containerized Controllers

Sang-Hun Lee [1], Jong-Seo Kim [2], Jong-Soo Seok [3] and Hyun-Wook Jin [4,*]

1 Hyundai Mobis Co., Ltd., Yongin-si, Gyeonggi-do 16891, Korea; sanghun@mobis.co.kr
2 LIG Nex1 Co., Ltd., Seongnam-si, Gyeonggi-do 13488, Korea; jongseo.kim@lignex1.com
3 Electronics and Telecommunications Research Institute, Daejeon 34129, Korea; jsseok@etri.re.kr
4 Department of Computer Science & Engineering, Konkuk University, Seoul 05029, Korea
* Correspondence: jinh@konkuk.ac.kr

Received: 8 September 2019; Accepted: 9 October 2019; Published: 11 October 2019

Abstract: The virtualization technology has a great potential to improve the manageability and scalability of industrial control systems, as it can host and consolidate computing resources very efficiently. There accordingly have been efforts to utilize the virtualization technology for industrial control systems, but the research for virtualization of traditional industrial real-time networks, such as Controller Area Network (CAN), has been done in a very limited scope. Those traditional fieldbuses have distinguished characteristics from well-studied Ethernet-based networks; thus, it is necessary to study how to support their inherent functions transparently and how to guarantee Quality-of-Service (QoS) in virtualized environments. In this paper, we suggest a lightweight CAN virtualization technology for virtual controllers to tackle both functionality and QoS issues. We particularly target the virtual controllers that are containerized with an operating-system(OS)-based virtualization technology. In the functionality aspect, our virtualization technology provides virtual CAN interfaces and virtual CAN buses at the device driver level. In the QoS perspective, we provide a hierarchical real-time scheduler and a simulator, which enable the adjustment of phase offsets of virtual controllers and tasks. The experiment results show that our CAN virtualization has lower overheads than an existing approach up to 20%. Moreover, we show that the worst-case end-to-end delay could be reduced up to 78.7% by adjusting the phase offsets of virtual controllers and tasks.

Keywords: virtualization; controller area network; fieldbus; real-time; container

---

1. Introduction

The contemporary industrial control systems comprise many sensors, actuators, and controllers connected through real-time networks. As the number of sensors and actuators in modern industrial plants increases drastically, the manageability of the controllers that directly interact with sensors and actuators in real-time becomes a serious concern. Accordingly, the demands for the flexibility with regard to hosting and consolidation of the controllers in large and complex industrial plants are constantly growing. For instance, to address the manageability and scalability in industrial control systems, there is an active movement to exploit cloud computing technologies in the infrastructure of smart manufacturing with the advent of the Industry 4.0 era [1–3].

In cloud computing, the virtualization is the key technology that provides the resource isolation and security between virtual machines [4]. Thus, it is expected that the industrial control software can also be efficiently deployed and executed in given computing resources by means of virtualization, while satisfying the requirements on security by preventing unauthorized resource access between virtual controllers. However, existing virtualization technologies in cloud computing do not support essential components of industrial control systems. It is particularly important to provide functional transparency and a real-time guarantee of industrial networks in virtualized environments. There were significant studies on network virtualization, but most of the existing research focused on the performance optimization of Transmission Control Protocol/Internet Protocol (TCP/IP) over Ethernet [5–10] or high-performance interconnects [11]. Although the Ethernet-based industrial networks, such as EtherCAT [12] and PROFINET [13], are emerging, traditional fieldbuses, such as Controller Area Network (CAN) [14], are still prevalent among the majority of control systems. CAN is a bus-based network standardized in ISO 11898. In order to support CAN in virtualized environments, we have to deal with following challenging issues:

- *Support for sharing of the network interface*: In order to allow several virtual controllers to share a physical CAN network interface in an isolated manner, the run-time support should be capable of multiplexing and demultiplexing the input/output (I/O) requests from multiple virtual controllers. However, the protocol stacks of CAN (i.e., CANopen [15]) implicitly assume that the CAN network interface can be dedicated to only one software controller.
- *Emulation of the media access control*: The characteristics of CAN are significantly different from general purpose networks. For example, the CAN message identifier is used in bus arbitration; that is, it is considered as a priority for bus arbitration. Thus, such characteristics have to be emulated in virtualized environments to preserve the behavior of controllers.
- *Low virtualization overheads*: As the traditional hypervisor-based virtualization (e.g., Xen [4], VMware [16], and VirtualBox [17]) adds significant run-time overheads, the operating-system (OS)-based virtualization (e.g., Container [18,19]) is emerging. Accordingly, we need a CAN virtualization technology that can be incorporated into the OS-based virtualization aiming to minimize the virtualization overheads.
- *Analysis of end-to-end delay*: In virtualized environments, multiple virtual controllers share the CPU resources; thus, the end-to-end delay of control loop highly depends on how the virtual controllers are scheduled. Therefore, we need a mechanism to analyze the worst-case end-to-end delay and minimize it to satisfy the requirements on real-time.

In this paper, we suggest a lightweight CAN virtualization technology for virtual controllers that are containerized with an OS-based virtualization. Our study mainly focuses on how to provide correct communication semantics and functionalities of industrial fieldbuses in OS-based virtualization, while providing low overheads and Quality-of-Service (QoS). The proposed scheme does not require any modifications of control applications and protocol stacks. There were also studies to virtualize CAN, but these were hardware-level approaches [20] or targeted the hypervisor-based virtualization [21]. We also suggest adjusting the phase offsets of virtual controllers and their tasks to minimize the end-to-end delay. By adjusting the execution point of tasks that perform communication over fieldbuses, we can improve the worst-case end-to-end delay. We implemented a simulation tool that finds a sub-optimal phase combination of virtual controllers and tasks. Although the phasing schemes were also discussed in other studies, these did not consider virtualized environments [22–25]. The performance measurement results show that our CAN virtualization technology hardly adds additional overheads and reports lower overheads than a hypervisor-based virtualization up to 20%. We also show that our phasing scheme can reduce the worst-case end-to-end delay by 47.0∼78.7%.

The rest of the paper is organized as follows: we discuss the related work in Section 2. We detail the suggested design of CAN virtualization and its implementation in Section 3. In this section, we present the device-driver-level CAN virtualization for containerized controllers and the simulation tool for optimal phasing of controllers and their tasks. The performance measurement results are presented in Section 4. Finally, we conclude this paper in Section 5.

2. Related Work

The virtualization technology provides multiple virtual platforms, each of which can run their own applications and system software on a single physical computing node. In legacy virtualization approaches, the software layer that provides the virtual machines is called hypervisor or Virtual Machine Monitor (VMM). A hypervisor can run on bare hardware (i.e., Type-1) or on top of an OS (i.e., Type-2). We call the OS running on the virtual machine as guest OS. In the Type-2 environment, the OS that hosts the hypervisor is called host OS. We can classify the virtualization technology into two: full-virtualization and para-virtualization. The full-virtualization allows the legacy software as either OS or applications to run in a virtual machine without any modifications. To do this, the hypervisors usually perform the binary translation and emulate every detail of physical hardware instruction sets. VMware [16] and VirtualBox [17] are examples of full-virtualization hypervisors. On the other hand, the para-virtualization requires modifications of guest OS in order to minimize the virtualization overhead. The hypervisors of para-virtualization provide guest OS with programming interfaces called hypercalls. Consequently, the para-virtualization presents better performance than full-virtualization. Xen [4] and XtratuM [26] are examples of para-virtualization hypervisors.

However, the para-virtualization also adds significant run-time overheads compared with the raw (i.e., non-virtualized) systems. The emerging OS-based virtualization [18] in which the OS instance is shared between the guest and the host domains does not induce significant overheads because the OS takes care of virtualization without extra software layers, such as hypervisor and multiple OS instances. In the OS-based virtualization, we call the guest domain as a container, and the OS has to guarantee the resource isolation between containers. The Linux kernel, for instance, provides *control group (cgroup)* and *name space* to guarantee the resource isolation with respect to resource usage and security, respectively [27].

The virtualization technology is mainly utilized in the cloud computing systems, but it also has very high potential of improving manageability and safety in industrial control systems. For example, the partitioning defined by ARINC-653 [28] and AUTOSAR [29] to provide temporal and spatial isolation between avionics and automobile applications can be ideally implemented by the virtualization technology [30]. In addition, there were efforts to host industrial control services in cloud infrastructures by virtualizing Programmable Logic Controllers (PLCs) and control networks [1–3]. These efforts correspond to the trend of running PLCs on open platforms such as PCs [31,32]. These studies have showed that the virtualization is a promising solution for providing infrastructure consolidation, manageability, resiliency, and security in industrial control systems. The existing studies, however, only targeted the Ethernet-based control networks, of which the virtualization technologies are already available; thus, there were no thorough discussions on how to host the traditional fieldbuses, such as CAN, in virtualized environments.

There has been significant research on network virtualization, which can be classified into network interface virtualization and Software-Defined Networking (SDN). Most of the existing network interface virtualization technologies focused on the performance optimization of TCP/IP over Ethernet interfaces [5–10] or high-performance interconnects such as InfiniBand [11]. A widely accepted approach is to provide multiple virtual network interfaces with the assistance of the network interface card. Since this approach requires the support from the network devices, it is not suitable to apply this to the fieldbus interface, which is not equipped with sufficient hardware resources to implement multiple virtual network interfaces. Though there was an architectural research on efficient network interface virtualization for CAN [20], it also highly depended on the assistance from a network interface card. To address the manageability and flexibility in the network architecture, the SDN technology that dissociates the control plane from data plane has been suggested [33]. There was a study to exploit SDN for control systems [34], but this also targeted only Ethernet because the existing SDN technology is limited primarily to IP-based networks.

Researchers also studied container scheduling for industrial IoT applications in cloud and fog computing environments [35,36]. However, they targeted soft real-time applications. There were studies to provide hard real-time scheduling in fog computing infrastructures [37,38], but these focused on task-level scheduling without consideration of container-level scheduling. In this paper, we study hierarchical CPU scheduling that performs task- and container-level scheduling for hard real-time control applications.

There is a lot of research on message scheduling [39–43] to meet the real-time constraints. Guaranteeing service-level real-time in a CAN-based networked control system is also studied [44]. There are several research works on synchronization between distributed controllers [45–48]. In this paper, we especially aim to adjust the phase offsets of virtual controllers and their tasks by means of the global clock and improve the worst-case end-to-end delay. Sung et al. [25] showed the possibility of synchronizing distributed control tasks by utilizing the global clocks of real-time control networks, but they considered only communication tasks while overlooking the preemption by higher-priority tasks. Kim and Kim [24], Kang et al. [23], and Lee at al. [22] exploited the global clock more positively to adjust task phases for isochronous control on EtherCAT. Craciunas et al. [49] tried to decide the optimal offset of tasks in terms of utility. However, previous studies did not consider the virtualized environments, where we have to deal with the phase of guest domains as well as the phase of tasks.

3. Virtualization of Controller Area Network

In this section, we suggest a lightweight virtualization of the CAN fieldbus for containerized virtual controllers. In addition, we implement a hierarchical real-time scheduler and a simulator to guarantee the real-time requirements of virtual controllers.

3.1. Design Issues

We can consider four design alternatives for virtualization of industrial network interfaces as shown in Figure 1. In the emulation scheme (Figure 1a), the hypervisor emulates the target network interface and provides a channel to access the physical network interface that can be different from the target interface. However, as mentioned in Section 2, the hypervisor-based virtualization increases run-time overheads. Thus, we target the OS-based virtualization and consider virtual controllers as containerized instances. In the relay scheme (Figure 1b), a daemon process manages actual data transmission. The virtual controllers have to communicate with this daemon process through Inter-Process Communication (IPC) channels to send and receive messages. This solution is the only way to make the virtual controllers share the CAN interface without modifications of underlying system software or hardware. However, this approach not only requires modifications of control applications, but also induces a significant overhead due to IPC. The network interface capable of self-virtualization (Figure 1c) provides the virtual interfaces by itself. However, since each virtual interface is dedicated to a virtual controller, this alternative adds memory and computation overheads onto the network interface. The industrial network interfaces are usually equipped with a low-speed processing unit and low memory space; thus, many interfaces are not capable of accepting this design choice. The driver-level virtualization (Figure 1d) is somewhat similar with self-virtualization but virtual interfaces are provided by the device driver. Compared with relay and self-virtualization approaches, the driver-level virtualization is superior in both performance and resource requirements. We will describe the details of the driver-level virtualization in Section 3.2.

To satisfy the real-time requirements of the virtual controllers, we have to pay careful attention to CPU scheduling. In virtualized environments, CPU scheduling is performed in a hierarchical manner [50]; first, the scheduler assigns the CPU resources to a container; then, the tasks that belong to the container are scheduled within the limit of the CPU resources assigned to the container. Legacy OS in cloud systems, however, focus on limiting the resource usage of containers rather than guaranteeing resources [51]. For instance, Linux keeps track of the resource usage of containers and throttles a container's resource usage if that container exceeds the limit. That is, Linux does not guarantee resources but limits those. Thus, it is difficult to guarantee the deadlines of real-time applications. Our hierarchical scheduler provides the resource reservation based on a periodic execution model and guarantees the deadlines of hard real-time tasks. We will describe the implementation of the hierarchical real-time scheduler in Section 3.3. The end-to-end delay denoted as D_{e2e} in Figure 2 is defined as the time from the beginning of the task that performs sensing and control to the completion of the task that performs actuation. For the sake of simplicity, we assume that the sensing and control operations are performed by a single task [24]. The sensing and control task sends control messages periodically to the actuation task. Since the tasks of the virtual controllers communicate through control networks, it is especially critical to decide when the tasks are scheduled and take part in communication. The simple examples in Figure 3 show how CPU scheduling impacts end-to-end delay. In these examples, $Node_i$ runs two virtual controllers denoted as VC_n. VC_0 comprises two tasks, τ_0 and τ_1. τ_0 senses the plant, decides the control, and sends a control message to the actuation task running on $Node_j$. τ_1 can be a Human–Machine Interface (HMI) task, for example. We assume that the sensing and control task, τ_0, sends a message to the network at the finish time and the actuation task receives a message at the start time. In Case 1 (Figure 3a), the CPU resource of $Node_i$ is assigned to VC_0 in time slot 0 and two tasks are executed in succession. However, the actuation task on $Node_j$ runs long after the message arrives. Thus, the end-to-end delay becomes large. If τ_0 decides a precise control based on the current situation of the plant, the control executed at $Node_j$ after a significant delay may not correspond to the situation at that time. In Case 2 (Figure 3b), if the period of VC_0 of $Node_i$ is two time slots, we can delay the execution of tasks in VC_0 of $Node_i$ to time slot 1 in which the release point of τ_0 is delayed even more. This results in a less end-to-end delay than Case 1, while still guaranteeing the deadlines of containers and tasks. In this paper, we define the *phase offset* as an intentional delay of execution of containers and tasks. We can also adjust the phase offset of the actuation container as shown in Case 3 (Figure 3c), where the release point of the actuation task is advanced from time slot 2 to 1. To find a sub-optimal combination of phase offsets, we implement a simulator, which will be detailed in Section 3.4.

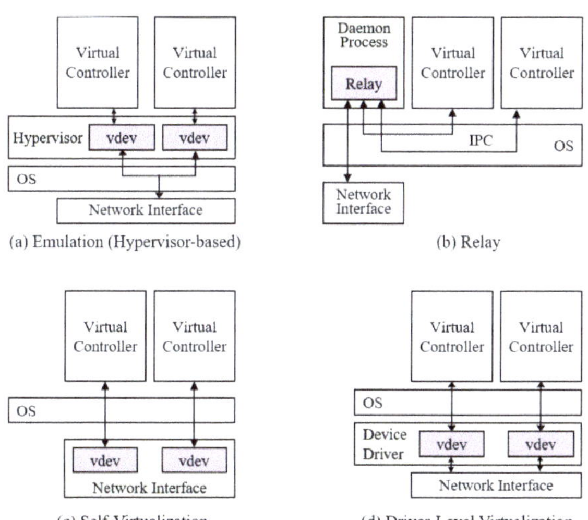

Figure 1. Alternatives for network interface virtualization.

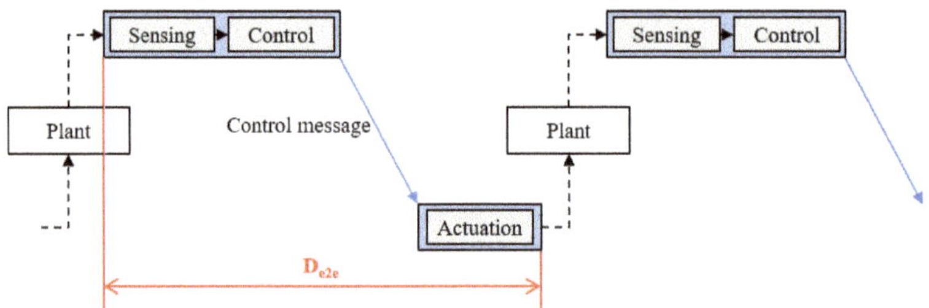

Figure 2. End-to-end delay of control loop.

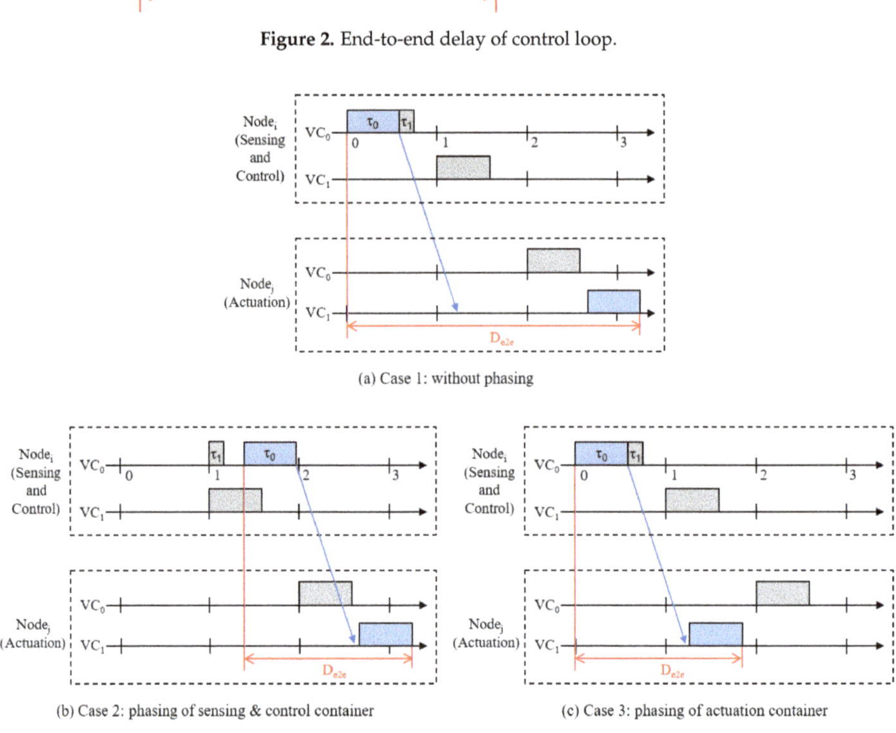

Figure 3. End-to-end delay and phase offsets.

3.2. Driver-Level CAN Virtualization

As discussed in Section 3.1, the driver-level virtualization is more beneficial than the other design alternatives when considering transparency, performance, and resource requirements all together. Therefore, in this paper, we suggest the driver-level virtualization of CAN and study its implementation issues in detail. Figure 4 shows the suggested design. In the device driver, there are two main components to provide the CAN virtualization: *virtual CAN interface* that emulates the behavior of the CAN network interface and *virtual CAN bus* that emulates the media access control. A virtual controller owns exclusively an instance of the virtual CAN interface and is connected to a virtual CAN bus.

To provide functional transparency for virtual controllers, we have to emulate the inherent features and characteristics of CAN in the virtualized environments. The header of the CAN message includes the 11-bit message identifier, which specifies the class of information the CAN message represents (e.g., speed or torque of motors). The information that a specific identifier represents can vary from system to system and is determined at the system design phase. The controllers broadcast CAN messages tagging a message identifier according to its assignment rule and receive messages by specifying interesting message identifiers. It is to be noted that a CAN message specifies neither source nor destination and is simply broadcast to all nodes in the same bus. The CAN device drivers and CANopen allow only one task to receive messages of a specific identifier. If different tasks running on the same node wish to receive the CAN messages of the same identifier, it is not guaranteed that all tasks receive the messages properly. Only one task that issues the receiving operations before others can receive the messages. To overcome this limitation and host several virtual controllers, we provide the virtual CAN interfaces, which are created dynamically at the run time on demand and assigned exclusively to a virtual controller. A virtual CAN interface has a pair of send and receive queues and data structures for locking of the queues.

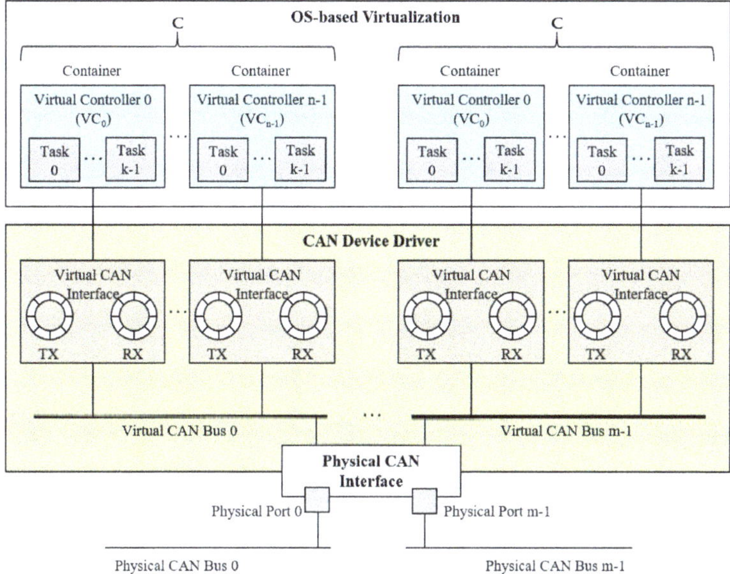

Figure 4. Driver-level Controller Area Network (CAN) virtualization.

The CAN message identifier is also used in bus arbitration. When several controllers try to access the CAN bus simultaneously, the controller that tries to send the message with the lowest value of message identifier gains bus access. This means that the message identifier is considered as a priority for bus arbitration, where the lower identifier value, the higher priority. Since the virtual controllers are considered as separate CAN nodes in the virtualized environments, we need to arbitrate the bus access among the virtual controllers. We emulate the behavior of media access control in the virtualized environments by introducing the virtual CAN bus in the device driver as shown in Figure 4. When the physical CAN interface is able to send a message, the virtual CAN bus searches the send queues of all virtual CAN interfaces connected to the virtual bus, chooses the message that has the lowest identifier value, and sends it to the physical CAN bus. Moreover, the virtual CAN bus emulates the broadcast media. If a virtual controller sends a CAN message, it is sent out to the physical CAN bus but also delivered to the other virtual controllers connected to the same virtual bus. In non-virtualized environments, the propagation delay of messages on physical CAN bus is very small; thus, the geographical order of controllers on the bus may not be a critical issue. However, the virtual CAN bus copies messages to multiple virtual CAN interfaces to emulate the broadcasting media. The overhead of this copy operation can expand the time differences between message arrival points at local virtual CAN interfaces and remote CAN interfaces. Though we cannot provide a comparable latency to the physical bus, to mitigate the side effect of the copy overhead, the virtual CAN bus copies the message to the local virtual CAN interfaces and then sends the message to physical CAN bus. In a similar way, when a message is received from the physical CAN bus, the virtual CAN bus inserts the message to the receive queues of all virtual interfaces connected to the virtual bus. Since filtering of interesting messages is performed by the upper layers, we do not consider it at the virtual CAN bus.

The physical CAN interface in Figure 4 has multiple ports. The CAN interfaces in sensors or actuators usually have a single port, but we generalize our design and implementation for multiple ports because the computing nodes in cloud system can be equipped with a multi-port CAN interface. This allows a single computing node to host multiple sets of virtual controllers that use different CAN buses.

3.3. Hierarchical Real-Time Scheduling

As described in Section 3.1, we implement a hierarchical real-time scheduler for containerized controllers on Linux. The scheduler uses a fixed-priority scheduling algorithm (e.g., rate monotonic (RM) scheduling [52]) for both virtual controllers and tasks. We consider a set of periodic virtual controllers (i.e., $\mathbb{C} = \{VC_0, VC_1, \cdots, VC_{n-1}\}$) for each processor. Each virtual controller VC_i is containerized with a different set of periodic tasks and uses a separate name space. VC_a has a higher priority than VC_b if $a < b$. We denote a virtual controller by $VC_i = (\Pi_i, \Theta_i, \Delta_i, \mathbb{T}_i)$, where Π_i is the period, Θ_i is the time duration reserved for each period, $\Delta_i \mid 0 \leq \Delta_i < \Pi_i$ is the phase offset (i.e., temporal offset from a certain reference time), and $\mathbb{T}_i$ is a set of periodic tasks, i.e., $\mathbb{T}_i = \{\tau_0^i, \tau_1^i, \cdots, \tau_{k-1}^i\}$. Task τ_a^i has a higher priority than task τ_b^i if $a < b$. A task is denoted as $\tau_j^i = (p_j^i, e_j^i, \delta_j^i)$, where p_j^i is the period, e_j^i is a range of execution time, and $\delta_j^i \mid 0 \leq \delta_j^i < p_j^i$ is the task-level phase offset. We assume that the relative deadline of each task is equal to its period. For the sake of simplicity, we also assume that a processor and a physical bus are dedicated to a set of VCs (i.e., $\mathbb{C}$) launched by a tenant. If a processor is shared between multiple tenants that submit their $\mathbb{C}$s at arbitrary time points, it is difficult to analyze and guarantee the end-to-end delay on the fly. In addition, if a physical bus (and a virtual bus) is shared by different tenants, there can be conflicts between different definitions of message identifiers of disparate $\mathbb{C}$s, which results in malfunctions. Once a $\mathbb{C}$ finishes, it releases the processor and bus resources occupied so that another following $\mathbb{C}$ can be used. The applications at the plant level (e.g., Supervisory Control and Data Acquisition (SCADA) [53]) may consist of multiple $\mathbb{C}$s.

The hierarchical real-time scheduler is implemented as a daemon process. The scheduler suspends and resumes tasks by using signals. Once the system initialization is completed, the scheduler starts the timer. We use a global timer synchronized across distributed nodes. Although the fieldbuses, such as TTCAN [54] and EtherCAT [12], provide a global clock in distributed systems, we additionally implement a global clock in software for cases where a hardware global clock is not supported. Our software global clock is synchronized by using IEEE 1588 [55]. The scheduler releases a virtual controller VC_i at Δ_i. Then, the scheduler assigns the CPU resources to VC_i for the duration Θ_i at every period Π_i. The task τ_j^i is released periodically after $\Delta_i + \delta_j^i$. The scheduler runs tasks based on their period p_j^i and execution time e_j^i within the CPU utilization of VC_i (i.e., Θ_i/Π_i).

Figure 5 shows how the phases Δ_i and δ_j^i decide the release point of a task. In this example, two virtual controllers (i.e., VC_0 and VC_1), each of which has two tasks, run on a single CPU. The first periods of VC_0 and VC_1 starts at Δ_0 and Δ_1, respectively. We assume that the highest-priority task of each virtual controller in this example has the zero phase offset (i.e., $\delta_0^0 = 0$ and $\delta_0^1 = 0$); thus, τ_0^0 and τ_0^1 are released as soon as the first period of each VC begins, whereas τ_1^0 and τ_1^1 start being released after $\Delta_0 + \delta_1^0$ and $\Delta_1 + \delta_1^1$, respectively. Since VC_0 has a higher priority than VC_1, the tasks of VC_1 are preempted by the tasks of VC_0. For example, we can see that the execution of τ_0^1 is delayed because it is preempted by τ_1^0. We will discuss the adjustment of phases in more detail in the next subsection.

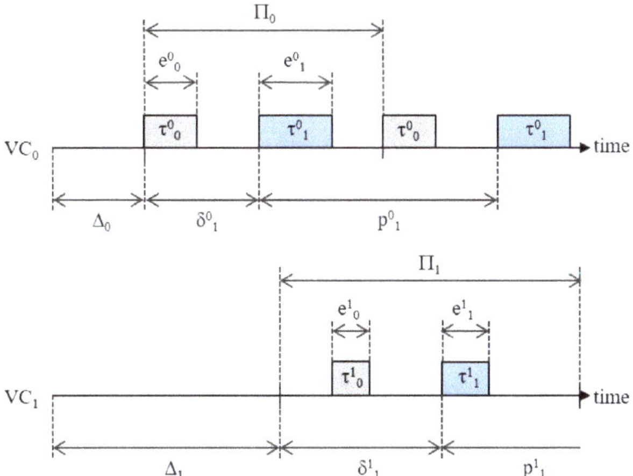

Figure 5. Hierarchical real-time scheduling.

3.4. Phasing of Virtual Controllers and Tasks

The worst-case end-to-end delay is an important metric of QoS for industrial control applications [56]. As we have discussed above, the worst-case end-to-end delay in virtualized environments is strongly influenced by the phase offsets of virtual controllers and tasks (i.e., Δ_i and δ^i_j). Researchers tried to find an optimal combinations of phase offsets by suggesting either an online algorithm [23] or a simulation-based offline approach [22]. However, they did not consider the virtualized environments. In this paper, we implement a simulator that performs a discrete-event simulation and provides a sub-optimal phase combination for virtual controllers and tasks. It is to be noted that the phase combination suggested by the simulator does not hinder the deadline guarantee of containers and tasks.

The simulator consists of configuration manager, node objects, simulator kernel, phase search manager, and log manager as shown in Figure 6. The configuration manager provides the user interfaces to configure simulation parameters, such as attributes of virtual controllers, tasks, and target fieldbus. The parameters are specified as an XML format and parsed by the configuration manager at the initialization phase. The node objects are created according to the simulation parameters. Each node object emulates a $\mathbb{C}$. The simulation kernel emulates the run-time behavior of overall system by performing hierarchical CPU scheduling and message transmission. Since the execution time of tasks varies for every period in real systems, the exec-time generator emulates this by generating time values in the range of e^i_j with a uniform distribution. In addition, the event handler and fieldbus interface components emulate system overheads, such as interrupt handling and Direct Memory Access (DMA). The simulation results are gathered by the IPC module and saved into text files by the log manager.

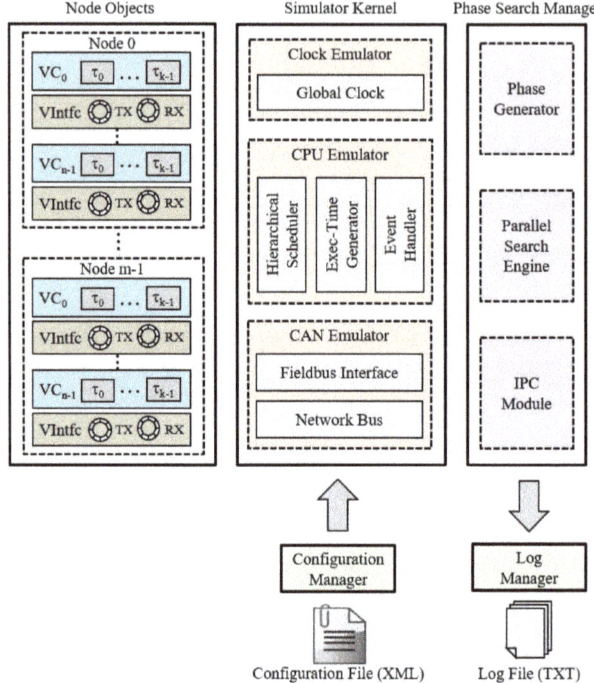

Figure 6. Simulator.

3.5. Implementation

We implemented the CAN virtualization at the Linux device driver of the PEAK-System CAN interface. In Linux, each port of the PEAK-System CAN interface is registered as a character device file. The device file is an abstraction implemented by OS to provide basic user-level operations on I/O devices. The applications and CANopen request the send and receive operations through the file I/O system calls, and the virtual file system of Linux internally calls the I/O functions provided by the device driver. In our implementation, a virtual CAN interface is created dynamically and destroyed at the run time when an application or CANopen calls the `open()` and `close()` system calls, respectively. A virtual CAN interface is exclusively assigned to a virtual controller. The send and receive operations are commenced by the `ioctl()` system call. In the case of send operation, the device driver first inserts the message to the send queue of the virtual interface, then seeks the message that has the highest priority (i.e., lowest message identifier number) across the send queues of the virtual interfaces, and not only sends the highest-priority message to the physical CAN interface, but also copies it into the receive queues of the other virtual interfaces. Regarding the receive operation, the interrupt handler in the device driver copies the received message into every receive queue of the virtual interfaces that are connected to the same virtual CAN bus. The `ioctl()` system call with receive command returns the message of desired identifier from the receive queue of the corresponding virtual CAN interface. We identify the corresponding virtual CAN interface by using the file descriptor passed by the system calls.

To force the container scheduler to behave like the RM scheduler, we set the `period` and `runtime` attributes of cgroup into Π_i and Θ_i, respectively. The Linux cgroup provides these attributes for a container that consists of the tasks scheduled by a Linux real-time scheduler (i.e., `SCHED_FIFO` or `SCHED_RR`). Then, it is guaranteed that the CPU usage of the container does not exceed the specified `runtime` for every `period`. We set the priority of tasks into the priority of the container to which the tasks belong. In addition, since the Linux real-time schedulers do not support periodic task scheduling, we implemented an overlay scheduler on Linux so that task scheduling can be performed with the RM algorithm. The overlay scheduler maintains a list of the container control blocks, each of which includes a list of task control blocks (i.e., $\mathbb{T}_i$). A task control block includes the attributes of the task, such as period p_j^i and worst-case execution time $Max(e_j^i)$. The overlay scheduler selects a task to run from the current container based on the RM algorithm and suspends/resumes tasks by using signals, such as `SIGSTOP` and `SIGCONT`.

The current implementation of the simulation tool performs an exhaustive search to find a sub-optimal phase combination; that is, it investigates all possible combinations of phase offsets by shifting the phase offsets of containers and tasks by a given time unit. The simulation tool analyzes the worst-case end-to-end delay of a phase combination by simulating the control loops for a given number of iterations. Thus, the total simulation time and the accuracy of analysis highly depend on the time unit generating phase combinations and the number of iterations to simulate. Although the current implementation is enough to show the benefits of phasing for virtual controllers, further study is needed to reduce the number of phase combinations investigated and reduce the total simulation time without sacrificing the analysis accuracy. The simulator creates multiple processes as many as CPU cores to run simulations for different phase offsets in parallel. Once the phase offsets of containers and tasks are decided by the simulator, we set the parameters (i.e., Δ_i and δ_j^i) of the hierarchical scheduler to the phase offset combination suggested.

3.6. Summary

In this subsection, we summarize how the suggested design and implementation address four challenges listed in Section 1. First, to support sharing of the CAN interface between virtual controllers, we proposed the virtual CAN interface and the virtual CAN bus as described in Section 3.2. A virtual controller is assigned a separate virtual CAN interface and allowed to use only the assigned virtual interface; that is, the other virtual interfaces of different virtual controllers are invisible as described in Section 3.5. The virtual CAN interfaces of virtual controllers in the same $\mathbb{C}$ are connected through a virtual CAN bus that takes care of multiplexing and demultiplexing of accessing to/from the physical CAN interface. Thus, we can isolate the communication of each virtual controller with respect to functionality, while allowing for sharing of the physical CAN interface.

Secondly, to emulate the media access control of CAN, the virtual CAN bus implements the bus arbitration and message broadcasting as described in Section 3.2. In CAN, the message identifier is used as the priority of messages in bus arbitration. As described in Section 3.5, the virtual CAN bus chooses the message that has the lowest identifier number from the send queues of the virtual CAN interfaces and sends it first. In addition, the virtual CAN bus implements message broadcasting by copying the message sent from a virtual controller or received from the physical CAN interface into the receive queues of the virtual interfaces connected to the same virtual CAN bus.

Thirdly, our driver-level CAN virtualization and hierarchical CPU scheduler support the OS-based virtualization, aiming for low virtualization overheads. As we have discussed earlier, the OS-based virtualization has lower overheads than the hypervisor-based virtualization. The CAN virtualization suggested in Section 3.2 is implemented at the CAN device driver; thus, it is transparent to the upper layers (i.e., OS kernel, CANopen, and applications) and harmonizes well with the OS-based virtualization. Moreover, the hierarchical CPU scheduler suggested in Section 3.3 is implemented for OS-based virtualization, targeting particularly containers and their tasks.

Finally, to analyze the end-to-end delay and enhance the worst-case end-to-end delay, we suggested a simulation tool and hierarchical real-time scheduling in Sections 3.4 and 3.3, respectively. The simulation tool analyzes the end-to-end delay with different phase offsets of containers and tasks and suggests a phase offset combination that can provide a sub-optimal worst-case end-to-end delay. The hierarchical real-time scheduler implements a global timer to synchronize between distributed virtual controllers and can adjust phase offsets with the scheduling of periodic containers and tasks.

4. Experimental Results

In this section, we analyze the overheads of our driver-level CAN virtualization. In addition, we show how phasing of virtual controllers and tasks can improve the worst-case end-to-end delay in the virtualized environments.

4.1. Comparisons with Hypervisor-Based Virtualization

Our CAN virtualization targets the OS-based virtualization aiming for less overheads. To analyze the virtualization overheads, we measured the Round-Trip-Time (RTT) between two different physical nodes connected through a CAN bus. Each node was equipped with an Intel i5 processor and a PEAK-System CAN interface and installed the Linux operating system. We measured RTT between two nodes that sent and received the same size messages in a ping-pong manner repeatedly for a given number of iterations. In the experiments, we considered the message size only up to 8-byte because a CAN frame can convey 8-byte payload in maximum. We drew a comparison between three cases: (i) original setup without virtualization, (ii) OS-based virtualized environment, and (iii) hypervisor-based virtualized environment. The original setup shows the base performance to be compared with. The OS-based virtualization shows the performance of our design suggested in Section 3. To measure the CAN virtualization overheads in hypervisor-based virtualization, we used the implementation suggested by Kim et al. [21]. It is to be noted that we applied the hypervisor-based virtualization to only one node, while measuring RTT on the other node that is not virtualized because the hypervisor-based virtualization does not provide an accurate timer to guest domains. Similarly, we also applied the OS-based virtualization to only one node for fairness. In addition, in the experiments, the virtual controllers did not follow the periodic execution models described in Section 3.3 to measure pure communication overheads, removing the impact of phase offsets and variable execution time of tasks. We will analyze the impact of different phase offsets in the next subsection.

Figure 7 shows the average RTT for different message sizes. As we can see, the OS-based virtualization hardly adds additional overheads compared with the original setup without virtualization, whereas the hypervisor-based virtualization shows higher overheads up to 20%. The low overheads of the OS-based virtualization is due to not only the absence of a hypervisor, but also our lightweight driver-level CAN virtualization.

In addition, we represented the distribution of RTTs measured in Cumulative Distribution Function (CDF) plots as shown in Figures 8–10. These graphs show the RTTs of 1, 4, and 8-byte messages. As we have discussed, the OS-based virtualization shows a comparable performance to a non-virtualized environment, while the hypervisor-based virtualization shows higher overheads. Moreover, we can observe that the jitters (i.e., difference between maximum and minimum RTTs) in OS-based virtualization are much less than those in hypervisor-based virtualization for all message sizes. The jitters for 4-byte messages, for example, were 6 μs with OS-based virtualization and 72 μs with hypervisor-based virtualization.

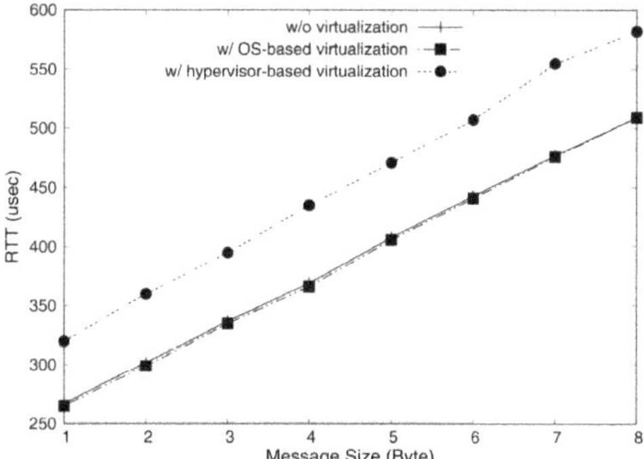

Figure 7. Average Round-Trip-Time (RTT) of different message sizes.

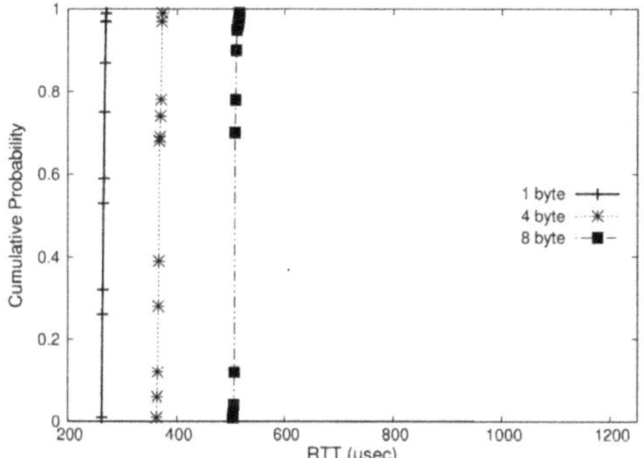

Figure 8. Cumulative Distribution Function (CDF) of RTT in a non-virtualized case.

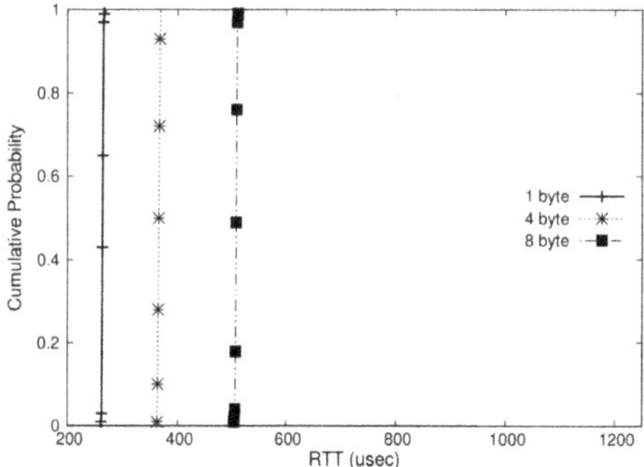

Figure 9. CDF of RTT in an operating-system(OS)-based virtualization case.

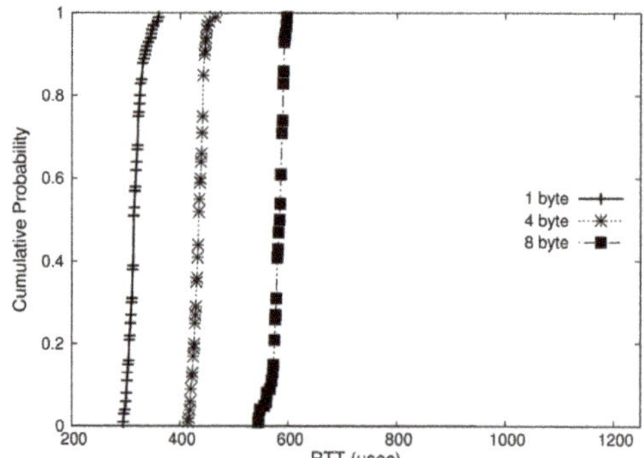

Figure 10. CDF of RTT in a hypervisor-based virtualization case.

4.2. Analysis of Worst-Case End-to-End Delay

In Section 3, we suggested a hierarchical real-time scheduler and a simulator, which can improve the worst-case end-to-end delay by adjusting the phase offsets of virtual controllers and tasks. To measure the end-to-end delay, we ran virtual controllers with our driver-level CAN virtualization on a master node equipped with an Intel i5 processor and a two-port PEAK-System CAN interface. Two worker nodes equipped with an Intel i3 processor were connected to the different CAN ports of the master node. In the experiments, we ran two virtual controllers, VC_0 and VC_1 on the master node, each of which communicates to a worker node through a different CAN bus. Each worker node runs only a single virtual controller (VC_0 or VC_1). We consider the scenario in which we run the virtual controllers on general-purpose computing nodes provided by a cloud and have a capable of adjusting phase offsets on both master and worker nodes. As defined in Section 3.1, the end-to-end delay (D_{e2e}) is the time from the beginning of the sensing and control task at the master to the completion of the actuation task at the worker. Our measurements were done in three steps. First, we randomly generated three test sets as shown in Table 1. In each test set, we assumed that τ_1 of VC_0 and τ_0 of VC_1 performed either sensing and control or actuation (denoted as (sc) and (a) in Table 1, respectively). Then, we ran simulations for each test set to find a sub-optimal combination of phase offsets. The simulation parameters are shown in Table 2. Finally, we applied the sub-optimal combinations suggested by the simulator to our experimental system and measured the actual end-to-end delays of 1000 messages.

Figures 11 and 12 show the end-to-end delays of VC_0 and VC_1 of test set 0, respectively. These graphs show the worst-case (denoted as MAX) and the best-case (denoted as MIN) delays and compare the values actually measured on a real system with those simulated to show that the errors by simulation are marginal (less than 2.5% for the worst-case end-to-end delay). As we can observe, phasing reduces the worst-case end-to-end delay by 59.2% for VC_0 and 47.0% for VC_1, respectively. Figures 13 and 14 show the end-to-end delay of every iteration. Although the error is often seen for some specific iteration because the simulation environment (e.g., execution time of tasks) and the actual system environment may be different at those points, it can be seen that the overall maximum and the minimum values of the simulation results and the actual measurement results are similar.

Table 1. Test sets (time unit: µs).

Test Sets	Sets of VCs	Virtual Controllers	Period (Π)	Duration (Θ)	Tasks	Period (p)	Execution Time (e)
0	$\mathbb{C}_0^{master}$	VC_0	2760	1000	τ_0	5530	330
					$\tau_1(sc)$	8800	580
					τ_2	9650	390
		VC_1	2970	1050	$\tau_0(sc)$	5950	310
					τ_1	7140	490
					τ_2	9980	550
	$\mathbb{C}_0^{worker0}$	VC_0	100,000	95,000	τ_0	5530	330
					$\tau_1(a)$	8800	580
					τ_2	9650	390
	$\mathbb{C}_0^{worker1}$	VC_1	100,000	95,000	$\tau_0(a)$	5950	310
					τ_1	7140	490
					τ_2	9980	550
1	$\mathbb{C}_1^{master}$	VC_0	4000	1500	τ_0	2000	60
					$\tau_1(sc)$	2000	100
					τ_2	4000	110
		VC_1	4000	1500	$\tau_0(sc)$	2000	50
					τ_1	2000	140
					τ_2	4000	120
	$\mathbb{C}_1^{worker0}$	VC_0	100,000	95,000	τ_0	2000	50
					$\tau_1(a)$	2000	100
					τ_2	4000	100
	$\mathbb{C}_1^{worker1}$	VC_1	100,000	950,000	$\tau_0(a)$	2000	50
					τ_1	2000	100
					τ_2	4000	100
2	$\mathbb{C}_2^{master}$	VC_0	2610	1100	τ_0	5230	660
					$\tau_1(sc)$	8840	470
					τ_2	9610	510
		VC_1	2840	1000	$\tau_0(sc)$	5680	440
					τ_1	10,580	650
					τ_2	11,090	420
	$\mathbb{C}_2^{worker0}$	VC_0	100,000	95,000	τ_0	5230	660
					$\tau_1(a)$	8840	470
					τ_2	9610	510
	$\mathbb{C}_2^{worker1}$	VC_1	100,000	95,000	$\tau_0(a)$	5680	440
					τ_1	10,580	650
					τ_2	11,090	420

Table 2. Simulation parameters.

Parameters	Value
Phasing resolution	50 µs
Simulation iterations	1000
Simulation resolution	10 µs
Interrupt handling overhead	20 µs
Tx and Rx queue size	10
Fieldbus bandwidth	1 Mbps
Fieldbus forwarding delay	1 µs
Message size	8 bytes
Direct Memory Access (DMA) overhead	10 µs

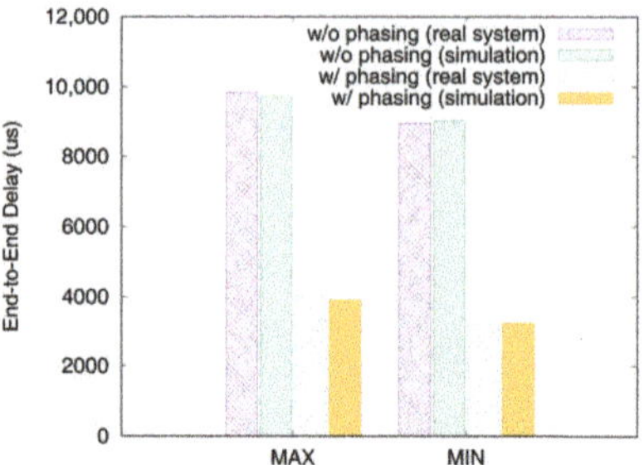

Figure 11. Max and min D_{e2e} of VC_0 in test set 0.

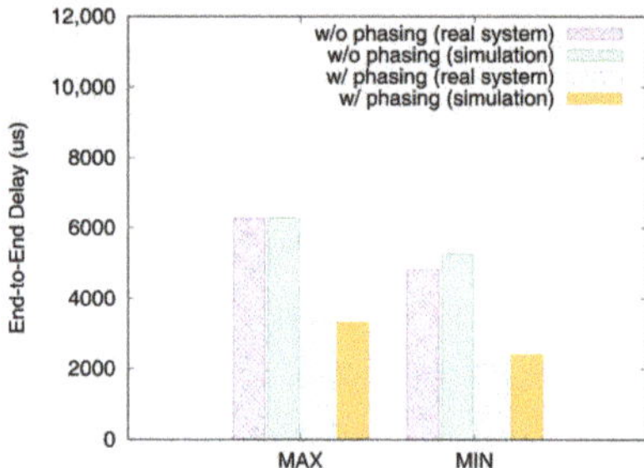

Figure 12. Max and min D_{e2e} of VC_1 in test set 0.

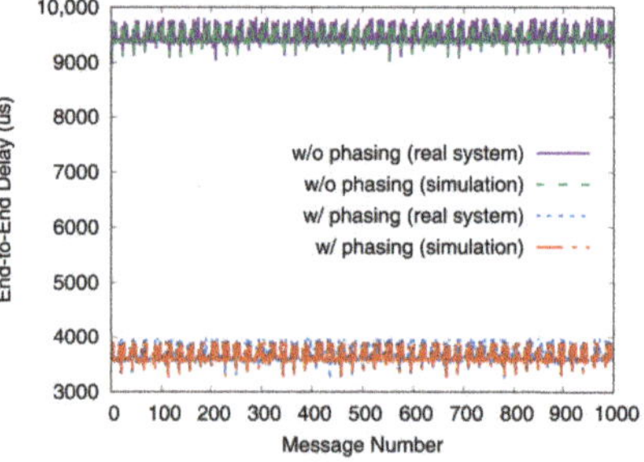

Figure 13. D_{e2e} distribution of VC_0 in test set 0.

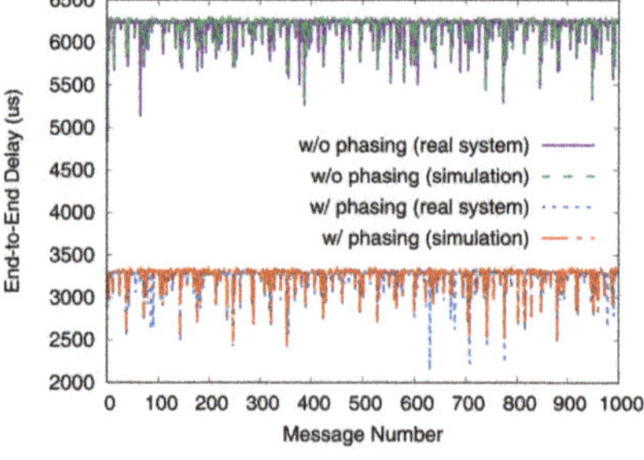

Figure 14. D_{e2e} distribution of VC_1 in test set 0.

Figures 15–22 show the measurement results for the test sets 1 and 2. We again see that the simulator can predict the worst-case end-to-end delays accurately and provide a sub-optimal phase combination successfully. In these experiments, the phasing scheme reduced the worst-case end-to-end delay on a real system up to 78.7% with test set 1 and 58.0% with test set 2, respectively.

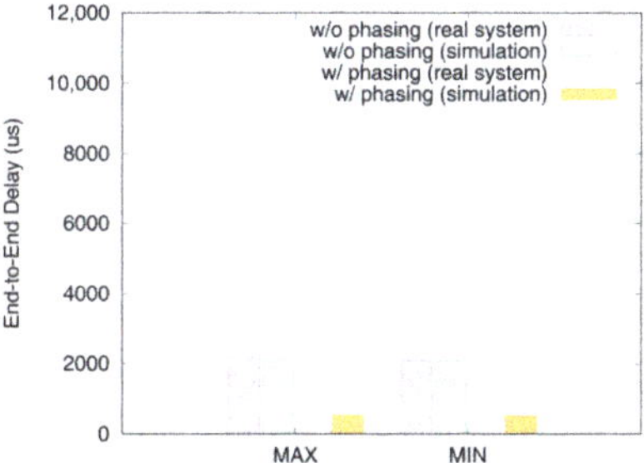

Figure 15. Max and min D_{e2e} of VC_0 in test set 1.

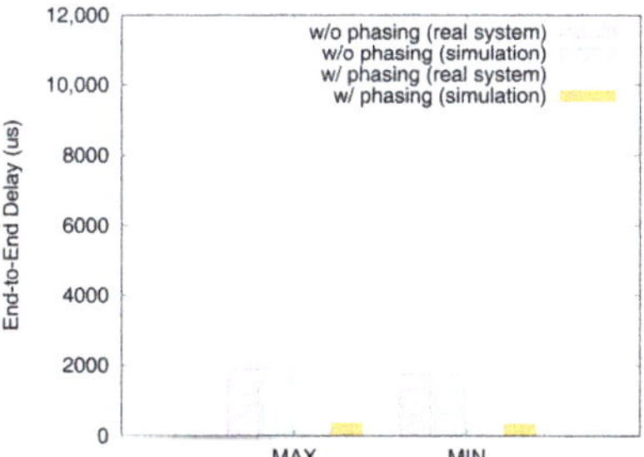

Figure 16. Max and min D_{e2e} of VC_1 in test set 1.

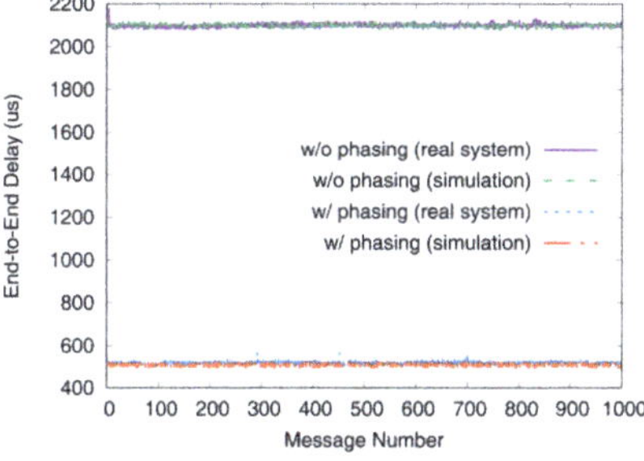

Figure 17. D_{e2e} distribution of VC_0 in test set 1.

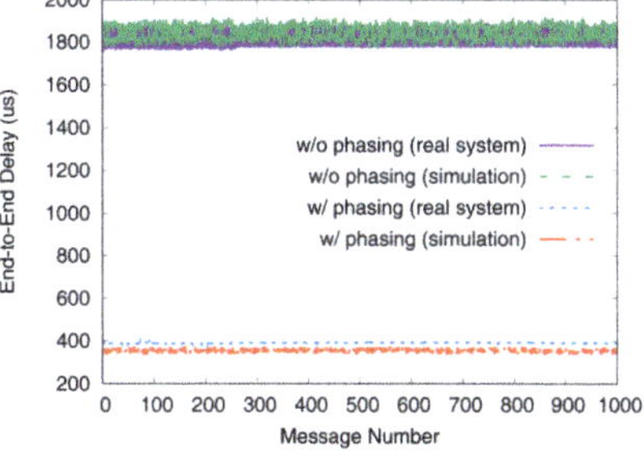

Figure 18. D_{e2e} distribution of VC_1 in test set 1.

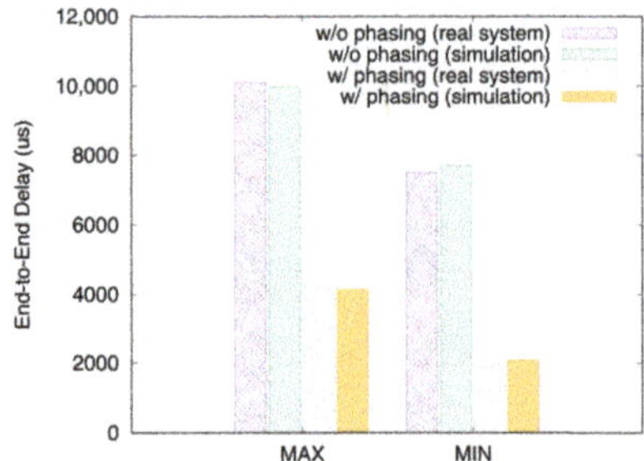

Figure 19. Max and min D_{e2e} of VC_0 in test set 2.

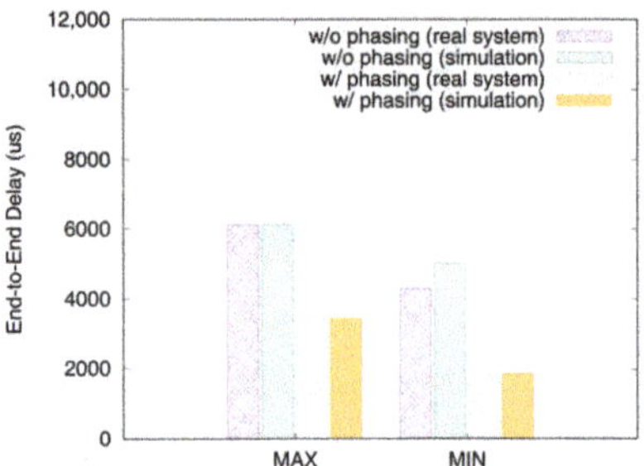

Figure 20. Max and min D_{e2e} of VC_1 in test set 2.

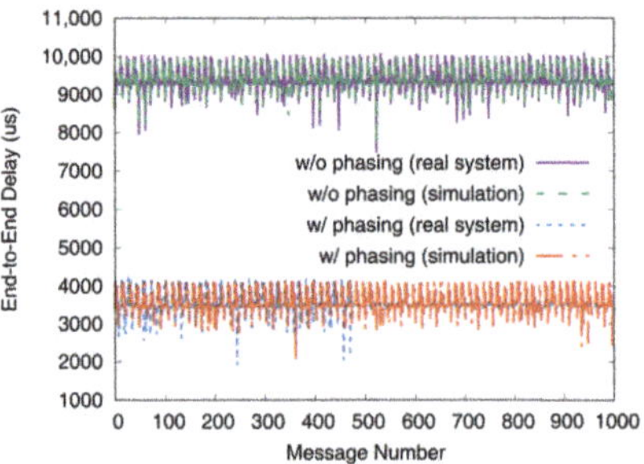

Figure 21. D_{e2e} distribution of VC_0 in test set 2.

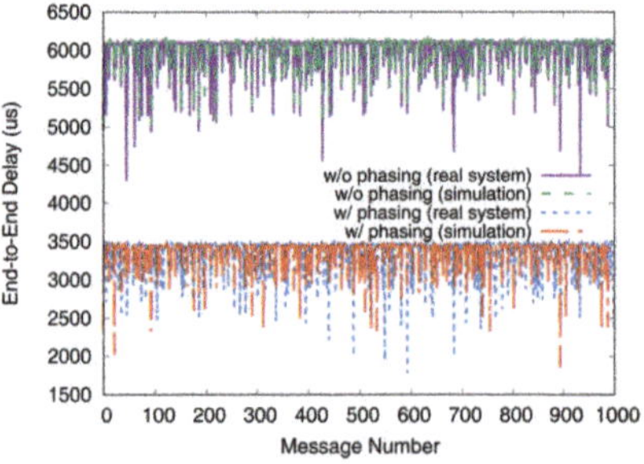

Figure 22. D_{e2e} distribution of VC_1 in test set 2.

5. Conclusions

The virtualization technologies can provide efficient hosting and consolidation of computing resources. Thus, several researchers tried to utilize the virtualization technologies in industrial control systems and showed benefits with respect to manageability and scalability. However, the support for traditional fieldbuses, such as CAN, in virtualized environments has not been studied thoroughly. In this paper, to tackle both functionality and QoS issues of CAN in virtualized environments, we suggested the lightweight CAN virtualization technology for containerized controllers. In the functionality aspect, our driver-level virtualization technology provided the abstractions for virtual CAN interfaces and virtual CAN buses, while preserving the transparency to system software and applications. In the QoS perspective, we provided a hierarchical real-time scheduler and phasing of virtual controllers and tasks. To provide a sub-optimal phase combination, we implemented a simulator. The experiment results showed that our CAN virtualization that targeted the OS-based virtualization had significantly lower overheads and less jitters compared with a hypervisor-based virtualized environment. In addition, we showed that the worst-case end-to-end delay could be reduced up to 78.7% by adjusting the phase offsets of virtual controllers and their tasks. As future work, we plan to apply our virtualization technology to a large-scale system that consists of more virtual controllers. To do this, we have to optimize the simulator, which currently takes several hours to generate a sub-optimal phase combination for the test cases discussed in this paper.

Author Contributions: Conceptualization, S.-H.L., J.-S.K., and H.-W.J.; methodology, S.-H.L., J.-S.K., J.-S.S., and H.-W.J.; software, S.-H.L., J.-S.K., and J.-S.S.; validation, S.-H.L. and H.-W.J.; formal analysis, S.-H.L. and H.-W.J.; investigation, S.-H.L., J.-S.K., and J.-S.S.; resources, S.-H.L., J.-S.K., and J.-S.S.; data curation, S.-H.L., J.-S.K., and J.-S.S.; writing—original draft preparation, S.-H.L. and H.-W.J.; writing—review and editing, H.-W.J.; visualization, S.-H.L.; supervision, H.-W.J.; project administration, H.-W.J.; funding acquisition, H.-W.J.

Funding: This work was supported by the National Research Foundation of Korea (NRF) grant funded by the Korea government (MSIT) (No. 2017R1A2B4012759).

Conflicts of Interest: The authors declare no conflict of interest

Abbreviations

The following abbreviations are used in this manuscript:

CAN	Controller Area Network
CFD	Cumulative Distribution Function
cgroup	Control Group
DMA	Direct Memory Access
HMI	Human–Machine Interface
I/O	Input/Output
OS	Operating System
PLC	Programmable Logic Controller
QoS	Quality-of-Service
RM	Rate Monotonic
RTT	Round-Trip-Time
SCADA	Supervisory Control and Data Acquisition
SDN	Software-Defined Networking
TCP/IP	Transmission Control Protocol/Internet Protocol
VC	Virtual Controller
VMM	Virtual Machine Monitor

References

1. Goldschmidt, T.; Murugaiah, M. K.; Sonntag, C.; Schlich, B.; Biallas, S.; Weber, P. Cloud-based control: A multi-tenant, horizontally scalable soft-PLC. In Proceedings of the 8th IEEE International Conference on Cloud Computing (CLOUD 2015), New York, NY, USA, 27 June–2 July 2015; pp. 909–916.

2. Givehchi, O.; Imtiaz, J.; Trsek, H.; Jasperneite, J. Control-as-a-service from the cloud: A case study for using virtualized PLCs. In Proceedings of the 10th IEEE Workshop on Factory Communication Systems (WFCS 2014), Toulouse, France, 5–7 May 2014.
3. Tasci, T.; Melcher, J.; Verl, A. A container-based architecture for real-time control applications. In Proceedings of the 2018 IEEE International Conference on Engineering, Technology and Innovation (ICE/ITMC), Stuttgart, Germany, 17–20 June 2018.
4. Barham, P.; Dragovic, B.; Fraser, K.; Hand, S.; Harris, T.; Ho, A.; Neugebauer, R.; Pratt, I.; Warfield, A. Xen and the art of virtualization. *ACM SIGOPS Oper. Syst. Rev.* **2003**, *37*, 164–177. [CrossRef]
5. Sugerman, J.; Venkitachalam, G.; Lim, B.-H. Virtualizing I/O devices on Vmware workstation's hosted virtual machine monitor. In Proceedings of the USENIX Annual Technical Conference, Boston, MA, USA, 25–30 June 2001.
6. Raj, H.; Schwan, K. High performance and scalable I/O virtualization via self-virtualized devices. In Proceedings of the International Symposium on High Performance Distributed Computing (HPDC 07), Monterey, CA, USA, 25–29 June 2007.
7. Santos, J. R.; Turner, Y.; Janakiraman, G.; Pratt, I. Bridging the gap between software and hardware techniques for I/O virtualization. In Proceedings of the USENIX Annual Technical Conference, Boston, MA, USA, 22–27 June 2008; pp. 22–27.
8. Russell, R. Virtio: Towards a de-facto standard for virtual I/O devices. *ACM SIGOPS Oper. Syst. Rev.* **2008**, *42*, 342–355. [CrossRef]
9. Ram, K.K.; Santos, J.R.; Turner, Y.; Cox, A.L.; Rixner, S. Achieving 10Gbps using safe and transparent network interface virtualization. In Proceedings of the International Conference on Virtual Execution Environments (VEE 09), Washington, DC, USA, 11–13 March 2009.
10. Li, J.; Xue, S.; Zhang, W.; Qi, Z. When I/O interrupt becomes system bottleneck: Efficiency and scalability enhancement for SR-IOV network virtualization. *IEEE Trans. Cloud Comput.* **2017**. [CrossRef]
11. Jose, J.; Li, M.; Lu, X.; Kandalla, K. C.; Arnold, M. D.; Panda, D. K. SR-IOV support for virtualization on Infiniband clusters: Early experience. In Proceedings of the 13th IEEE/ACM International Symposium on Cluster, Cloud, and Grid Computing (CCGrid 2013), Delft, The Netherlands, 13–16 May 2013; pp. 385–392.
12. Jansen, D.; Buttner, H. Real-time Ethernet: The EtherCAT solution. *Comput. Control Eng.* **2004**, *15*, 16–21. [CrossRef]
13. Feld, J. PROFINET-scalable factory communication for all applications. In Proceedings of the 2004 IEEE International Workshop on Factory Communication Systems (WFCS 2004), Vienna, Austria, 22–24 September 2004.
14. Farsi, M.; Ratcliff, K.; Barbosa, M. An overview of controller area network. *Comput. Control Eng.* **1999**, *10*, 113–120. [CrossRef]
15. Pfeiffer, O.; Ayre, A.; Keydel, C. The CANopen standard. In *Embedded Networking with CAN and CANopen*; Copperhill Technologies Corporation: Greenfield, MA, USA, 2008; pp. 39–112.
16. VMware. Available online: http://www.vmware.com (accessed on 2 September 2019).
17. VirtualBox. Available online: http://www.virtualbox.org (accessed on 2 September 2019).
18. Morabito, R.; Kjällman, J.; Komu, M. Hypervisors vs. lightweight virtualization: A performance comparison. In Proceedings of the 2015 IEEE International Conference on Cloud Engineering (ICCE 2015), Tempe, AZ, USA, 9–13 March 2015.
19. Morabito, R. Virtualization on internet of things edge devices with container technologies: A performance evaluation. *IEEE Access* **2017**, *5*, 8835–8850. [CrossRef]
20. Herber, C.; Richter, A.; Rauchfuss, H.; Herkersdorf, A. Spatial and temporal isolation of virtual CAN controllers. *ACM SIGBED Rev.* **2014**, *11*, 19–26. [CrossRef]
21. Kim, J.-S.; Lee, S.-H.; Jin, H.-W. Fieldbus virtualization for integrated modular avionics. In Proceedings of the 16th IEEE International Conference on Emerging Technologies and Factory Automation (ETFA 2011), Toulouse, France, 5–9 September 2011.
22. Lee, S.-H.; Jin, H.-W.; Kim, K.; Lee, S. Phasing of periodic tasks distributed over real-time fieldbus. *Int. J. Comput. Commun. Control* **2017**, *12*, 645–660. [CrossRef]
23. Kang, H.; Kim, K.; Jin, H.-W. Real-time software pipelining for multidomain motion controllers. *IEEE Trans. Ind. Inform.* **2016**, *12*, 705–715. [CrossRef]
24. Kim, I.; Kim, T. Guaranteeing isochronous control of networked motion control systems using phase offset adjustment. *Sensors* **2015**, *15*, 13945–13965. [CrossRef]

25. Sung, M.; Kim, I.; Kim, T. Toward a holistic delay analysis of EtherCAT synchronized control processes. *Int. J. Comput. Commun. Control* **2013**, *8*, 608–621. [CrossRef]
26. Carrascosa, E.; Coronel, J.; Masmano, M.; Balbastre, P.; Crespo, A. XtratuM hypervisor redesign for LEON4 multicore processor. *ACM SIGBED Rev.* **2014**, *11*, 27–31. [CrossRef]
27. Seyfried, S. Resource management in Linux with control groups. In Proceedings of the 17th International Linux System Technology Conference (Linux-Kongress 2010), Nuremberg, Germany, 17–18 April 2013.
28. Aeronautical Radio Inc. *Avionics Application Software Standard Interface Part 1—Required Services*; SAE-ITC: Bowie, MD, USA, 2015.
29. AUTOSAR. Available online: http://www.autosar.org (accessed on 2 September 2019).
30. Han, S.; Jin, H.-W. Resource partitioning for integrated modular avionics: Comparative study of implementation alternatives. *Softw. Pract. Exp.* **2014**, *44*, 1441–1466. [CrossRef]
31. Katz, R.; Min, B.-K.; Pasek, Z. Open architecture control technology trends. *ERC/RMS Rep.* **2000**, *35*, 1–26.
32. Hong, K.-S.; Choi, K.-H.; Kim, J.-G.; Lee, S. A PC-based open robot control system: PC-ORC. *Robot. Comput. Integr. Manuf.* **2001**, *17*, 355–365. [CrossRef]
33. Nunes, B.A.A.; Mendonca, M.; Nguyen, X.N.; Obraczka, K.; Turletti, T. A survey of software-defined networking: Past, present, and future of programmable networks. *IEEE Commun. Surv. Tutor.* **2014**, *16*, 1617–1634. [CrossRef]
34. Cruz, T.; Simões, P.; Monteiro, E. Virtualizing programmable logic controllers: Toward a convergent approach. *IEEE Embed. Syst. Lett.* **2016**, *8*, 69–72. [CrossRef]
35. Kaur, K.; Garg, S.; Kaddoum, G.; Ahmed, S.H.; Atiquzzaman, M. KEIDS: Kubernetes based energy and interference driven scheduler for industrial IoT in edge-cloud ecosystem. *IEEE Internet Things J.* **2019**. [CrossRef]
36. Hong, C.H.; Lee, K.; Kang, M.; Yoo, C. qCon: QoS-aware network resource management for fog computing. *Sensors* **2018**, *18*, 3444. [CrossRef]
37. Barzegaran, M.; Cervin, A.; Pop, P. Towards quality-of-control-aware scheduling of industrial applications on fog computing platforms, In Proceedings of the Workshop on Fog Computing and the IoT (Iot-Fog '19), Montreal, QC, Canada, 15 April 2019.
38. Yin, L.; Luo, J.; Luo, H. Tasks scheduling and resource allocation in fog computing based on containers for smart manufacturing. *IEEE Trans. Ind. Inform.* **2018**, *14*, 4712–4721. [CrossRef]
39. Martins, E.; Neves, P.; Fonseca, J. Architecture of a fieldbus message scheduler coprocessor based on the planning paradigm. *Microprocess. Microsyst.* **2002**, *26*, 97–106. [CrossRef]
40. Zheng, W.; Chong, J.; Pinello, C.; Kanajan, S.; Sangiovanni-Vincentelli, A. Extensible and scalable time triggered scheduling. In Proceedings of the International Conference on Application of Concurrency to System Design (ACSD), St. Malo, France, 6–9 June 2005.
41. Velasco, M.; Marti, P.; Yepez, J.; Villa, R.; Fuertes, J.M. Schedulability analysis for CAN-based networked control systems with dynamic bandwidth management. In Proceedings of the IEEE International Conference on Emerging Technologies and Factory Automation (ETFA), Palma de Mallorca, Spain, 22–25 September 2009.
42. Lukasiewycz, M.; Gla, M.; Milbredt, P.; Teich, J. FlexRay schedule optimization of the static segment. In Proceedings of the International Conference on Hardware/Software Codesign and System Synthesis (CODES+ISSS), Grenoble, France, 11–16 October 2009.
43. Schneider, R.; Bordoloi, U.; Goswami, D.; Chakraborty, S. Optimized schedule synthesis under real-time constraints for the dynamic segment of FlexRay. In Proceedings of the IEEE/IFIP International Conference on Embedded and Ubiquitous Computing (EUC 10), Hong Kong, China, 11–13 December 2010.
44. Zhang, H.; Shi, Y.; Wang, J.; Chen, H. A new delay-compensation scheme for networked control systems in controller area networks. *IEEE Trans. Ind. Electron.* **2018**, *65*, 7239-7247. [CrossRef]
45. Elmenreich, W. Time-triggered fieldbus networks—State of the art and future applications. In Proceedings of the IEEE Symposium on Object Oriented Real-Time Distributed Computing (ISORC 08), Orlando, FL, USA, 5–7 May 2008.
46. Cena, G.; Bertolotti, I.C.; Scanzio, S.; Valenzano, A.; Zunino, C. On the accuracy of the distributed clock mechanism in EtherCAT. In Proceedings of the IEEE International Workshop on Factory Communication Systems (WFCS), Nancy, France, 18–21 May 2010.
47. Felser, M. Fieldbus based isochronous automation application. In Proceedings of the IEEE International Conference on Emerging Technologies and Factory Automation (ETFA), Palma de Mallorca, Spain, 22–25 September 2008.

48. Marti, P.; Camacho, A.; Velasco, M.; Mares, P.; Fuertes, J.M. Synchronizing sampling and actuation in the absence of global time in networked control systems. In Proceedings of the IEEE International Conference on Emerging Technologies and Factory Automation, Bilbao, Spain, 13–16 September 2010.
49. Craciunas, S.S.; Oliver, R.S.; Ecker, V. Optimal static scheduling of real-time tasks on distributed time-triggered networked systems. In Proceedings of the IEEE International Conference on Emerging Technologies and Factory Automation (ETFA), Barcelona, Spain, 16–19 September 2014 .
50. Shin, I.; Lee, I. Compositional real-time scheduling framework with periodic model. *ACM Trans. Embed. Comput. Syst.* **2008**, *7*, 30. [CrossRef]
51. Yim, Y.-G.; Jo, H.-C.; Jin, H.-W.; Lee, S.-I. "Schedulability analysis of Linux task groups for hard real-time systems. In Proceedings of the 17th Real Time Linux Workshop (RTLWS 2015), Graz, Austria, 21–22 October 2015.
52. Liu, C.; Layland, J. Scheduling algorithms for multiprogramming in a hard-real-time environment. *J. ACM* **1973**, *20*, 46–61. [CrossRef]
53. Boyer, S.A. *SCADA: Supervisory Control and Data Acquisition*; International Society of Automation: Research Triangle, NC, USA, 2009.
54. Fuhrer, T.; Muller, B.; Dieterle, W.; Hartwich, F.; Huge, R.; Walther, M. Time triggered communication on CAN (time triggered CAN-TTCAN). In Proceedings of the International CAN Conference, Helsinki, Finland, 11–14 June 2000.
55. Eidson, J.C. A detailed analysis of IEEE 1588. In *Measurement, Control, and Communication Using IEEE 1588*; Springer: Palo Alto, CA, USA, 2006; pp. 61–132.
56. Kim, K.; Sung, M.; Jin, H.-W. Design and implementation of a delay-guaranteed motor drive for precision motion control. *IEEE Trans. Ind. Inform.* **2012**, *8*, 351–365. [CrossRef]

Article

A Design Approach to IoT Endpoint Security for Production Machinery Monitoring

Stefano Tedeschi [1], Christos Emmanouilidis [1,*], Jörn Mehnen [2] and Rajkumar Roy [3]

[1] Manufacturing Department, Cranfield University, Cranfield MK43 0AL, UK; s.tedeschi@cranfield.ac.uk
[2] Design, Manufacturing & Engineering Management Department, University of Strathclyde, Glasgow G1 1XJ, UK; jorn.mehnen@strath.ac.uk
[3] School of Mathematics, Computer Science & Engineering, City, University of London, London EC1V 0HB, UK; r.roy@city.ac.uk
* Correspondence: christosem@cranfield.ac.uk

Received: 9 April 2019; Accepted: 16 May 2019; Published: 22 May 2019

Abstract: The Internet of Things (IoT) has significant potential in upgrading legacy production machinery with monitoring capabilities to unlock new capabilities and bring economic benefits. However, the introduction of IoT at the shop floor layer exposes it to additional security risks with potentially significant adverse operational impact. This article addresses such fundamental new risks at their root by introducing a novel endpoint security-by-design approach. The approach is implemented on a widely applicable production-machinery-monitoring application by introducing real-time adaptation features for IoT device security through subsystem isolation and a dedicated lightweight authentication protocol. This paper establishes a novel viewpoint for the understanding of IoT endpoint security risks and relevant mitigation strategies and opens a new space of risk-averse designs that enable IoT benefits, while shielding operational integrity in industrial environments

Keywords: industrial IoT; security; legacy production machinery; real-time condition monitoring

---

1. Introduction

Industry 4.0 has a profound transformative effect on manufacturing environments, bringing in Internet of Things (IoT) connectivity to enable interaction that goes beyond basic machine-to-machine (M2M) communication. Such connectivity scales up the requirements of production data management and leads towards data-driven service innovation in manufacturing, wherein data analytics play a key role [1]. However, the potential arising from such enhanced connectivity is not sufficiently addressed in legacy production machinery, which is often poorly connected [2]. The connectivity capabilities of computer numerical controlled (CNC) machine tools remained constrained within the standardised CNC programming data exchanges, and further limited by a lack of versatile open application programming interfaces (APIs), making it difficult to monitor and control their functions within the whole production process [3]. CNC machine tools may already support a number of diagnostic services, which can be supplemented by additional sensors for direct or indirect monitoring. Such upgrades can be fitted within a networked factory environment through, making the machinery part of the Internet of Things (IoT) environment. IoT offers flexible means for connecting, as well as augmenting even modern machinery through advanced real-time data acquisition and monitoring services [4]. However, added connectivity brings in additional security and integrity risks for industrial environments. While security management has received extensive attention in the information security field, the functionality in production environments is delivered by the employed operational technology at the physical edge, and as such, its endpoint security deserves further attention. The potential operational impact that any security breaches may have on the integrity of industrial systems can be profound and need to be taken into account within the context of the targeted application domain at design stage.

An abstract view of the nature of threats relevant to legacy production machinery is illustrated in Figure 1, showing physical threats at the lower layer, human interaction ones are at the top, and various types of technical threats in between. Physical threats involve actual physical tampering and may have direct tangible impact, for example causing physical damage on machinery, production, and infrastructure, or harm nearby personnel. Advanced technical threats refer to technology enabled access to different network layers and may involve data and software tampering. Human interaction threats are relevant to human interaction with technical systems.

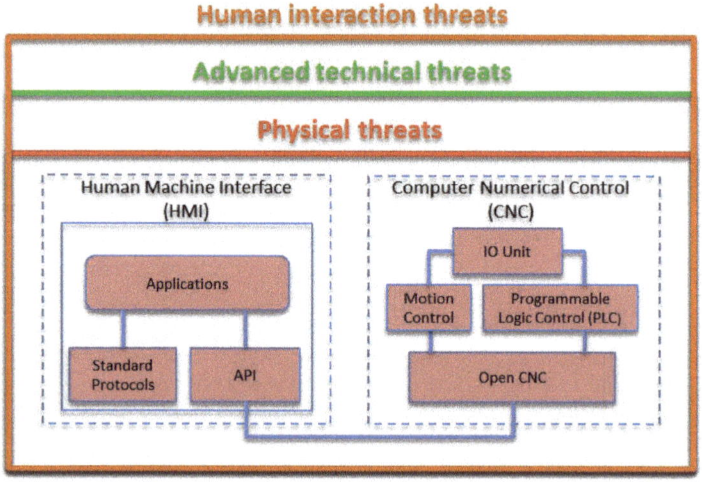

Figure 1. Abstract view of threats for legacy production machinery monitoring. API: application programming interfaces.

While connected and smart environments are gradually implemented in healthcare, industrial, military, and transportation applications, security and privacy issues are increasingly highlighted as the major sources of risks. Industrial systems, in particular, strongly depend on preserving their physical and functional integrity, in additional to typical trust, identity, access control, and data protection through mechanisms, and require due consideration at the design stage of any networking upgrade to offer protection against multiple potential threats [5–7]. For example, permitting the cloning of tags or signal replaying [8,9] may allow attackers to gain access to critical data, services, and facilities. Information can be indirectly extracted from network, hardware, and software components, as some IoT systems may be susceptible to reverse engineering [10]. Defence techniques to prevent such attacks include cryptography [11], secure authentication protocols [12], improved resistance to cloning [13], and automatic malware detection [14]. However, such countermeasures are not included by design in typical industrial IoT endpoint devices, often either due to their resource-constraint nature, or through lack of appropriate designs, allowing such devices to be exposed to threats targeting real-time machine data access, tampering [15] with production machinery, modifying machining software or machine code, cloning devices, as well as initiating denial of service (DoS) or reverse engineering processes.

Various security approaches have been proposed to address or mitigate potential threats [16]. Established methodologies, such as STRIDE [17], first documented internally at Microsoft, involve threat identification and modelling as key activities, while others, such as PASTA [18] take a comprehensive application-oriented and risk-based perspective. While these methodologies have been successfully applied in practice, they do not offer higher-level guidelines and do not sufficiently address risks introduced at the endpoint-level of the IoT stack, considering the real-time application context and integration with industrial control systems (ICSs), such as supervisory control and data acquisition (SCADA), distributed control systems (DCS), and programmable logic controllers (PLC). Non-internetworked industrial control and monitoring systems were not vulnerable to cyber-attacks and there was limited or no clear direct physical integration between them and higher-tier enterprise systems [19]. In contrast, in modern cyber physical systems (CPS) and cyber physical production systems (CPPS) [20], supervisory control systems increasingly employ IoT connectivity to enable ubiquitous real-time monitoring [21], thereby also increasing the risk of cyber-attacks [22,23]. The mitigation of such risks needs to be introduced at the design or runtime stages [24,25]. Unsecured system operation may result in a real loss of service or loss of industrial environment control [26], often facilitated by obsolete versions of operating systems. The integration of IoT with ICS supports the aggregation of factory data to feed into SCADA and enterprise systems, giving rise also to new security challenges. The difference with previous generation SCADA systems is highlighted in Reference [27] where the need to handle endpoint security at the IoT device level is emphasised, which can be considered from a systems viewpoint, such as via SySML modelling of CPS agents [28]. The increasing incorporation of IoT in CPPS has motivated the development of assessment and identification techniques for CPS vulnerabilities, such as in the two-phased approach of Reference [29]. Specifically, phase one involves representing various processes via an intersection mapping of cyber, physical, cyber-physical, and human entities; and phase two introduces a decision tree-based structure for intuitive risk-based vulnerability analysis (e.g., low, medium, and high risk) [30], in a way that bears similarity to STRIDE and part of the PASTA methodology, but without the risk mitigation phase, demonstrating the approach on an automotive manufacturer case. IoT-enabled (or to this effect hybrid) SCADA systems employing wireless sensors network (WSN) may be vulnerable to external attacks. The impact of such threats to SCADA components needs to be analysed to prioritise risks [31].

In contrast with conventional information technology (IT), ICS are operational technology (OT) [32], acting to afford reliable real-time operations with required execution and safety properties at real production time. Security incidents have raised safety concerns in CPPS. Manufacturing enterprises have been the target of different cyber-attacks, aiming to acquire and gain access to sensitive information [33,34], or have fallen victims to ransomware operations targeting to block the computer access [35,36]. "Stuxnet" [37] was a notable worm attack which hit industrial PLC and SCADA vulnerabilities of nuclear plants, by being capable of periodically changing the frequencies of variable frequency drives, affecting centrifuge normal condition operation [38], even if the centrifuges themselves were equipped with cyber and physical security systems. In order to detect unexpected changes, enterprises often use quality control (QC) systems to alert for abnormal quality variations, However, these need to be both robust and strong in covering a range of relevant variations, and to be effective in this context they require threat analysis to better understand relationships between QC, manufacturing, and cyber-physical systems at design stage [39]. It is therefore, important to contextualise security approaches to the nature of CPS and ICS [40] and provide an analysis of security threat types and vulnerabilities, with an outline of security methods for attack prevention, detection, and recovery [41]. For example, in physical attacks, physical accessibility to the target device is by itself the prime vulnerability; data tampering with IoT networking, software attacks exploiting vulnerabilities inside IoT applications; and encryption attacks, involving breaking the system encryption are among the possible threats [42,43]. No security approach in CPPS environments would be sufficient without securing also the human interaction not only with computing and communication devices, but also with physical production assets, to mitigate functional integrity risks.

Focusing on IoT endpoints, device security can be supported by authentication mechanisms. For instance, identity-based authentication based on software defined networking (SDN) can target the distributed nature of wireless sensing in IoT, while consuming reduced resources, compared to public key cryptography (PKC) approaches [44]. Alternative authentication protocols diversify their approach between resource-rich and resource-constrained nodes. An example is the two-stage PAuthKey protocol, with a registration stage for obtaining cryptography credentials and an authentication step for establishing the communication [45]. Only resources-rich nodes communicate with the registration authority and the communication with constrained-resource nodes is then authenticated via implicit certificates. In practice, end-to-end security is applicable to the application layer, while lower layers rely on media access control (MAC)-tier security [46]. Overall, such an approach delegates security to edge nodes, enhancing resources efficiency. A hardware authentication method is presented in Reference [47], wherein each device is equipped with a unique fingerprint, consisting of multiple features, such as location, transmitter state, or physical object state. Alternatively, the authentication scheme is linked to distributed denial of service (DDoS) attack prevention via an algorithm that collects information from nodes to detect an attacker so as to prevent the working node from serving the malicious attacks [48]. In Reference [49], a dual authentication based on certificates and using datagram transport layer security (DTLS) between constrained IoT devices is proposed. An alternative approach uses a lightweight key agreement protocol to ensure anonymity, data secrecy, and trust between wireless sensor network (WSN) nodes in the IoT network [50]. In another network-centric approach, privacy invasion targeting networking patterns can be mitigated through synthetic packet-injection to hide real network traffic [51].

Considering that the intended functionality of CPPS is determined at the design stage, the same should be the case for IoT security in production environments, taking into account the potential sources of attacks and system vulnerabilities [52]. Therefore, understanding the nature and functionality of industrial systems is a prerequisite to designing their IoT security. With this in mind, after analysing and synthesising requirements for industrial systems, the Industrial Internet Consortium (IIC) has put forward the Industrial Systems Security Framework (IISF) [53]. The key differentiating factor between IISF and other IT or nonindustrial IoT security approaches lies in the joint handling of IT and OT. Security is viewed upon from the perspective of the potential impact on the delivered functionality of industrial systems, i.e., overall industrial systems' trustworthiness. This is translated into a risk-based framework, directly linking security threats to risks arising from their impact on industrial systems trustworthiness. Recognising the ecosystem nature of IoT installations, IISF considers the whole system lifecycle and the permeation of trust across the system life-cycle phases and the system actors involved in them. IISF highlights the architecture view of IoT by considering security at the different layers of the IoT implementation stack, starting from the shop floor IoT end points. The shop floor end points include sensors, actuators, as well as connected production machines, which now become exposed to cyber-attacks, and therefore, lessons learned from IoT security need to be applied to develop strategies for networked production environments security. The IISF highlights the importance of the principle of isolation when securing IoT endpoints. This refers to process isolation within the operating system, container isolation implementing hardware or software-enforced boundaries, and virtual isolation protecting individual virtual instances of a trusted execution environment. However, endpoint security is still not sufficiently covered when upgrading legacy production equipment with IoT capabilities. This fundamental baseline of IoT endpoint security in industrial environment is, therefore, the target of the security thinking approach introduced in this paper, which includes:

- A novel risk-averse IoT endpoint security design thinking approach for industrial environments.
- An innovative IoT device security implementation of the design thinking approach, motivated by the isolation principle and applied at the interfaces between the key components of an IoT endpoint device and supported by a new lightweight authentication protocol with real-time features.
- Application of the above on a typical industrial case, that of production machinery monitoring.

The paper is organised as follows: Section 2 introduces the new endpoint security design thinking approach, comprising five stages. Section 3 analyses vulnerabilities when introducing IoT-enabled monitoring in manufacturing environments and introduces a relevant threats taxonomy, corresponding to the first two stages of the approach. Section 4 deals with stage 3 and employs an attack tree-modelling methodology to analyse security when introducing IoT in such environments, and presents steps taken to address them through adopting the subsystem isolation principle for an IoT data acquisition unit (DAQ). Section 5 implements stage 4 and applies the proposed approach on a legacy production machinery monitoring application. A representative implementation example case of the design solution, tested against a set of typical selected key attack types, is presented in Section 6, which corresponds to the final stage. Section 7 offers a discussion regarding limitations and further work, while Section 8 presents the conclusion.

2. Design Thinking for IoT Security in Industrial Environments

Production machinery real-time monitoring is a major application target when introducing IoT in industrial environments and IoT endpoint devices are a fundamental component for any IoT security approach designed for such monitoring. Consistent with relevant recommendations and standards [54] and taking into account the nature of the manufacturing domain, the present research proposes a design thinking approach that clearly takes into account the application context and the context-specific potential impact of security compromises, a process more aligned with PASTA rather than STRIDE. The introduced systematic design thinking approach for IoT device security includes five key stages (Figure 2). Feedback from each phase may reveal a need to reconsider analysis, modelling, design, and implementation choices of all earlier phases.

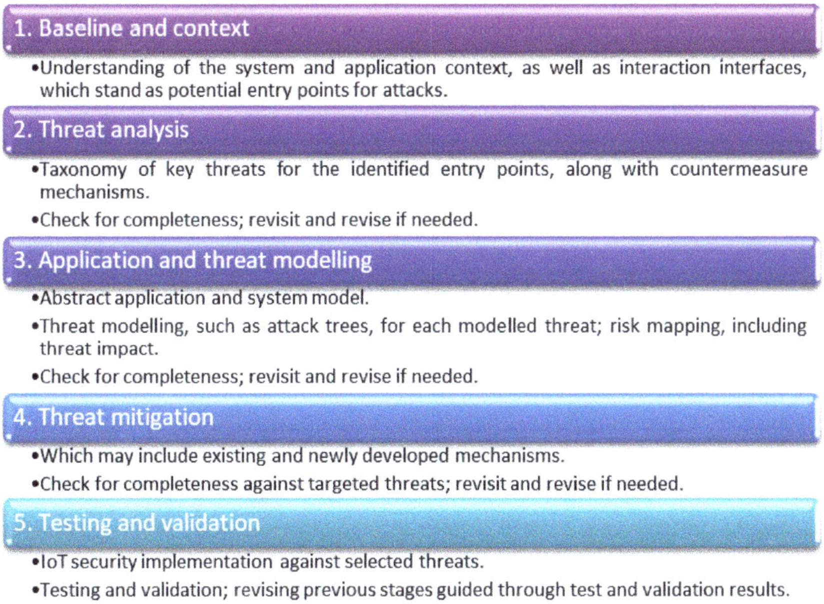

1. Baseline and context

- Understanding of the system and application context, as well as interaction interfaces, which stand as potential entry points for attacks.

2. Threat analysis

- Taxonomy of key threats for the identified entry points, along with countermeasure mechanisms.
- Check for completeness; revisit and revise if needed.

3. Application and threat modelling

- Abstract application and system model.
- Threat modelling, such as attack trees, for each modelled threat; risk mapping, including threat impact.
- Check for completeness; revisit and revise if needed.

4. Threat mitigation

- Which may include existing and newly developed mechanisms.
- Check for completeness against targeted threats; revisit and revise if needed.

5. Testing and validation

- IoT security implementation against selected threats.
- Testing and validation; revising previous stages guided through test and validation results.

Figure 2. Design thinking approach for Internet of Things (IoT) device security.

(a) **Baseline and context**: This stage involves analysis of current practices in production environments. An understanding of the application context and system component interfaces, which may be exposed to security threats, is necessary to apply proposed concepts to a specific application target.
(b) **Threat analysis**: Having identified the high-level system interfaces which pose security risks, this stage involves an analysis of key security issues and vulnerabilities related to the implementation of IoT inside a production environment. Each vulnerability exploited by a threat can create an adverse impact on system integrity. A taxonomy of threats is produced, classifying them under the broader categories of physical, human interaction, and advanced technical ones. For each threat, possible mitigation mechanisms are proposed, and impact risk assessment is performed. Risk is quantified in three categories (High, Medium, and Low), consistent with recommendations [55].
(c) **Application and threat modelling**: The third phase provides the application context needed for an effective approach. It produces a more detailed model of the targeted system, along with its interfaces and functionality. Modelling tools include data flow diagrams (DFD) [56] to understand the permeation of data trust between components, and systematic threat modelling via attack trees [57], which need to be checked for coverage of security threats.
(d) **Threat mitigation**: The fourth phase deals with design and implementation of security threats-mitigation mechanisms. In the present work, an instance of the overall process is created and applied to the real-time monitoring application relevant to production environments.
(e) **Testing and validation**: This includes testing and validation of the mitigation mechanisms against selected threats. Testing may include simulation and functional testing, while validation may be performed in a test or a controlled operational environment. Results from functional and penetration testing can be fed back to improve the mitigation effectiveness. The functional aim of the test in the selected application case is to deliver uninterrupted real-time monitoring.

Figure 3 shows a simplified flowchart for the proposed systematic approach applied to a real-time monitoring application relevant to production environments. For illustration purposes, this lists three types of attacks, namely network, system communication, and DAQ. These will be considered in more detail in the context of analysing the selected application case in Sections 3 and 4, dealing with stages a, b, and c of the approach. To demonstrate the application of the new approach, the implementation and testing of mitigation mechanisms against denial of service (DoS) [58] and clone attacks [59] are presented in Sections 5 and 6, corresponding with stages d and e. This involved the development of an innovative IoT endpoint device security implementation, introducing a new lightweight authentication protocol, consistent with the isolation principle and integrated in a prototype IoT DAQ device.

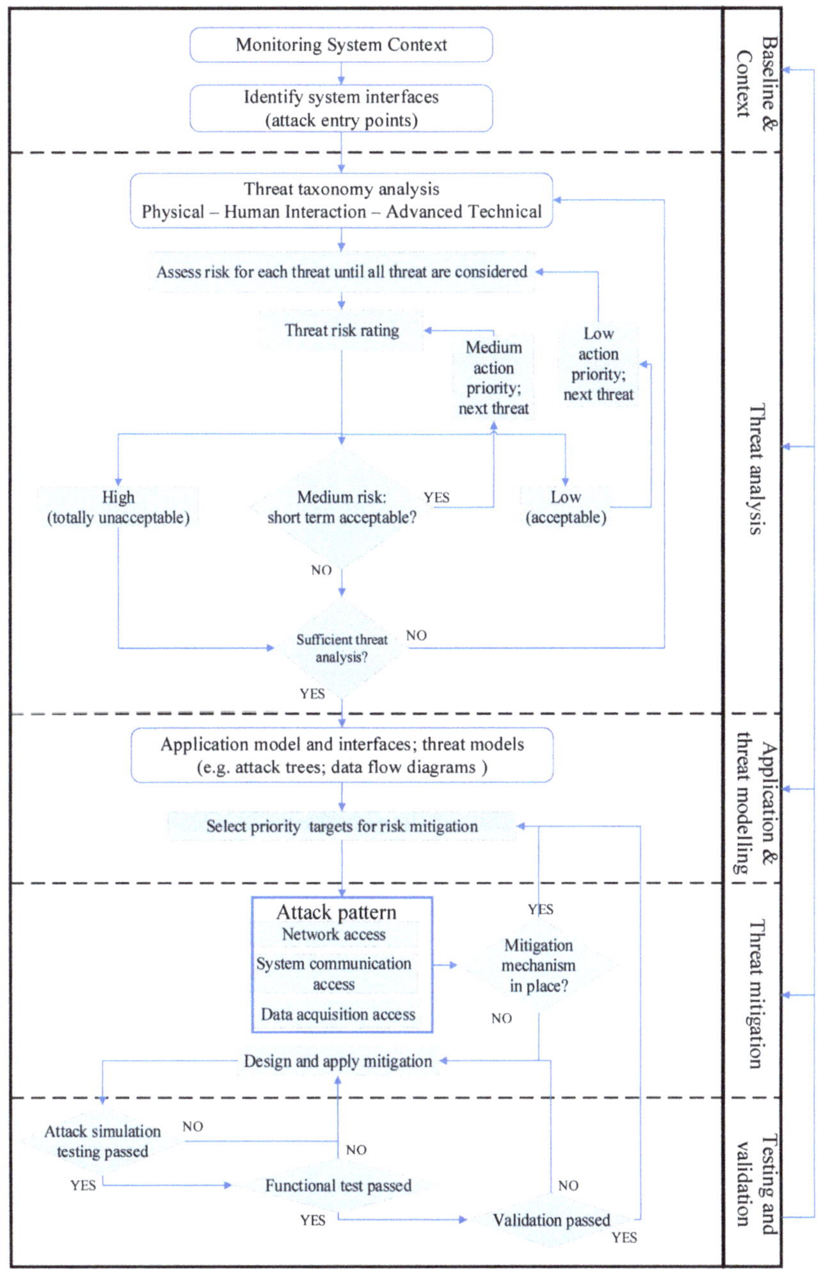

Figure 3. Design thinking for IoT device security in production environments.

3. Monitoring Systems Security in Industrial Environments

3.1. Baseline and Context

The proposed design thinking approach is applied on a widely employed application in industrial environments, namely real-time condition monitoring (CM). CM refers to data acquisition and processing to infer the state of a machine over time [60]. It enables the identification of recommended maintenance actions based on the actual condition of monitored assets, rather than at predetermined intervals, thus allowing a condition-based maintenance (CBM) strategy to be implemented [61]. The determination of an appropriate CM approach consistent with a CBM strategy involves cost–benefits analysis, equipment audits, reliability and criticality audits, monitoring methods selection, data acquisition and analysis, determination of appropriate maintenance actions, and review processes [62]. A typical real-time condition monitoring system for legacy production machinery comprises sensors, a DAQ unit or microprocessor, computing resources, and adequate software [63], which may also be compactly available as a data-logging device. Signals acquired via the DAQ are processed by dedicated software, enabling the machine health to be determined. More advanced condition monitoring may also involve prognostics, and maintenance action determination. In wireless sensing, measurements can be acquired through a DAQ equipped with connectivity. Remote monitoring systems (RMS) [63] may already employ network communication between monitored machinery and back-end systems, or may involve retrofitting monitored assets with a communication device. RMS are applicable to both production processes and products. Connected products are amongst the prime developments which contributed to the concept of closed-loop product lifecycle management (PLM). IoT technologies not only enable product connectivity but also create data flows that upgrade the value proposition of product usage in operating environments [64,65]. Including IoT connectivity in such products creates additional vulnerabilities and this applies to IoT-enabled production machinery too. Therefore, the integration of IoT on legacy production machinery requires a rethinking of their security design [52].

Figure 4 offers an abstract view of a machinery real-time monitoring system highlighting potential entry points for security attacks, assuming three standard communication types, namely wired or wireless device peer-to-peer (P2P), fieldbus, and Ethernet, as part of stage one of the approach. The networking enables data flows through sensors, PLCs, DCSs, programmable automation controllers (PAC), and human-machine interfaces (HMI), which in turn can drive recommendations for maintenance actions, and their planning and execution. This mapping can be looked upon from the viewpoint of the ISA-95 reference architecture, as adapted and mapped in five layers by the European Union Agency for Network and Information Security (ENISA) for the scope of smart manufacturing security [23]. Specifically, the field level of Figure 4 corresponds to Level 1, the control level to Level 2, the operator level to Level 3, and the upper-level refers to the application context, which in this case refers to interfaces exposed to devices accessing maintenance management and planning software and services, corresponding to Level 4. Unless the permeation of trust in such an architecture is duly considered, IoT-enabled industrial monitoring systems create increased security risks. Therefore, the additional focus is on the interfaces exposed to attacks, as per the first stage of the design approach of this paper. The next section provides an overview of threats analysis by threat type, applicable to industrial environments, relevant to stage 2 of the approach, while stage 3 in Section 4 studies in detail data interfaces and corresponding attack models for typical key security breaches in the studied problem, namely network, system communication, and DAQ access.

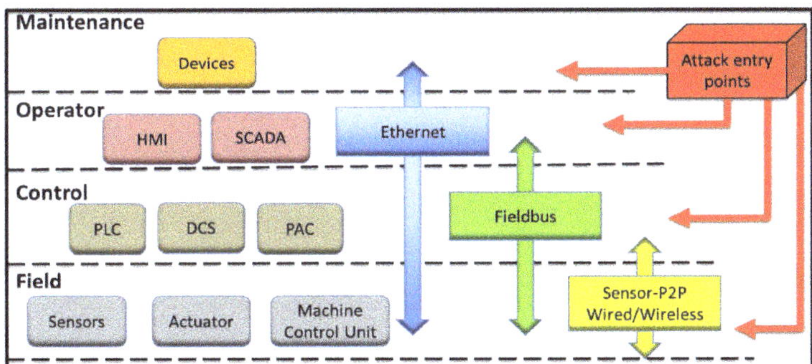

Figure 4. Legacy production machinery real-time monitoring system attack entry points. SCADA: Supervisory control and data acquisition; DCS: Distributed control systems; PAC: Programmable automation controllers; P2P: Peer-to-peer.

3.2. Threat Analysis

Threat analysis is the main activity in stage 2 of the process. The ISO27000 family of standards offers a broadly adopted framework for information security, including recommendations for information security management systems (ISMS, ISO/IEC 27001 [53]), where threat identification, as part of security risk assessment (ISO27005) [66], is central to devising a security approach. Vulnerabilities can be exploited by attack events to trigger security breaches, which, depending on the resulting sequence of events, may cause adverse impacts. Such impacts need to be translated into security risk mapping and quantification [67]. However, such recommendations are not specific enough to cover monitoring architectures for legacy production machinery. Approaches relevant to cloud security risk management [68] and lessons from other domains, such as finance, wherein cyber-attacks were already the prime sources of money loss, highlight the need to perform a domain-specific threat analysis to prevent adverse impacts [69]. Threat analysis is incomplete if it does not deal with application domain considerations. ENISA has produced a threat taxonomy for Industry 4.0 [23], which classifies threats into (a) nefarious activity or abuse; (b) eavesdropping, interception, or hacking; (c) physical attack; (d) unintentional or accidental; (e) failures or malfunctions; (f) outages; (g) legal; and (h) disaster. In this study, legacy production machinery and their monitoring systems define the application domain scope. Considering the focus on operational technology, physical types of threat are of prime concern [70]. Furthermore, considering the roles of personnel in production operations, a second key category would need to concentrate on human interactions. Finally, as advanced technology is involved in such manufacturing environments, compared to legacy production ones, technical threats is a natural third broad category. Therefore, the paper proposes that an appropriate high-level threat taxonomy should analyse human interaction (HIT), advanced technical (ATT), and physical threats (PT). Automatic operations are excluded from human–machine interactions and all operations that require human intervention (semiautomatic and manual) are included within human interactions. The software and network entry points are hard to enumerate and are subject to change. Entry points for hardware attacks are fewer and moderately well determined but attack targets can be diverse, targeting for example, information leakage [71], tampering [72], denial of service (DoS) [73], or cloning [59]. For each threat type, threat analysis needs to identify and describe activities which may allow the relevant vulnerability to be exploited (Table 1).

Table 1. An example of threat analysis. HIT: Human interaction; ATT: Advanced technical; PT: Physical threats.

Activity Description	Impact Examples	Countermeasure Mechanisms	Threat Types		
			HIT	ATT	PT
1. Negligence Errors and vulnerabilities linked to the launch of a new network within a production environment.	Network delays or errors lead to poor control or loss of control over certain production processes	Operations procedures to be followed by personnel installing or using the network	x		
2. Social Engineering Uses the human behaviour to gain security access without the victim realising the manipulation.	Could cause system integrity loss (e.g., data loss or tampering, system process malfunctions, poor product quality, health and safety issues).	Training about the social engineering threat, company policy, and procedures.	x		
3. Denial of service (DoS) Channel is flooded with data, exhausting bandwidth.	Breakdown of network control, causing loss of production monitoring and control capabilities	Network traffic analysis and detection systems		x	

The potential harm that a threat may cause when exploiting vulnerabilities is assessed by rating the impact in categories such as those recommended in the National Institute of Standards and Technology (NIST) standard [55] (Table 2). Risk impact is linked to the functionality and integrity of the installation, and so risk analysis needs to consider its specific context. Risk levels can be adapted for a finer risk granularity if needed to serve specific application needs. The likelihood of the identified risks is then assessed (Table 3) and the final risk impact is quantified as the product of risk impact and likelihood (Table 4). IoT-enabled production assets create enhanced production data flows and therefore, DFD is a fitting model to study security vulnerabilities of key system entities. DFDs employ symbols for key processes and entities:

- ✓ External entities (EE), considered as end-point of a system;
- ✓ Processes (P), such as system or unit functionality;
- ✓ Data flows (DF), i.e., ways to transfer data;
- ✓ Data storage (DS), such as database or files for recorded information.

Table 2. Impact rating.

High (H)	The Threat Is Unacceptable and Immediate Measures Are Needed to Reduce It to Preserve Data or System Integrity.
Medium (M)	The threat may be acceptable over the short term but countermeasures to reduce the risk should be implemented.
Low (L)	The risks are acceptable. Measures to reduce risk can be taken in conjunction with other actions, for example, during upgrades.

Table 3. Risk likelihood (chance rating).

High (H)	A Highly Motivated and Sufficiently Capable Threat-Source; Protection Countermeasures Are Ineffective.
Moderate (M)	The source of the threat is motivated and capable, but some countermeasures in the short term could hinder the success of attacks.
Low (L)	Limited motivation and capability of threat-source; the countermeasures are sufficient to prevent the hazard.

Table 4. Score rating (SR) = Impact rating × Chance rating.

		Impact →→		
		Low (L)	**Moderate (M)**	**High (H)**
Chance →→	**High (H)**	L × H = M	M × H = H	H × H = H
	Moderate (M)	L × M = L	M × M = M	H × M = H
	Low (L)	L × L = L	M × L = L	H × L = M

Finally, Table 5 offers a threat classification scheme along with risk impact quantification and applicable DFD modelling entities. Risks with high chance and impact are likely to occur, will have a significant impact, and should be given priority for mitigation. Risk quantification in the tables is indicative and actual risks in a specific implementation are likely to differ. An expert view of risk quantification in such industrial settings is available by ENISA [23]. This type of analysis is a necessary step to establish a sound baseline for designing a security approach to reduce risks for remote monitoring when upgrading legacy equipment with IoT devices. Having concluded with the stage 2 of the proposed approach, stage 3 aims to produce more detailed threat analysis for the targeted application domain, as described in the next section.

Table 5. Threats classification for IoT-enabled production environments.

Activity	**Threat Types**			**External Entity (EE)**	**Data Flow (DF)**	**Data Store (DS)**	**Process (P)**	**Impact Rating**	**Chance Rating**	**Score Rating**
	HIT	**ATT**	**PT**							
Negligence	X						X	M	M	M
Social Engineering	X			X			X	H	L	M
Tampering			X		X	X	X	H	L	M
Physical Intrusions	X		X		X	X	X	H	L	M
User Misuse	X			X			X	H	L	M
Unauthorised remote accesses	X						X	H	L	M
External hardware	X						X	H	L	M
Physical destruction	X		X				X	H	L	M
Command injection		X			X		X	M	L	L
Denial of Service (DoS)		X			X	X	X	H	M	**H**
Signal replaying		X			X	X	X	M	L	L
Cloning		X			X	X	X	H	M	**H**
Remote switch off		X			X		X	H	L	M
Signal blocking or jamming		X			X		X	H	L	M
Reverse engineering		X	X		X	X	X	H	L	M
Side-channel		X	X		X	X	X	H	L	M
Wireless zapping		X			X		X	M	L	L
Software compromise	X	X			X	X	X	H	L	M
Electromagnetic interference			X				X	M	L	L
Cable cuts			X	X			X	H	L	M
Power fluctuation			X				X	M	M	M
Voltage spikes			X				X	H	L	M
Installation errors			X				X	M	L	L
Takeover of an authorised session			X	X	X	X	X	H	L	M

4. Application and Threat Modelling

4.1. Application Model and Data Interfaces

In the first part of stage 3 of the proposed approach, the application model considers key components of a machinery monitoring architecture and their data interfaces (links), to enable studying IoT security requirements in more detail. Following a representation similar to Reference [74], a simplified mapping of data exchanges is shown in Figure 5.

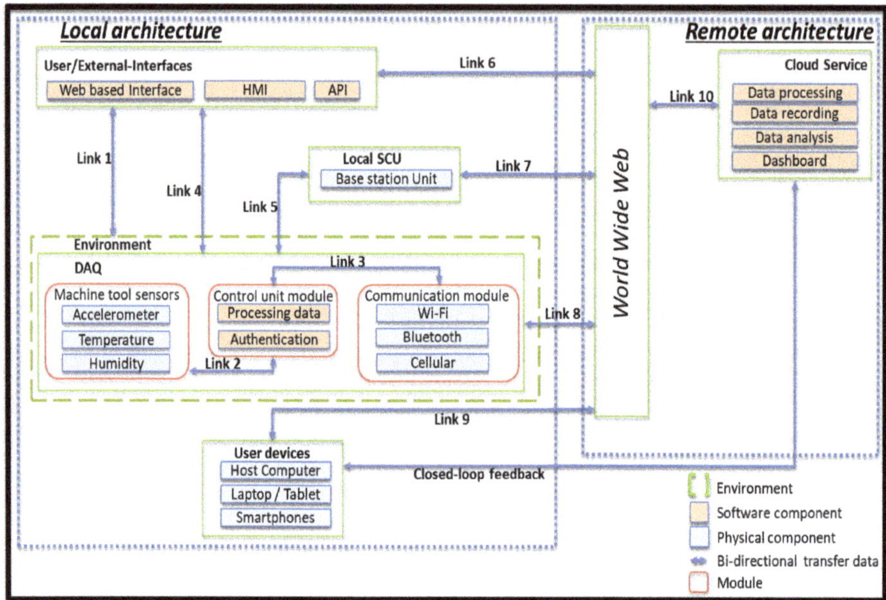

Figure 5. Abstract application model of the connected legacy production machinery. SCU: System control unit.

The local architecture includes the workplace with a legacy production machine and the IoT-enabled DAQ. The DAQ comprises three modules. The sensors module is physically attached to the machine. Collected real-time data are processed in the control unit module and can be passed to external or visual user interfaces. Transmitted data are sent to cloud-based systems, to the system control unit (SCU), or a mixture of both via the communication module. The remote architecture stores, manages, analyses, and visualises data on a dashboard to aid future actions. Such functionality is offered through the cloud to end-user devices, which can reside inside the local architecture. The data flows across the links are:

- Link 1: The environment includes the legacy production machinery, the DAQ modules, with access to configuration and management web services.
- Link 2: Data acquired from the sensor module are sent to the control module.
- Link 3: The control module manages the authentication process and passes data to the communication module.
- Link 4: The DAQ provides a user-interface to manage and visualise the data acquisition in real-time, residing within the monitored facility.
- Link 5: The DAQ and the SCU exchange data between the sensors and the local architecture.
- Link 6: Interfaces offer data visualisation and support or trigger appropriate actions.
- Link 7: The SCU employs cloud access to offer machine data management to users.
- Link 8: The DAQ communicates with cloud services via the internet.
- Link 9: User devices are communicating with the cloud or server through the internet, exchanging information relevant monitoring information.
- Link 10: Data management and visualisation services are made available to the user.

The mapped links are likely attack-entry points for the manufacturing environment. Link 1 is an entry point for physical threats, which can compromise the integrity of hardware devices, sensors, systems, and data. An attack can occur through connection to web-based interface, which is an entry point for software threats (e.g., viruses and trojans), as well as DoS and remote access control attack. Data interfaces require physical access to components, which can be exposed to cloning, side-channel, and reverse engineering attacks, but may also malfunction due to electromagnetic interference, voltage spike, and power fluctuation. Links 2 and 3 are entry points for physical intrusion and tampering, as well as cloning, side-channel, and reverse engineering. Links 4 and 5 are entry points for command injection, software attacks, DoS, and cloning, as well as unauthorised remote access. Links 6, 7, 8, and 9 are entry points for attacks causing a network breakdown or system process malfunctions through ATTs, such as DoS, command injections, reverse and social engineering attacks. Link 10 requires access credentials and is an entry point for error and omission, unauthorised remote access, social engineering, command injection, DoS, and software attacks. HITs are relevant to all interfaces and components and may cause data loss, process malfunctions, and network breakdown. Having available an abstract application and data exchange model helps towards application-specific threat modelling. DFDs between subsystems create an understanding of the permeation of trust between boundaries. Figure 6 shows a simplified DFD for IoT-enabled monitoring. Dotted line rectangles denote trust boundaries of subsystems; solid line rectangles represent external subsystems; arrows indicate data flows; interfaces with external entities and storage are marked with a solid coloured rectangle (not part of DFD). Data flows and trust boundaries constitute intermediate attach goals, and their modelling is the subject of the next section.

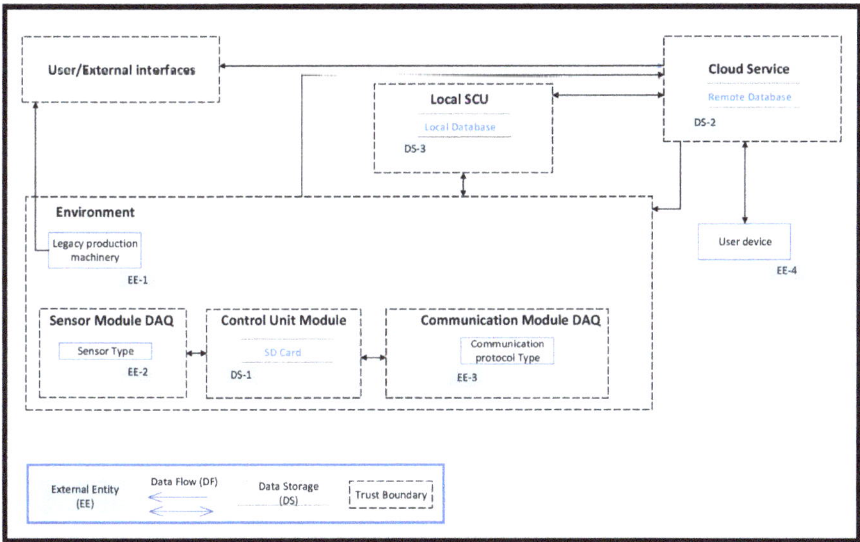

Figure 6. Data flow diagram of the connected legacy production machinery. DAQ: Data acquisition.

4.2. Threat Goals Modelling

The second part of stage 3 of the proposed approach deals with detailed threat modelling. In order to devise mitigation mechanisms, it is of interest to further understand specific goals that an attacker may set in pursuing attack targets. Focusing on the two high-risk priority attack threats (Table 5), namely DoS and cloning, it is of interest to study potential attack intentions and consequences. The main goals are: Gaining network or access, communication access to the supervisory and control architecture [75], and modifying the DAQ [76]. The potential impacts of these goals are analysed in Table 6, consistent with the reliability-oriented approach FMEA (failure mode and effects analysis). Specifically, impacts could affect different functions, which in the case of a production machine could be stated as [62]: P: Primary, affecting functions required to fulfil the machinery intended output (e.g., production of an item); S: Secondary, supporting the primary function (e.g., managing coolant in a machine tool); C: Control and protective, affecting the ability to control a process (e.g., adjusting feed rate in machining) or protecting workers, equipment, or the environment (e.g., stopping machining after tool breakage); I: Information, affecting ability to provide monitoring information for a function (e.g., failure to provide or display temperature reading); and U: Interface, affecting the interaction interface between two items. This makes the understanding of the potential consequences of an attack more tangible and aids the design and development of impact mitigation. The attack goals are next modelled, for example, via attack tree modelling, which is a common structured approach to illustrate in a logical way the main goals of an attacker. The top tree node is a key attack target. Lower level goals and individual malicious activities, which may contribute to reaching that goal, are located below the main node. Steps between the lower nodes and the top node depict intermediate states or attacker subgoals. This modelling is now applied for the machinery monitoring application, defining attack trees for the identified threats (e.g., Tables 1 and 5) and specifically for each of the attack goals of Table 6.

Table 6. Attack goals and impacts. P: Primary; S: Secondary; I: Information; U: Interface.

Attack Goal	Impact Description	Function Code
Network access	Inability to communicate with the DAQ	I, U
	Inability to communicate with the Cloud	U
	Inability to communicate with the SCU	C, U
	Inability to communicate with User Devices	I, U
	Inability to upgrade firmware	I
System communication access	Inability to use the HMI	I, U, P
	Inability to use the DAQ modules	I, C
	Inability to use the legacy production machinery	P, S
DAQ access	Inability to collect correct sensor data	C, I
	Inability to protect sensor data	C, I
	Inability to send data correctly	C, I, U

4.2.1. Network Access

Gaining access to the network wherein the monitoring system operates, an attacker can use malicious or fraudulent actions to gain access to data devices or server systems connected with the network. Figure 7 shows the attack tree that models the network access threat goals. Typically, enterprises may have a private and public network. Subject to access rights, these are exposed to personnel, customers, partners, or suppliers. Within the intranet there may be parts of the architecture which can be modified by access to the hardware for upgrading firmware, updating software, and replacing components, whereas other entities do not require physical access to the architecture and are only modified remotely. Upon gaining physical access to the hardware, the attacker can further access the network through devices, cables or ports, radio interference, or wireless and wired networking means. Without physical access, network access can be achieved via social engineering [77]. While encryption and a media access control (MAC) filter can be applied as security measures, spoofing attacks [78] can still be used to gain access to the network. An attacker can bypass strong encryption methods (such as Pretty Good Privacy (PGP) or Advanced Encryption Standard (AES)), by obtaining the encryption password mostly through social engineering, installing some malware for reading the password, or by breaking into specific network devices via a side-channel method. If the system devices are equipped with a weak encryption method, it may be easily broken with cryptography attacks. On the extranet side, the system can be equipped with password authentication. An attacker can use the dictionary method to guess the password, then bypass the firewall and gain access to the local network.

4.2.2. System Communication Access

Remote access applications allow ubiquitous supervision and control through networked devices, whilst HMIs allow enable control via a front machine panel. An attack may seek to gain access to the communication system to compromise supervisory systems and modify machine or process parameters. An attack tree analysis for this threat goal is shown in Figure 8. If there is no authentication requirement, an attack can easily succeed in gaining access. When authentication is enforced, an attack may guess the access key by the dictionary method [79], or bypass the password using a backdoor secret method, such as chipset, cryptosystem, and an algorithmic structured query language (SQL) code injection [80]. When encryption is employed, the attack can obtain the key through a social engineering method or malware injection. Systems without encryption are susceptible to man-in-the-middle (MITM) method, where the attacker can spoof the system identity, waiting for a user to login and then save the credentials for future access. If physical access to the HMI is gained, the attacker can use an infected USB dongle to compromise the control system or employ reverse engineering to gain communication or achieve this without physical access via social engineering.

4.2.3. Data Acquisition (DAQ) Access

Figure 9 displays the attack tree to acquire access to the DAQ. The side-channel method is one of the simplest physical access methods, allowing DAQ access to make modifications, such as install new firmware or patch, or replace hardware components. Using the network, the attacker can use SQL injection [81] to gain access to user devices or gain authentication to infect the DAQ with malware, and through the replay attack to spoof data. An attack can target DAQ access after remotely logging in with credentials to launch a DoS attack and flood available bandwidth. Accessing the sensors, the attacker can compromise hardware or software components to affect normal DAQ operation. When sensor authentication is not employed, an attack can gain DAQ access using log files to spoof data. From the extranet, an attack can gain DAQ access via the MITM method, SQL injection, or spoofing sensor information, replay attack method, or flood its connection via DoS. Alternatively, an attacker can remotely gain authentication to the cloud service and control the DAQ from there.

Attack tree modelling is a structured methodology for analysing security to drive the design and implementation of appropriate mitigation mechanisms. The next section takes into account such analysis to develop and test IoT endpoint device security for legacy production machinery monitoring.

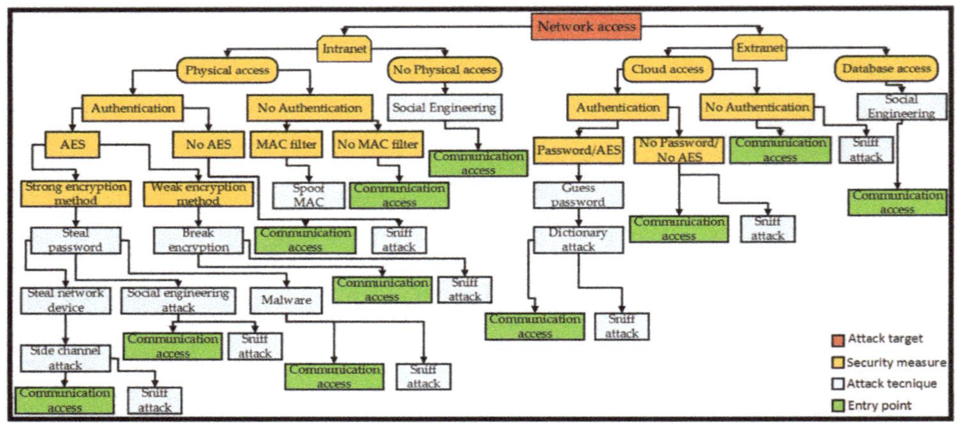

Figure 7. Network access attack tree. MAC: Media access control; AES: Advanced Encryption Standard.

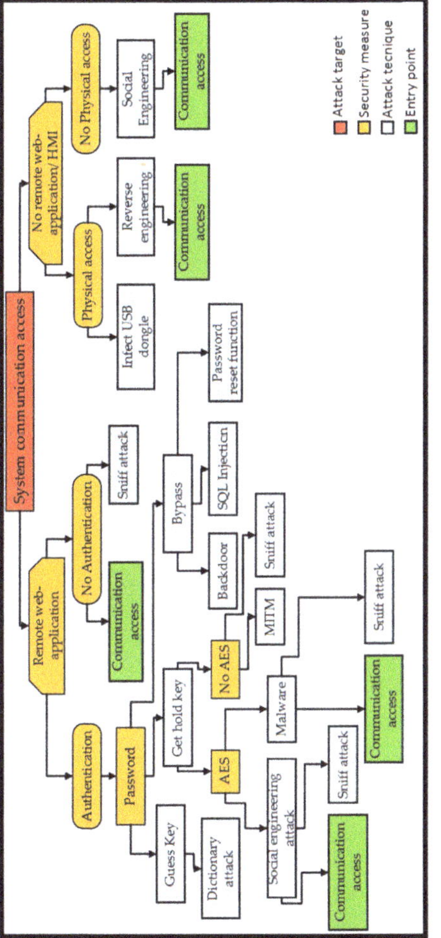

Figure 8. System communication access attack tree. MITM: Man-in-the-middle.

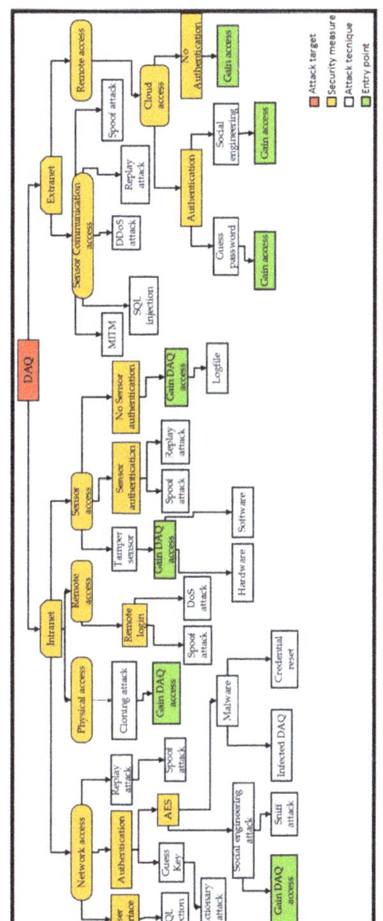

Figure 9. DAQ access attack tree. SQL: Structured query language.

5. Threat Mitigation for IoT-Enabled Production Machinery

After the steps of Section 4 for threat analysis and application-specific modelling, stage 4 of the proposed approach introduces mitigation mechanisms for priority risks. Retrofitting monitoring solutions on machinery typically involves devices that integrate acquisition, processing, and transmission of data. Such units are compact but may have security shortcomings. In some cases, they employ a single communication protocol for real-time data transmission, which can be restrictive in the sense that if a single communication protocol is compromised, the whole process integrity might be so too. However, increasingly, IoT devices offer multiconnectivity options, which add more flexibility but still the choice of protocol is preset and fixed in most cases. A typical IoT device includes I/O ports for sensing and actuation (1st Module), CPU and memory (2nd Module), communications (3rd Module), and powering options [82]. Each of them in order may be considered to extend the functionality of the previous one, but in integrated IoT devices, their trust boundary encompasses them all together (Figure 10). Such a device can be compromised if any of the three modules is compromised, for example, through cloning. IoT endpoint security can benefit from the IISF principle of component or subsystem isolation and this is adopted here. In contrast to monolithic devices, the proposed design choice is for a modular security approach, by decomposing the overall trust boundary to create a separate trust boundary for each component and implementing security mechanisms in the communication between them.

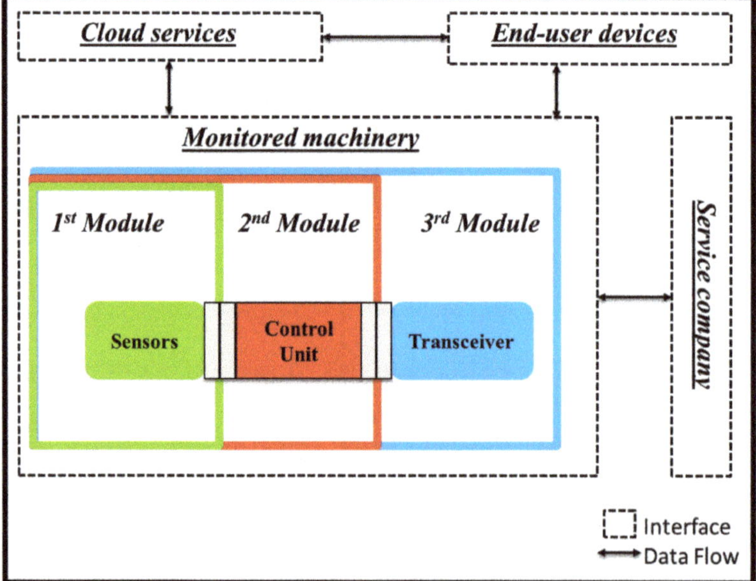

Figure 10. The IoT DAQ for legacy production machinery.

The Authentication Protocol

The authentication protocol for the modular IoT DAQ is illustrated in Figures 11 and 12. Specifically, the flowchart in Figure 11 shows the process flow, while the DFD in Figure 12 depicts the data flow. The protocol comprises four steps: Log identity authentication, encrypted communication, secure connection, and authentication, and will be referred to as LCCA. All LCCA phases employ AES cryptography. The phases are serially executed and failure to execute one as specified, results in issuing a security alert. The LCCA protocol includes a set of keycodes, passwords, baud rates, and frequency values as part of its mechanism to progress through the four phases and can be applied for the communication between the control module and the two other modules. The LCCA flow is described next for the communication between the sensor and the control unit module (Figure 11).

Phase 0: Start

- The system initialises the set of keycodes, passwords, baud rate, and frequency relationships to be used, and then enters a sleep mode (START), waiting for the first connection.

Phase 1: Log identity authentication

- This phase handles the log identity between the modules. Specifically, if the control unit module recognises the sensor module log identity, the protocol proceeds, otherwise the process freezes. In our example log identities (and to the same purpose baud rate, frequency values, and passwords) are prestored in a dictionary embedded in each module, but algorithmic approaches to dynamically create them could be employed instead.

Phase 2: Encrypted communication

- This step sets an agreed value for the data transfer rate (baud rate) between the control unit and sensor modules. In this way, the two modules engage in a handshake process. The LCCA algorithm sets the initial rate (baud rate 1) and at real-time every fixed time period (in this example, 3 ms) the algorithm changes the control unit rate with a new rate value (baud rate 2), according to (based on frequency x in Figure 11) a formula known in advance between the modules. Upon agreement, data exchange progresses, and all data transfers are encrypted. Any mismatch between the two, which may arise as a result of a security breach, will pause communication and set the system to sleep mode, issuing an alert. Once encrypted communication is established, the process advances to the next stage, otherwise, the connection is closed and returns to phase 1.

Phase 3: Secured connection

- This phase covers the connection between the control unit and the sensor module. Once encrypted communication is established, the control module will expect to receive a frequency value from the sensor module to set a new connection rate at predetermined intervals (set here every 3 milliseconds). If the frequency value is recognised by the control unit module, the protocol continues to the next phase, otherwise will pause communication and set the system to sleep mode, issuing an alert. The modules establish connection, and the control module sends the new frequency in a continuous loop employing the baud rate agreed in phase 2.

Phase 4: Authentication

- In this phase, the sensor module alphanumeric password is checked by the control unit. An admissible alphanumeric password is a combination of a minimum of eight characters, including lowercase and uppercase, numbers, and symbols. Additional measures prevent using the same password twice; dictionary words, or sequences; usernames or information that might become publicly associated with the user. If the control unit module does not recognise the password, authentication ends unsuccessfully, and the process moves back to step 3.

The DFD of Figure 12 is a detailed version of Figure 6 to illustrate the data flow through the trust boundaries when the IoT device is equipped with the added security provisions. Instead of the single trust boundary around the IoT device, there are now three trust boundaries, one for each module, and an overall boundary is highlighted for the whole machine equipped with the IoT device. Next, an implementation instance of the LCCA mitigation mechanism and its testing are presented.

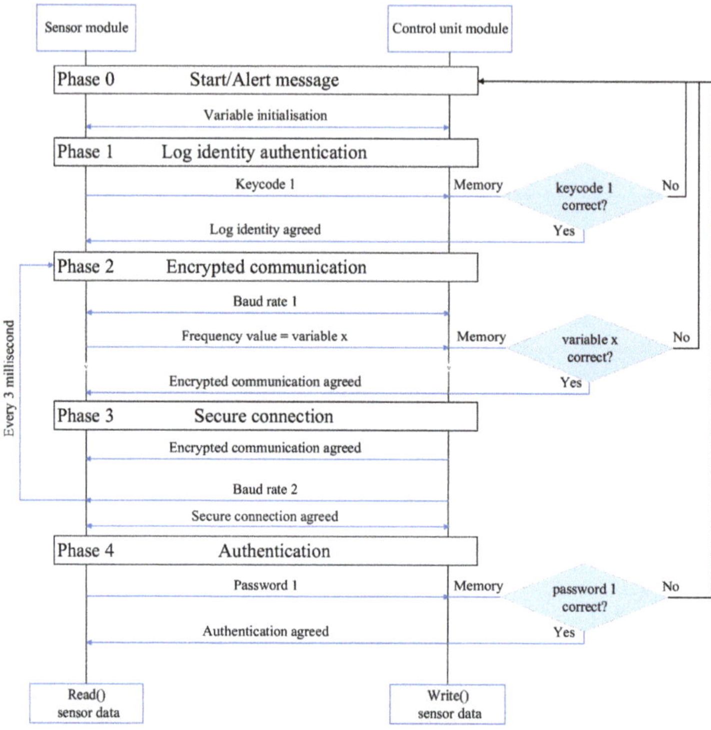

Figure 11. Log identity authentication, encrypted communication, secure connection, and authentication (LCCA) protocol.

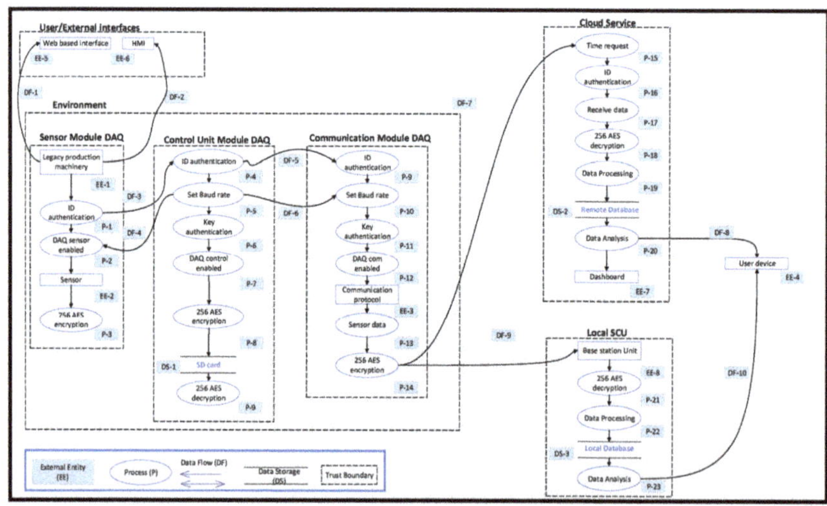

Figure 12. Data flow diagram of the connected legacy production machinery.

6. Pilot Implementation and Testing

This section describes the final stage of our security design approach and presents an implementation instance focusing on mitigation mechanisms for DoS and clone attacks, which are considered typical threats, are relevant to production environments, and were marked with a high impact score in the earlier analysis. The hypothesis of the experiments is that of a DoS or clone attack succeeding. This could be achieved through a number of intermediate goals, as shown in the relevant attack tree in Section 5. Once successful, the attacks aim to deprive the IoT device of vital resources and to compromise monitoring data. The mitigation mechanisms follow the principle of modularity and the LCCA protocol described in the previous section. The objective of the testing is to assess the ability of the implemented approach to avert these two types of attacks. An industrial DMG NTX 1000 CNC Mill Turn Centre (twin-spindle turning centre with five-axes milling capability) was employed for the experiments. The remainder of this section describes the physical instantiation of the IoT DAQ unit and the mitigation mechanisms testing DoS and cloning attacks.

6.1. DoS Attack

This section describes the implementation of the mitigation mechanism for the DoS attack. The test was run during the warm-up phase of the machine tool operation. The functional objective was to introduce the IoT DAQ for real-time monitoring of signals, such as acceleration and temperature from the machine spindle, then send the encrypted sensor data to a server (ownCloud) integrated into a raspberry pi 2 model band gain authentication to access and visualise data. The IoT modules are emulated through Arduino Uno units. The attack goal was to generate a DoS situation to jam the IoT device, affecting its battery life and communications, or gain access to monitored machine parameters, such as the spindle temperature and acceleration. The attack tree in Figure 9, shows attack paths that can lead to achieving the target. The modular IoT DAQ is shown in Figure 13. The control unit module is equipped with a 32 GB SD card to store data and its CPU runs the authentication protocol. The sensor module comprises a bottom layer that includes the sensors, the CPU and memory of the control module and the battery to supply the entire IoT unit during data acquisition and protocol execution; and a top layer that includes a relay board to manage data acquisition and apply the mitigation mechanism.

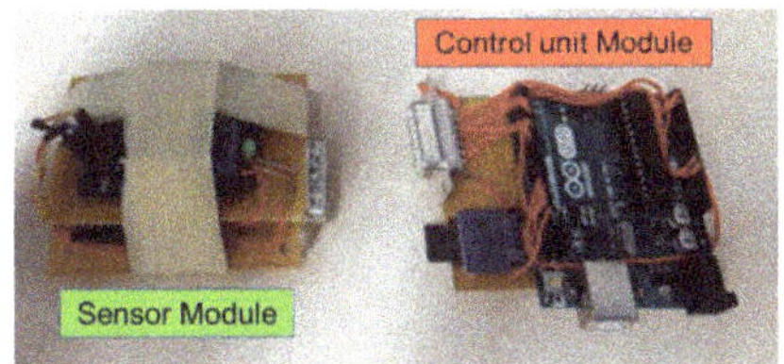

(**a**): Sensor and control unit modules

(**b**): Sensor and control unit attached to machine

(**c**): Data acquisition on a mobile device

Figure 13. Prototype of modular IoT DAQ.

The control unit module is equipped with code to calculate CPU and RAM usage. If the control unit does not identify correct credentials, i.e., valid keycode between the modules, the data acquisition and transmission processes are interrupted, sending an alert message to the user device. A snapshot of the user device screen during monitoring real-time data is shown in Figure 13c, where current data are shared with end-user devices and are visualised. The web-server, cloud, or end-user device are attack points, exposing the monitoring device to a DoS attack aiming to take down its operational capacity.

A DoS attack emulation scenario was set up (Figure 14) and includes:

1. The machine tool equipped with the sensor module on the spindle;
2. A hub for a monitoring service provider equipped with an API to make available, through the local network, the machine tool state and performance;
3. End-user devices used to monitor the machine tool anywhere and anytime;
4. The cloud service for processing, analysing, and planning maintenance interventions.

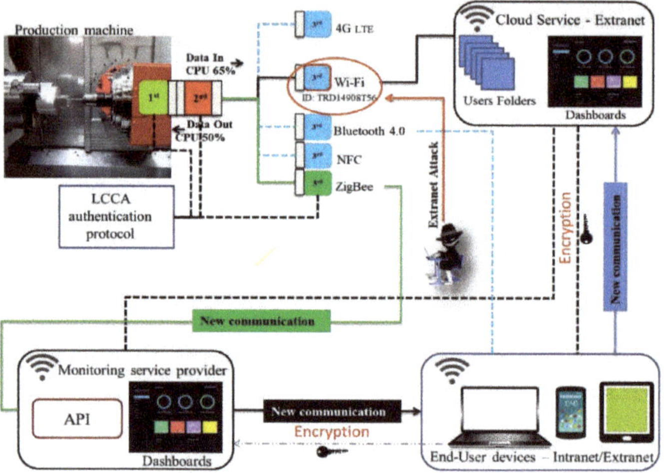

Figure 14. An external DoS attack.

All communications apply AES 256 encryption. The test aimed to simulate a denial of service (DoS) [73] via the network. The attacker gained network access through any of the earlier mentioned methods and is ready to generate connection requests to the communication module using its source address rather than the attack target. In this way, the communication module will respond affirmatively to the connection request not by the attacker but to the target of the attack. The result is a vicious circle that will quickly exhaust the targeted resources and flood the network with traffic (Figure 15). The attack generates an infinite request for access after spoofing the IP of the system through a fake source address and bypassing the firewall. At the same time, the targeted systems attempt to access the data when the sensor module seeks to exchange condition monitoring data with the control module. The large number of responses from the control module causes bandwidth exhaustion and hence a crash. An Arduino Uno was used for generating a connection request to the communication module, so as to affect the targeted device. The control unit module is connected to the sensor module via COM3 port and the communication module through the COM10 port. The attacker, after gaining the network authentication, could take down the capacity of the control module generating an autonomous function, able to generate infinite access requests, delay services, and reduce the battery life.

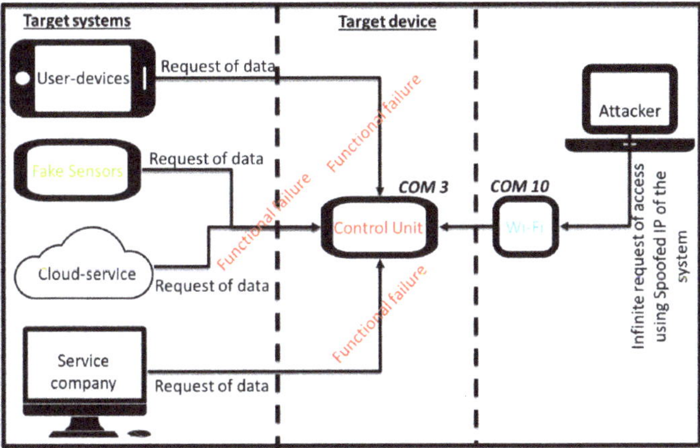

Figure 15. A DoS attack process.

As a result of this attack, the control unit will output the estimated ratio of CPU usage as shown in Figure 16 for the control unit module:

$$CPU\ utilisation\ (\%) = \frac{CPU\ Idlecountintime\ frame\ \times\ 100}{Maximum\ number\ of\ CPU\ Idlecountintime\ frame}$$

The CPU utilisation procedure includes two phases:

- Phase 1: The real-time operating clock (RTOC) is used to estimate CPU/core utilization. The scheduler system tick is used for this purpose, as it is based on timer interrupt, which is considered as a relatively accurately measure of elapsed time.
- Phase 2: Counting maximum idle count; an estimation is obtained through observing idle counts during a measurement period. If no task is performed (besides the timer interrupt) this represents the maximum number of idle counts and corresponds to 0% utilisation. Estimation accuracy errors tend to become insignificant when the CPU utilization measurement period is sufficiently large. After calculation of maximum idle counts, no code or task can be added to the idle task.

The CPU utilisation is contrasted against the expected average value for this device, which in this case was known to be 71% without any attacks. The initialisation stage when starting the CPU generates a level of 22% usage and this is due to a delay of the function printer at the screen. Reaching 100% is a strong indication that the CPU is under attack. In our case, the DoS attack materialises by running our application via a host computer on the intranet. The control unit and transmitter module exchange information using the LCCA protocol (Figure 11) to detect significant deviations from the expected standard operation. If the DoS attack occurs on the current available channel for exchanging data (for example, on the Wi-Fi module circled red in Figure 14), the control module recognises the attack and shuts off the current communication path. The scope of this test was to perform an end to end functional testing without fully emulating any kind of DoS attack or their formal mitigation mechanisms. The aim was to illustrate how the isolation principle is applied through the LCCA protocol to reduce relevant security risks. The simple detection technique can nonetheless be replaced by a more sophisticated mechanism, while following a similar isolation principle in the communication between modules.

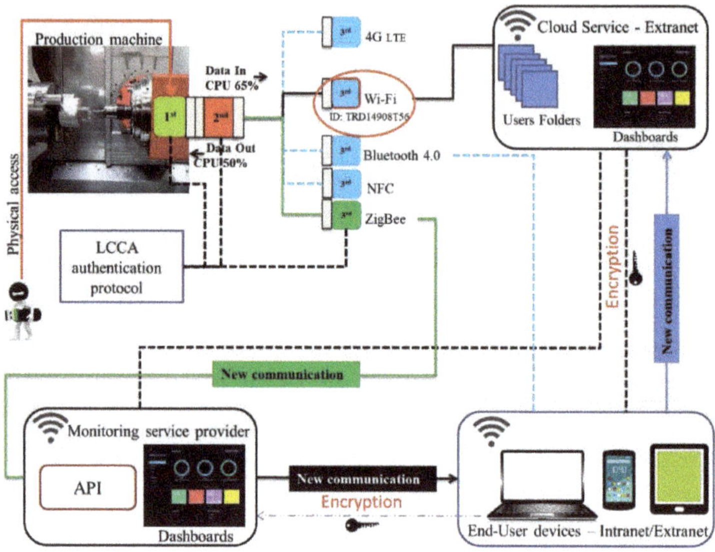

Figure 16. A DoS attack initiated via infected USB dongle.

6.2. Cloning Attack

In the second test, the attacker gains access to the communication module by the social engineering method or via bypassing the firewall. The mitigation mechanism is to use the control unit module to read the authentication key and detect abnormal states. When some of the hardware or software parameters (e.g., voltage, current, system memory, CPU usage, connection buses, or IDs) change in unexpected ways, the control unit module closes the current connection with the malicious hardware or software components and initiates an alternative way for exchanging data with the target. In the scenario of Figure 16, the attacker gains physical access to the sensor module and is able to clone it using fake modules equipped with reprogrammed the firmware. Figure 17 shows the authentication process between two modules of the modular IoT DAQ. For each connection between the modules, the control unit module generates a new unique authentication key (Step 1). The key is stored within the sensor module in a buffer of characters under a private class that does not allow modifications by other users (Step 2). The last phase (Step 3) checks the sensor unique key and compares it to the one in the control unit module buffer. If the sensor unique key matches the key inside the control buffer unit, the sensor module gets access to phase 2 of the authentication protocol (Figure 11). Upon guessing the authentication key, the attacker gains access to the target device and initiates the DoS attack. Figure 18 shows the DoS attack when an infected USB dongle is employed for upgrading an infected kernel inside of the machine [83]. Such an attack may employ multiple attacking nodes, which together form a botnet. A botnet is a network controlled by a master bot and is made up of devices infected by specialised malware, known as bots or zombies [84]. In a cloning attack of a wireless sensor network architecture, once a sensor node is compromised, the adversaries can easily capture other sensor nodes and deploy several clones that have legitimate access to the network (legitimate IDs, passwords, and other security credentials) [85]. The cloning attack affects the mobile communication protocol as well. Subscriber identity module (SIM) cloning by physical access is a simple process and the attacker must have a software program, a SIM reader, and a SIM chip writer [86]. Such examples highlight the risk of cloning attacks, which can be addressed by cryptography or physically unclonable functions (PUFs) [87]. In the IoT DAQ the control unit module is the master that controls all operations and requires protection. The cloning attack involves tampering with the sensor and communication modules, aiming to compromise the architecture integrity and modify the behaviour of the modules.

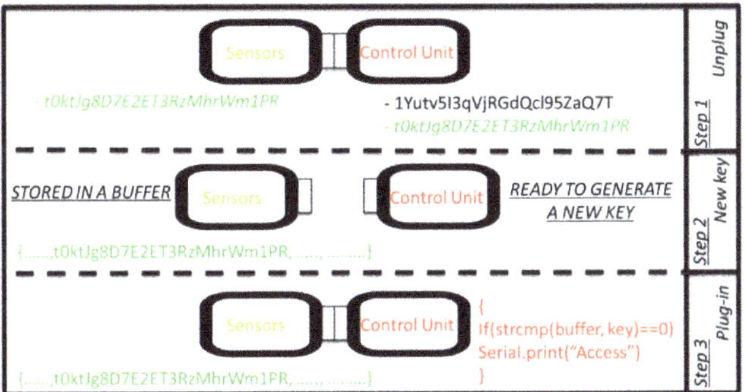

Figure 17. Control unit and sensor; communication module authentication process.

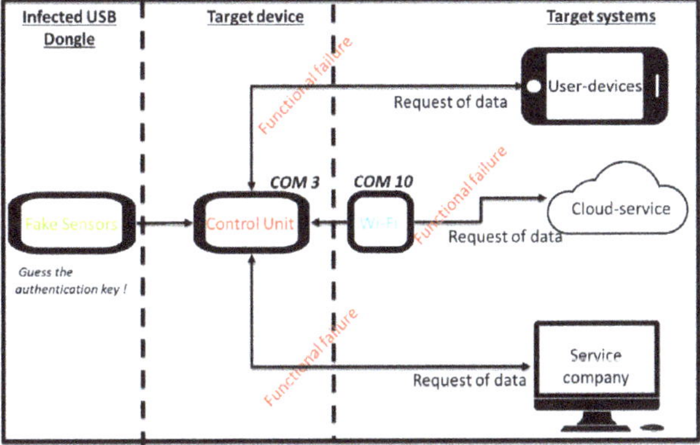

Figure 18. Authentication method vs. the distributed denial of service (DDoS) attack with infected USB dongle.

Figure 18 illustrates also the mitigation approach effect for the cloning attack in case the attacker guesses the key. The control module analyses data from the sensor and communication modules to detect deviation between the original and the clone sensor behaviour. When the clone sensor is identified, the control module disables all communication with the clone sensor. Figure 19 shows two different cases of sensor communication. In the first case, the control module is connected with the original sensor (green module); reading parameters, such as ID, password, CPU usage, static RAM (SRAM) byte sketch size; and hardware parameters through the INA219 sensor (power supply and current). At the second case (bottom), the clone sensor (amber module) shows the same hardware and software of the original sensor but the malicious code for compromising the monitoring system is also included. To detect signs of a cloning attack, the control module monitors changes in CPU usage, power supply and current, comparing them against typical values. In addition, the control unit reads the sketch byte size to understand the credibility of the sensor module. The sketch byte size is stored into the microcontroller SRAM and show the unique value of the sketch. If the adversary seeks to modify the code to add the malicious part and leave the rest of the sensor module the same as the original, the control module can recognise it as a clone module and will not share any information with it because of the deviation of sketch byte size and level of usage of SRAM. Physical parameters can help to single out unexpected changes to hardware parameters. The authentication ID control mechanism brings the probability of successful cloning threat events to a lower level, reducing the impact score rating.

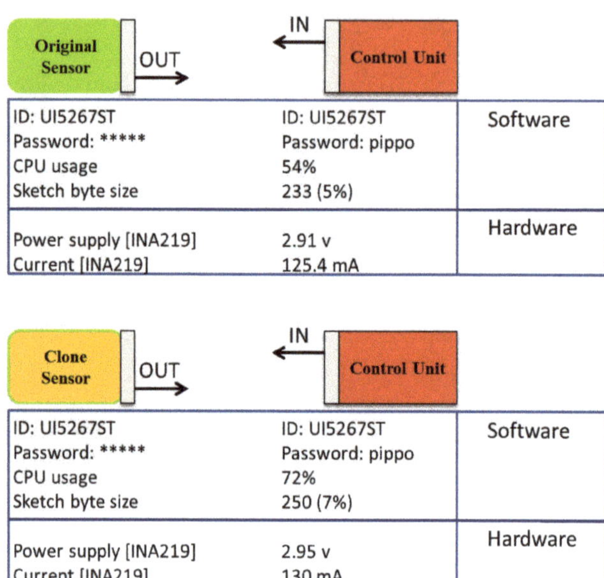

Figure 19. Reading byte between two modules.

7. Discussion and Related Future Work

The large amount of interconnected things for advance manufacturing brings new cyber risks. Production environments are strongly characterised by jointly involving OT and IT and the potential impact of any security breaches on the integrity of industrial systems can be very tangible and highly critical. Cyber security must therefore be a vital part of the design, operations, and strategy processes, and should be considered from the very beginning of any new connected Industry 4.0 driven initiative.

The reported work introduced a systematic design thinking approach to attack the new risks arising with the Industry 4.0 connectivity. The new approach draws parallels with previous and ongoing activities (e.g., PASTA, IISF, ENISA) but is positioned towards the concrete context of retrofitting legacy production machinery with IoT-enabled monitoring capabilities all the way from the study of requirements, and through threat and application modelling, all the way to threat mitigation design, implementation, and testing with prime focus on IoT endpoint security. As an exemplar of dealing with this challenge, this paper introduced a new security hardware IoT device for remote monitoring application in a production environment, which is managed through a flexible but strong lightweight authentication protocol and mechanisms for isolation between the key subsystems of an IoT endpoint device. This was tested through a real-world case study where security flaws were deliberately introduced, a qualitative risk assessment was applied, and relevant risks were mitigated.

Overall, the main paper contribution is in the overall design thinking approach, while several additional contributions which were included, such as the new authentication protocol implementing in effect the isolation principle at the IoT endpoint subsystem level. However, the principle of subsystem isolation goes well beyond physical or subsystem interfaces isolation. IoT deployments now make extensive use of containerisation technologies and IoT devices can themselves be put under this context through container engines and container APIs [88], which can link endpoint devices to an extended system of IoT-enabled application services [89]. Significant ongoing research currently targets the extension of IoT resources orchestration to jointly include both edge nodes and the cloud [90]. Based on the above, the research presented in this paper opens up several threads for further work:

- Comprehensive mitigation mechanisms for the range of identified threats. While the reported work presented an implementation example of the design thinking approach, which included specific instances of mitigation mechanisms relevant to preventing DoS and clone attack threats, any alternative and more comprehensive mechanism can be employed instead but would still need to be included within the context of an overall design approach for IoT security.
- Eventually, any introduced mitigation mechanism needs to be scrutinised for effective protection against attacks. The reported work intention was to present the multistage design thinking approach, with mitigation being a concrete step within this. Any final deployment of adopted solutions needs to be preceded by extensive and systematic testing against attacks. Such a testing will need to consider simulation or indeed emulation of attacks, as well as mechanisms for their systematic generation [90].
- The reported work includes risk assessment and mitigation as part of the five-stage systematic approach. However, risk quantification was only indicative and of qualitative nature. Further work is needed in the direction of systematic risk quantification, including approaches for data- and evidence-driven risk quantification [91]. While this is highly important for IoT endpoint devices, overall IoT network security is only as good as its weakest link and a weak node may have scalable negative impacts to the whole IoT network. Further work needs to put into such a context any risk-based approach to security and duly take into account complexity considerations.
- The isolation principle in IoT is effectively applied through virtualisation and containerisation technologies, as expressed, for example, by the IISF. While such technologies were more applicable to cloud services, they are increasingly expanded and implemented at the edge node level. IoT endpoint device security can strongly benefit via joint physical and virtual isolation, and future research need to align relevant research with such IoT architecture patterns [89,90].
- Organisations seeking to adopt security-by-design approaches would benefit from methodologies and tools that assist in appropriate prioritisation of any upgrades related to security. It is futile to implement the most sophisticated approach for part one aspect of security, when others are left too weak. Maturity assessment methods and tools are helpful to this end. Future work would need to look how to best place a design thinking approach, such as the one presented in this paper, within the context of overall organisational security maturity management [92].

8. Conclusions

This paper introduced a novel endpoint security design approach to address security issues when upgrading production machinery with IoT connectivity to deliver real-time condition monitoring for legacy production machinery. The approach considers best practice and guidelines to formulate a new domain-specific approach, contributing to bridging the gap between introducing IoT connectivity at the shop floor and shielding system and operational integrity. The main concepts of the new approach are the application-aware viewpoint, as opposed to generic security measures, the adoption of the principle of subsystem isolation, and the development of a new multistage but lightweight authentication protocols, which are all contributing to increasing the required complexity of any attack approach to achieve compromising the IoT device and associated monitoring and production processes. The concrete implementation of this approach was demonstrated through two industrial legacy machinery attack scenarios based on different attack entry points, for DoS and cloning attacks. The approach enables the mapping and prioritisation of threats and risks in a domain-specific application-oriented way, which, in turn, allows the identification of priorities for intervening with mitigation approach and lowers integrity risks.

While the new approach and its implementation focuses on the key design aspects, rather than on any single sophisticated detection mechanism, it is worth noticing that the employed mechanisms can be upgraded to introduce stronger detection, and therefore, response capabilities. Future research needs to target such capabilities but will need to develop a systematic approach for testing. The risk-based part of the methodology needs to evolve further from qualitative to quantitative, and be linked to the results of the testing phase to improve security performance. Production environments are considerably different from others due to the dominant presence of OT, which may imply significant operational impacts. It is for this reason, that dedicated testbeds and domain-specific security metrics need to be developed and employed in a systematic testing and evaluation process, while for the detection and response mechanism, other sophisticated algorithmic and other approaches could be used as part of the overall methodology for IoT device endpoint security protection. Overall, this paper included a discussion with leads to further research (Section 7), pointing out the need for further research in the direction of (a) comprehensive mitigation mechanisms; (b) systematic test generation and validation of solutions; (c) automated and data-driven risk assessment; (d) impact of endpoint vulnerabilities on overall IoT network security; (e) virtual isolation, IoT edge node containerisation and virtual–physical nodes orchestration; (f) systematic maturity assessment and management for IoT security.

Author Contributions: S.T. performed the reported research, the design, and development of the modular IoT DAQ and the authentication protocol, performed the experiments, and authored the manuscript. C.E. defined the structure, co-authored the manuscript, supervised the research, and provided guidance on the research methodology. J.M. initiated and co-supervised the research, provided discussion and feedback, as well as advice on the design of the modular IoT DAQ. R.R. co-initiated the research and provided discussion and feedback.

Funding: This research was co-funded by EPSRC, grant number EP/I033246/1, and Kennametal Inc and was conducted in the EPSRC Centre for Innovative Manufacturing in Through-life Engineering Services.

Acknowledgments: The contribution of Michael Farnsworth, now at the University of Sheffield, towards the simulation code in the testing experiments is gratefully acknowledged.

Conflicts of Interest: The authors declare no conflict of interest.

Data Access: Data underlying the testing can be accessed at https://doi.org/10.17862/cranfield.rd.8143340.

Abbreviations

AES	Advanced Encryption Standard
CBM	Condition-Based Maintenance
CM	Condition Monitoring
CNC	Computer Numerical Control
CPS	Cyber-Physical Systems
CPPS	Cyber Physical Production Systems
CPU	Central Processing Unit
DAQ	Data Acquisition
DFD	Data Flow Diagram
DoS	Denial of Service
DTLS	Datagram Transport Layer Security
HMI	Human Machine Interface
ICS	Industrial Control Systems
IT/OT	Information Technology/Operational Technology
IoT	Internet of Things
MAC	Medium Access Control
MITM	Man-in-the-Middle
LCCA	Log identity, Communication, Connection, Authentication
OT	Operational Technology
P2P	Peer-to-Peer
PAC	Programmable Automation Controller
PLC	Programmable Logic Controller
PLM	Product Lifecycle Management
RMS	Remote Monitoring System
SCADA	Supervisory Control and Data Acquisition
SCU	System Control Unit
SQL	Structured Query Language
WSN	Wireless Sensor Network

References

1. Lee, J.; Kao, H.; Yang, S. Service innovation and smart analytics for Industry 4.0 and big data environment. In Proceedings of the 6th CIRP Conference on Industrial Product Service Systems, Windsor, ON, Canada, 1 May 2014; pp. 3–8.
2. Deshpande, A.; Pieper, R. Legacy Machine Monitoring using power signal analysis. In Proceedings of the ASME 2011 International Manufacturing Science and Engineering Conference MSEC2011, Corvallis, OR, USA, 13–17 June 2011; pp. 207–214.
3. Hascoet, J.Y.; Rauch, M. Enabling Advanced CNC Programming with openNC Controllers for HSM Machines Tools. *High Speed Mach.* **2016**, *2*, 1–14. [CrossRef]
4. Elghazel, W.; Bahi, J.; Guyeux, C.; Hakem, M.; Medjaher, K.; Zerhouni, N. Dependability of wireless sensor networks for industrial prognostics and health management. *Comput. Ind.* **2015**, *68*, 1–15. [CrossRef]
5. Sadeghi, A.R.; Wachsmann, C.; Waidner, M. Security and privacy challenges in industrial Internet of Things. In Proceedings of the 52nd ACM/EDAC/IEEE Design Automation Conference (DAC), San Francisco, CA, USA, 8–12 June 2015; pp. 1–6.
6. Yang, Y.; Wu, L.; Yin, G.; Li, L.; Zhao, H. A Survey on Security and Privacy Issues in Internet-of-Things. *IEEE Internet Things* **2017**, *4*, 1250–1258. [CrossRef]
7. Zhou, L.; Yeh, K.H.; Hancke, G.; Liu, Z.; Su, C. Security and Privacy for the Industrial Internet of Things: An Overview of Approaches to Safeguarding Endpoints. *IEEE Signal Proc. Mag.* **2018**, *35*, 76–87. [CrossRef]
8. Murakami, K.; Suemitsu, H.; Matsuo, T. Classification of repeated replay-attacks and its detection monitor. In Proceedings of the 6th IEEE Global Conference on Consumer Electronics (GCCE), Nagoya, Japan, 24–27 October 2017; pp. 1–2.
9. Patil, D.S.; Patil, S.C. A Novel Algorithm for Detecting Node Clone Attack in Wireless Sensor Networks. In Proceedings of the International Conference on Computing, Communication, Control and Automation (ICCUBEA), Pune, India, 17–18 August 2017; pp. 1–4.
10. Fernandez-Carames, T.M.; Fraga-Lamas, P.; Suarez-Albela, M.; Castedo, L. Reverse Engineering and Security Evaluation of Commercial Tags for RFID-Based IoT Applications. *Sensors* **2017**, *17*, 28. [CrossRef]
11. Ogiela, L.; Ogiela, M.R. Insider Threats and Cryptographic Techniques in Secure Information Management. *IEEE Syst. J.* **2017**, *11*, 405–414. [CrossRef]
12. Iqbal, M.A.; Bayoumi, M. A Novel Authentication and Key Agreement Protocol for Internet of Things Based Resource-constrained Body Area Sensors. In Proceedings of the 4th International Conference on Future Internet of Things and Cloud Workshops, W-Fi Cloud 2016, Vienna, Austria, 22–24 August 2016; pp. 315–320.
13. Abawajy, J. Enhancing RFID tag resistance against cloning attack. In Proceedings of the 3rd International Conference on Network and System Security, Goal Coast, Australia, 19–21 October 2009; pp. 18–23.
14. Ray, B.; Huda, S.; Chowdhury, M.U. Smart RFID reader protocol for malware detection. In Proceedings of the 12th ACIS International Conference on Software Engineering, Artificial Intelligence, Networking and Parallel/Distributed Computing, Sydney, Australia, 6–8 July 2011; pp. 64–69.
15. Desai, A.R.; Hsiao, M.S.; Wang, C.; Nazhandali, L.; Hall, S. Interlocking obfuscation for anti-tamper hardware. In Proceedings of the 8th Annual Cyber Security and Information Intelligence Research Workshop: Federal Cybersecurity R and D Program Thrusts, CSIIRW 2013, Oak Ridge, TN, USA, 8 January 2013; pp. 1–4.
16. Shostack, A. *Threat Modelling: Design for Security*; Wiley: Indianapolis, IN, USA, 2014; pp. 29–351.
17. Khan, R.; McLaughlin, K.; Laverty, D.; Sezer, S. STRIDE-based threat modelling for cyber-physical systems. In Proceedings of the IEEE PES Innovative Smart Grid Technologies Conference Europe (ISGT-Europe), Torino, Italy, 26–29 September 2017; pp. 1–6.
18. Ucedavelez, T.; Morana, M.M. *Risk Centric Threat Modeling: Process for Attack Simulation and Threat Analysis*; Wiley: Hoboken, NJ, USA, 2015; pp. 317–342.
19. Wang, E.K.; Ye, Y.; Xu, X.; Yiu, S.M.; Hui, L.C.K.; Chow, K.P. Security issues and challenges for cyber-physical systems. In Proceedings of the IEEE/ACM Int'l Conference on Green Computing and Communications & Int'l Conference on Cyber, Physical and Social Computing, Washington, DC, USA, 18–20 December 2010; pp. 733–738.
20. Monostori, L. Cyber-physical production systems: Roots, expectations and R&D challenges. In Proceedings of the 47th CIRP Conference on Manufacturing Systems, Windsor, ON, Canada, 28–30 April 2014; pp. 9–13.

21. Alexandru, A.M.; De Mauro, A.; Fiasché, M.; Sisca, F.G.; Taisch, M.; Fasanotti, L.; Grasseni, P. A smart web-based maintenance system for a smart manufacturing environment. In Proceedings of the 1st IEEE International Forum on Research and Technologies for Society and Industry Leveraging a Better Tomorrow (RTSI), Torino, Italy, 16–18 September 2015; pp. 1–6.
22. Weippl, E.; Kirseberg, P. Security in cyber-physical production systems: A roadmap to improving IT-security in the production system lifecycle. In Proceedings of the AEIT International Annual Conference, Cagliari, Italy, 20–22 September 2017; pp. 1–6.
23. ENISA. Good Practices for Security of Internet of Things in the Context of Smart Manufacturing, November 2018. Available online: https://www.enisa.europa.eu/publications/good-practices-for-security-of-iot (accessed on 13 May 2019).
24. Wolf, M.; Serpanos, D. Safety and Security in Cyber-Physical Systems and Internet-of-Things Systems. *Proc. IEEE* **2018**, *106*, 9–20. [CrossRef]
25. Koopman, P.; Wagner, M. Transportation CPS Safety Challenges. In Proceedings of the NSF workshop Transportation Cyber-Physical Systems, Pittsburgh, PA, USA, 23–24 January 2014; pp. 1–3.
26. Fidler, D.P. Was Stuxnet an Act of War? Decoding a Cyberattack. *IEEE Secur. Priv.* **2011**, *9*, 56–59. [CrossRef]
27. Johnson, C. Securing the participation of safety-critical SCADA systems in the industrial internet of things. In Proceedings of the 11th International Conference on System Safety and Cyber Security (SSCS), London, UK, 11–13 October 2016; pp. 11–13.
28. Hehenberger, P.; Vogel-Heuser, B.; Bradley, D.; Eynard, B.; Tomiyama, T.; Achiche, S. Design, modelling, simulation and integration of cyber physical systems: Methods and applications. *Comput. Ind.* **2016**, *82*, 273–289. [CrossRef]
29. DeSmita, Z.; Elhabashy, A.E.; Wells, J.L.; Jaime, A.C. An approach to cyber-physical vulnerability assessment for intelligent manufacturing systems. *J. Manuf. Syst.* **2017**, *43*, 339–351. [CrossRef]
30. Hutchins, M.J.; Bhinge, R.; Maxwell, K.M.; Robinson, S.L.; Sutherland, J.W.; Dornfeld, D. Framework for Identifying Cybersecurity Risks in Manufacturing. *Procedia Manuf.* **2015**, *1*, 47–63. [CrossRef]
31. Finogeev, A.G.; Finogeev, A.A. Information attacks and security in wireless sensor networks of industrial SCADA systems. *J. Ind. Inf. Integr.* **2017**, *5*, 6–16. [CrossRef]
32. Conklin, A.W. IT vs. OT Security: A Time to Consider a Change in CIA to Include Resilience. In Proceedings of the 49th Hawaii International Conference System Sciences (HICSS), Koloa, HI, USA, 5–8 January 2016; pp. 2642–2647.
33. Global Cyber Executive Briefing-Manufacturing. Available online: https://www2.deloitte.com/global/en/pages/risk/articles/Manufacturing.html (accessed on 13 May 2019).
34. ISTR Internet Security Threat Report 2018; Volume 23. Available online: https://www.symantec.com/content/dam/symantec/docs/reports/istr-23-2018-en.pdf (accessed on 13 May 2019).
35. The Verge. Available online: https://www.theverge.com/2017/5/14/15637472/renault-nissan-shut-down-french-uk-factories-wannacry-cyberattack (accessed on 13 May 2019).
36. Ylmaz, E.N.; Ciylan, B.; Gönen, S.; Sindiren, E.; Karacayılmaz, G. Cyber security in industrial control systems: Analysis of DoS attacks against PLCs and the insider effect. In Proceedings of the 6th International Istanbul Smart Grids and Cities Congress and Fair (ICSG), Istanbul, Turkey, 25–26 April 2018; pp. 81–85.
37. Langner, R. Stuxnet: Dissecting a Cyberwarfare Weapon. *IEEE Secur. Priv.* **2011**, *9*, 49–51. [CrossRef]
38. Chen, T.M.; Abu-Nimeh, S. Lessons from Stuxnet. *Computer* **2011**, *44*, 91–93. [CrossRef]
39. Elhabashy, A.; Lee, J.W.; Camelio, J.; Woodall, W.H. A Cyber-physical Attack Taxonomy for Production Systems: A Quality Control Perspective. *J. Intell. Manuf.* **2016**, 1–16. [CrossRef]
40. Krotofil, M.; Gollmann, D. Industrial control systems security: What is happening. In Proceedings of the 11th IEEE International Conference on Industrial Informatics (INDIN), Bochum, Germany, 29–31 July 2013; pp. 664–669.
41. Kim, H.; Kang, E.; Broman, D.; Lee, E.A. An architectural mechanism for resilient IoT services. In Proceedings of the 1st ACM Workshop on Internet of Safe Things, Delft, The Netherlands, 5 November 2017; pp. 8–13.
42. Do, Q.; Martini, B.; Choo, K.-K.R. Is the data on your wearable device secure? An Android Wear smartwatch case study. *Softw. Pract. Exp.* **2016**, *47*, 391–403. [CrossRef]
43. Ibrahim, A.; Sadeghi, A.R.; Tsudik, G.; Zeitouni, S. DARPA: Device Attestation Resilient to Physical Attacks. In Proceedings of the 9th ACM Conference on Security & Privacy in Wireless and Mobile Networks, Darmstadt, Germany, 18–20 July 2016; pp. 171–182.

44. Salmam, O.; Abdallah, S.; Elhaji, H.I.; Chehab, A.; Kayssi, A. Identity-based authentication scheme for the internet of things. In Proceedings of the Symposium on Computers and Communication (ISCC), Messina, Italy, 27–30 June 2016; pp. 1109–1111.
45. Porambage, P.; Schmitt, C.; Kumar, P.; Gurton, A.; Ylianttila, M. Pauthkey: A pervasive authentication protocol and key establishment scheme for wireless sensor networks in distributed applications. *Int. J. Distrib. Sens. Netw.* **2014**, *10*, 1–14. [CrossRef]
46. Haythem Ahmad, B.S.; Mohammed, F.D. Event-driven hybrid MAC protocol for a two-tier cognitive wireless sensor network: Design and implementation. *Int. J. High Perform. Comput. Netw.* **2016**, *9*, 271–280. [CrossRef]
47. Sharaf-Dabbagh, Y.; Saad, W. On the authentication of devices in the internet of things. In Proceedings of the 17th IEEE Symposium on A World of Wireless, Mobile and Multimedia Networks (WoWMoM), Coimbra, Portugal, 21–24 June 2016; pp. 1–3.
48. Zhang, C.; Green, R. Communication security in internet of things: Preventive measure and avoid ddos attack over iot network. In Proceedings of the 18th Symposium on Communication & Networking, Alexandria, VA, USA, 12–15 April 2015; pp. 8–15.
49. Dos Santos, L.G.; Guimaraes, T.V.; Da Cunha, R.G.; Granville, Z.L.; Tarouco, R.M. A dtls-based security architecture for the internet of things. In Proceedings of the IEEE Symposium on Computers and Communication (ISCC), Larnaca, Cyprus, 6–9 July 2015; pp. 809–815.
50. Jebri, S.; Abis, M.; Bouallegue, A. An Efficient scheme for anonymous communication in iot. In Proceedings of the 11th International Conference on Information Assurance and Security (IAS), Marrakech, Morocco, 14–16 December 2015; pp. 7–12.
51. Yoshigoe, K.; Dai, W.; Abramson, M.; Jacobs, A. Overcoming invasion of privacy in smart home environment with synthetic packet injection. In Proceedings of the TRON Symposium (TRONSHOW), Tokyo, Japan, 9–10 December 2014; pp. 1–7.
52. Mehnen, J.; He, H.; Tedeschi, S.; Tapoglou, N. *Practical Security Aspects of the Internet of Things. Cybersecurity for Industry 4.0*; Thames, L., Schaefer, D., Eds.; Springer Nature: Birmingham, UK, 2017; pp. 225–242.
53. Industrial Internet Security Framework Technical Report. Available online: https://www.iiconsortium.org/IISF.htm (accessed on 13 May 2019).
54. ISO/IEC 27001. Available online: https://www.iso.org/isoiec-27001-information-security.html (accessed on 13 May 2019).
55. Information Security. Guide for Conducting Risk Assessment. NIST Special Publication 800-30. Available online: https://nvlpubs.nist.gov/nistpubs/Legacy/SP/nistspecialpublication800-30r1.pdf (accessed on 13 May 2019).
56. Schmalenberg, F.; Vandenhouten, R. An advanced data processing environment based on data flow diagrams with a flexible triggering and execution model. In Proceedings of the 14th International Symposium on Applied Machine Intelligence and Informatics (SAMI), Herlany, Slovakia, 21–23 January 2016; pp. 159–164.
57. Hui, X.; Jun, S.; Xinlu, Z.; Lingyu, Y. Attack identification for software-defined networking based on attack trees and extension innovation methods. In Proceedings of the 9th IEEE International Conference on Intelligent Data Acquisition and Advanced Computing Systems: Technology and Applications (IDAACS), Bucharest, Romania, 21–23 September 2017; pp. 485–489.
58. Maslina, D.; Rajah, R.; Mary, G.; David, A.; Abdul Fuad, A.R.; Azni, A.H. Denial of service: (DoS) Impact on sensors. In Proceedings of the 4th International Conference on Information Management (ICIM), Oxford, UK, 25–27 May 2018; pp. 270–274.
59. Georges, A.K.; Niki, P.; Iyengar, S.S.; Beltran, J.; Kamhoua, C.; Hernandez, B.L.; Njilla, L.; Makki, A.P. Preventing Colluding Identity Clone Attacks in Online Social Networks. In Proceedings of the 37th International Conference on Distributed Computing Systems Workshops (ICDCSW), Atlanta, GA, USA, 5–8 June 2017; pp. 187–192.
60. ISO 13372:2012. *Condition Monitoring and Diagnostics of Machines—Vocabulary*; ISO: Geneva, Switzerland, 2012.
61. CEN-EN 13306:2017. *Maintenance—Maintenance Terminology*; CEN: Brussels, Belgium, 2017.
62. Crespo-Márquez, A. *The Maintenance Management Framework: Models and Methods for Complex Systems Maintenance*; Springer: Berlin/Heidelberg, Germany, 2007.
63. Mori, M.; Fujishima, M.; Komatsu, M.; Zhao, B.; Liu, Y. Development of remote monitoring and maintenance system for machine tools. *CIRP Ann. Manuf. Technol.* **2008**, *57*, 433–436. [CrossRef]

64. Emmanouilidis, C.; Bertoncelj, L.; Bevilacqua, M.; Tedeschi, S.; Ruiz Carcel, C. Internet of Things—Enabled Visual Analytics for Linked Maintenance and Product Lifecycle Management. In Proceedings of the 16th IFAC Symposium on Information Control Problems in Manufacturing, Bergamo, Italy, 11–13 June 2018; pp. 435–440.
65. Kiritsis, D. Closed-loop PLM for intelligent products in the era of the Internet of things. *Comput. Aided Des.* **2011**, *43*, 479–501. [CrossRef]
66. ISO/IEC 27005:2018. *Information Technology—Security Techniques—Information Security Risk Management*, 3rd ed.; ISO: Geneva, Switzerland, 2018.
67. Information Security. National Institute of Standards and Technology (NIST). Available online: https://nvlpubs.nist.gov/nistpubs/Legacy/SP/nistspecialpublication800-30r1.pdf (accessed on 13 May 2019).
68. Zhang, X.; Wuwong, N.; Li, H.; Zhang, X. Information Security Risk Management Framework for the Cloud Computing Environments. In Proceedings of the 10th International Conference on Computer and Information Technology (CIT 2010), Bradford, UK, 29 June–1 July 2010; pp. 1328–1334.
69. Risk Assessment Report Based on the 2014 Data Branch. University if Washington, IMT 552. Available online: https://www.slideshare.net/DivyaKothari1/jpmorgan-chase-co-risk-assessment-report-62798131 (accessed on 28 March 2016).
70. Cyber Security: Challenges Ahead. Available online: http://www.nexusacademicpublishers.com/uploads/portals/Cyber_Security_Challenged_Ahead.pdf (accessed on 13 May 2019).
71. Abhishek, D.; Gokham, M.; Joseph, Z. Detecting/preventing information leakage on the memory bus due to malicious hardware. In Proceedings of the Design, Automation & test in Europe Conference & Exhibition (DATE 2010), Dresden, Germany, 8–12 March 2010; pp. 1–6.
72. Becher, A.; Benenson, Z.; Dornseif, M. Tampering with Motes: Real-World Physical Attacks on Wireless Sensor Networks. In Proceedings of the Third International Conference, Security in Pervasive Computing (SPC), York, UK, 18–21 April 2006; pp. 104–118.
73. Fachkha, C.; Bou-Harb, E.; Debbabi, M. Inferring distributed reflection denial of service attacks from darknet. *Comput. Commun.* **2015**, *62*, 59–71. [CrossRef]
74. Meyer, D.; Haase, J.; Eckert, M.; Klauer, B. A Threat-Model for Building and Home Automation. In Proceedings of the IEEE 14th International Conference on Industrial Informatics (INDIN), Poitiers, France, 19–21 July 2016; pp. 860–866.
75. Kang, D.J.; Lee, J.J.; Kim, S.J.; Park, J.H. Analysis on cyber threats to SCADA systems. In Proceedings of the IEEE Transmission & Distribution Conference & Exposition, Seoul, Korea, 26–30 October 2009.
76. Young, N. Cyber security for automatic test equipment. In Proceedings of the IEEE AUTOTESTCON, Schaumburg, IL, USA, 9–15 September 2017.
77. SANS: The Threats of Social Engineering and Your Defense Against It. Available online: https://www.sans.org/reading-room/whitepapers/engineering/threat-social-engineering-defense-1232 (accessed on 13 May 2019).
78. Rai, K.K.; Asawa, K. Impact analysis of rank attack with spoofed IP on routing in 6LoWPAN network. In Proceedings of the 10th International Conference on Contemporary Computing (IC3), Noida, India, 10–12 August 2017; pp. 1–5.
79. Nakhila, O.; Zou, C. Parallel Active Dictionary Attack on IEEE 802.11 Enterprise Networks. In Proceedings of the Military Communications Conference, MILCOM 2016-IEEE, Baltimore, MD, USA, 1–3 November 2016; pp. 1–6.
80. Maraj, A.; Rogova, E.; Jakupi, G.; Grajqevci, X. Testing techniques and analysis of SQL injection attack. In Proceedings of the 2nd International Conference on Knowledge Engineering and Applications, London, UK, 21–23 October 2017; pp. 1–5.
81. Khanna, S.; Verma, A.K. Classification of SQL Injection Attacks Using Fuzzy Tainting. In *Progress in Intelligent Computing Techniques: Theory, Practice, and Applications*; Springer: Singapore, 2018; pp. 463–469.
82. Tedeschi, S.; Emmanouilidis, C.; Farnsworth, M.; Mehnen, J.; Roy, R. New threats for old manufacturing problems: Secure IoT-Enabled monitoring of legacy production machinery. In Proceedings of the APMS 2017: IFIP International Conference on Advances in Production Management Systems: The Path to Intelligent, Collaborative, and Sustainable Manufacturing, Hamburg, Germany, 3–7 September 2017; pp. 391–398.

83. Oliveira, J.; Frade, M.; Pinto, P. System Protection Agent Against Unauthorized Activities via USB Devices. In Proceedings of the 3rd International Conference on Internet of Things, Big Data and Security (IoTBDS) Funchal, Madeira, Portugal, 19–21 March 2018; pp. 237–243.
84. Bertino, E.; Islam, N. Botnets and internet of things security. *Computer.* **2017**, *50*, 76–79. [CrossRef]
85. Aansari, H.M.; Vakili, T.V. Detection of clone node attack in mobile wireless sensor network with optimised cost function. *Int. J. Sens. Netw.* **2017**, *24*, 149–159. [CrossRef]
86. Quirke, J. Security in the GSM System. Available online: https://pdfs.semanticscholar.org/b0c8/493e0c6b6e5e08d870a1b318401236e07e82.pdf (accessed on 13 May 2019).
87. Helfmeier, C.; Boit, C.; Nedospasov, D.; Seifert, J.P. Cloning Physically Unclonable Functions. In Proceedings of the IEEE International Symposium on Hardware-Oriented Security and Trust (HOST), Austin, TX, USA, 2–3 June 2013; pp. 1–6.
88. Großmann, M.; Illig, S.; Matějka, C.L. SensIoT: An Extensible and General Internet of Things Monitoring Framework. *Wirel. Commun. Mob. Comput.* **2019**, 1–15. [CrossRef]
89. Morabito, R.; Petrolo, R.; Loscrì, V.; Mitton, N. Reprint of: LEGIoT: A Lightweight Edge Gateway for the Internet of Things. *Futur. Gener. Comput. Syst.* **2019**, *92*, 1157–1171. [CrossRef]
90. Teixeira, F.A.; Pereira, F.M.Q.; Wong, H.; Nogueira, J.M.S.; Oliveira, L.B. SIoT: Securing Internet of Things through distributed systems analysis. *Futur. Gener. Comput. Syst.* **2019**, *92*, 1172–1186. [CrossRef]
91. Matheu-García, S.N.; Hernández-Ramos, J.L.; Skarmeta, A.F.; Baldini, G. Risk-based automated assessment and testing for the cybersecurity certification and labelling of IoT devices. *Comput. Stand. Interfaces* **2019**, *62*, 64–83. [CrossRef]
92. ENISA, Threat Landscape Report 2018, ETL 2018|1.0, January 2019. Available online: https://www.enisa.europa.eu/publications/enisa-threat-landscape-report-2018 (accessed on 13 May 2019).

Article

Anomaly Detection Trusted Hardware Sensors for Critical Infrastructure Legacy Devices

Apostolos P. Fournaris [1,*,†], Charis Dimopoulos [2,†], Konstantinos Lampropoulos [2] and Odysseas Koufopavlou [2]

1 Industrial Systems Institute, R.C. ATHENA, Patras Science Park, 26504 Platani-Patras, Greece
2 Electrical and Computer Engineering Department, University of Patras, Rion Campus, 26504 Rion-Patras, Greece; c.dimopoulos@upnet.gr (C.D.); klamprop@ece.upatras.gr (K.L.); odysseas@ece.upatras.gr (O.K.)
* Correspondence: fournaris@isi.gr
† These authors contributed equally to this work.

Received: 15 February 2020; Accepted: 25 May 2020; Published: 30 May 2020

Abstract: Critical infrastructures and associated real time Informational systems need some security protection mechanisms that will be able to detect and respond to possible attacks. For this reason, Anomaly Detection Systems (ADS), as part of a Security Information and Event Management (SIEM) system, are needed for constantly monitoring and identifying potential threats inside an Information Technology (IT) system. Typically, ADS collect information from various sources within a CI system using security sensors or agents and correlate that information so as to identify anomaly events. Such sensors though in a CI setting (factories, power plants, remote locations) may be placed in open areas and left unattended, thus becoming targets themselves of security attacks. They can be tampering and malicious manipulated so that they provide false data that may lead an ADS or SIEM system to falsely comprehend the CI current security status. In this paper, we describe existing approaches on security monitoring in critical infrastructures and focus on how to collect security sensor–agent information in a secure and trusted way. We then introduce the concept of hardware assisted security sensor information collection that improves the level of trust (by hardware means) and also increases the responsiveness of the sensor. Thus, we propose a Hardware Security Token (HST) that when connected to a CI host, it acts as a secure anchor for security agent information collection. We describe the HST functionality, its association with a host device, its expected role and its log monitoring mechanism. We also provide information on how security can be established between the host device and the HST. Then, we introduce and describe the necessary host components that need to be established in order to guarantee a high security level and correct HST functionality. We also provide a realization–implementation of the HST overall concept in a FPGA SoC evaluation board and describe how the HST implementation can be controlled. In addition, in the paper, two case studies where the HST has been used in practice and its functionality have been validated (one case study on a real critical infrastructure test site and another where a critical industrial infrastructure was emulated in our lab) are described. Finally, results taken from these two case studies are presented, showing actual measurements for the in-field HST usage.

Keywords: security; hardware design; trust; cryptography; anomaly detection

---

1. Introduction

In recent years, many critical infrastructures (CIs) around the world adopted various Information and Communication Technologies (ICT) advances, in an effort to become more flexible and cost effective. However, this adaptation was not made carefully and with a thorough evaluation on the implications it introduced to their security. Numerous new devices with advanced computation power and connectivity capabilities are constantly been installed in CIs and the—once closed and isolated CI systems—are becoming more and more vulnerable to new types of threats and attacks. Various reports and research teams have proved how dangerous this situation is. For example, a team called "SCADA StrangeLove team" was able, back in 2013, to get full control of various industrial infrastructures (energy, oil and gas, chemical and transportation CIs). They claimed to have found more than 60,000 online control systems that were exposed. Furthermore, even though nowadays CIs are more secure, the number and sophistication of cyberattacks is still increasing [1]. There is a critical need to fortify CIs to the maximum possible, since a major cyberattack in one of them may cause severe problems not only at a technical level but also in the economy, public safety, etc. [2].

On the other hand, protecting CI systems is a highly complicated task. A large number of diverse security systems and protection mechanisms must collaborate [3,4]. Solutions like Anomaly Detection Systems (ADS), Intrusion Detection (IDS), Antivirus tools for Malwares and Ransomwares, DDoS protection, Endpoint security, Hardware, protection for CI devices, Access control, etc., are just few of the technical tools that can be used to fortify such complex environments [5]. In addition, apart from the technical parts, a CI security must also include human training (personnel, users, etc.), Private-Public Partnerships, Assessments, Vulnerability Analyses, etc.

Usually the above security tools and solutions are integrated inside a Unified Threat Management (UTM) system or a Security Information and Event Management (SIEM) system. The difference between a SIEM and a UTM is that the SIEM does not exactly integrate security components but only collects reporting information (e.g., logs, reports, events, etc.) and combines it with input from other sources in order to "assemble a puzzle" which would eventually identify a possible security risk.

Inside this wide area of security solutions, this work examines innovations on a very specific aspect of CI protection—the design of trusted sensors for Anomaly Detection Systems (ADS) and SIEMs. An ADS can be described as a solution which extends the functionality of an Intrusion Detection System (IDS). In particular, the ADS not only monitors very specific, predetermined network metrics but also collects information from multiple other sources to estimate the security status of an IT system. Such sources can be distributed sensors inside a CI that generate logs which are collected by a centralized ADS or SIEM analyzer. The success rate of an ADS detection process (small false positives or negatives number) heavily relies on the quality ADS analyzer algorithm (Machine Learning techniques are also used nowadays) and the accuracy of the collected data [6,7]. Obviously, data from maliciously manipulated sensors can lead an ADS in producing false results and keep CI administration ignorant or falsely alert on a possible cybersecurity attack [8,9].

In this paper, we review the option of using hardware means in order to secure sensors' ADS/SIEM transmitted data, instilling trust in the overall process. In addition, extending the work in [2], we propose a Hardware Security Token to be physically connected to legacy CI devices and act as a trusted ADS sensor for failed access attempts as well as a mechanism for providing authentication and integrity to sensor's collected data. In the paper, we analyze the HST architecture and approach and we describe how it can achieve a level of trust in the associated host device using an appropriate security protocol. We also describe the main HST functionality achievable through the use of a dedicated host software program for accessing the HST as well as the HST log reporting mechanism on the ADS monitoring system. Then we describe a realization–implementation of the HST using an FPGA SoC evaluation board and we show how the HST services can be accessed. Finally, in the paper, two case studies using the HST in practice are described (one case study on a real critical infrastructure test site and another where a critical industrial infrastructure was emulated in our lab) and results are presented showing the HST capabilities in practice. The rest of the paper is organized as follows. In Section 2, an overview of a CI ADS sensors is made, security issues that may arise are described and a relevant threat model is presented. In Section 3, mechanisms to create trusted ADS sensors are described and in Section 4 the hardware assisted sensor approach, architecture, and functionality are proposed. Section 5 provides a realization of the HST along with use case scenarios of its usage and Section 6 concludes the paper.

2. Critical Infrastructure Security Monitoring System Anomaly Detection Sensors

Considering the security threats and challenges that many critical infrastructures have, there is a considerable need to continuously monitor such infrastructures during their regular operation for security anomalies that can be linked to some security attack [10–13]. Typical IT systems have a series of well-developed tools that, using a wide range of technologies and methods, can detect, respond and mitigate security attacks. The generic category of run-time monitoring systems may comprise of various components like intrusion detection systems (IDS), zero-vulnerability malware detectors and anomaly detectors that are all interconnected under a Security Information and Event Management (SIEM) system [7,14]. The Runtime monitoring Anomaly Detection mechanism is usually responsible for the correlation between various events and logs to extract security alerts and make attack mitigation suggestions. However, CI runtime security monitoring must consider the CI specificities that differ from those of a typical IT system [15,16].

CI systems (CIS) have a close association with the physical world (they monitor and respond to physical processes), thus they constitute an ideal realization of cyberphysical systems (or system of systems) and should be approached in that way in terms of security. According to [6], there are four basic characteristics that distinguish a CIS from typical IT systems in terms of runtime security intrusion detection. Due to their connection between the cyber and the physical world, the CIS devices measure physical phenomena and perform physical processes that are governed by the laws of physics. Thus, a CIS security monitoring system must perform physical process monitoring. Furthermore, typical CIS use many OT components, thus they are highly focused on automation and time driven processes that realize closed control loops operating with no human intervention (and its associated unpredictability). This kind of behavior focuses on Machine to Machine communications, increases the regularity and predictability of the CIS activities and makes them attractive to attackers [3]. Thus, the CIS security monitoring system should be able to monitor regularly closed control loops. Thirdly, the attack surface of a CIS is considerably broader than that of an IT system. CISs consist of many heterogeneous subsystems, including IT and OT devices. They follow a broad range of different, not IT related, control protocols like ISA 100, Modbus, CAN, etc. Some of these devices and protocols have proprietary software or standards that may make IT countermeasures unfitting [2,5]. This reason along with the fact that a successful CIS attack has high impact and thus high payoff, attracts very skilled attackers that can mount very sophisticated attacks on CPSs and CIS [13]. Such attacks are usually very hard to discover and document. Attackers exploit CIS zero-day vulnerabilities which would render many IT security monitoring toolsets useless (e.g., knowledge-based toolsets [6]).

Lastly, many CIS consist of legacy hardware that is difficult to modify or physically access. Such components may be partially analog, have very limited installed software resources and can be dictated by physical processes. The biggest challenge in such legacy devices is how to install security monitoring sensors on them and how to predict/model their behavior correctly in order to detect possible anomalies. Since legacy devices do not have many computational resources, it becomes hard for the monitoring system to retain its real-time responsiveness when collecting security metrics from them.

Runtime Security monitoring in the CIS domain, considering the above specificities, can take various forms. Monitoring relies on two core functions—the collection of data from various CIS sources and the analysis of data in a dedicated runtime security monitoring subsystem. To achieve appropriate data collection, the security monitoring system must deploy security agent software on the monitored CIS devices or introduce virtual entities (Virtual Machines) for data collection within the CIS infrastructure. All collected data are analyzed in the CIS runtime security monitoring system that uses data mining, machine learning, pattern recognition or statistical data analysis to extract metrics on security issues that may take place inside the CIS at runtime. Such issues may be possible incidents, threats that can be binary characterized as bad/good or continuously characterized by a specific significance grade. The performance of the security monitoring system is measured by the False Positive Rate (FPR), the False Negative Rate (FNR) and the True Positive Rate (TPR). The system is also measured in terms of incident detection latency and consumed resources number, computational overhead, excessive network traffic and power consumption [6].

The functionality and services SIEM and ADS applications provide can be considered a necessity and an integral component of a CIS security monitoring system. The basis of their functionality lies in the collection of various metrics reporting the health of a computing system and its network done by a broad network of local and remote sensor entities. These programs vary from honeypots, package analyzers, port scanner to antivirus or antimalware solutions. As a valuable source of ADS input data, one can consider the various OS activities, such as successful or failed authentication and authorization, collected by the OS' system log manager. The purpose of the above is for those sensors to track specific activities, log them and provide the necessary information to the ADS when some prerequisites defined by the security administrator are met. The collection of those events can be in real time or near real time, aiding the security administrators with visual cues in their attempt to monitor the overall system's security status and informing them of abnormalities that may lead to a compromised system. ADS sensors typically can be deployed as a software program installed on a host machine.

The heterogeneity CIS exhibit is directly reflected in the diverse nature of the sensors necessary to collect and log information for a CIS ADS. This variety of sensors, coupled with the critical nature of the overall system and the exposure various CI devices have on the CI premises (remote locations, power plants, factories) greatly increases the risk of successful device tampering or manipulation by an adversary. This type of compromise (often executed on the hardware level) is mostly ignored by the device's software ADS sensor and often leads to data manipulation of the ADS logging mechanism. As an implication, the ADS is provided with fabricated data, leading to either the suppression of real or creation of false anomaly events on the security monitoring mechanism.

2.1. Threat Model

There exist only a few works that attempt to ensure trust in the information collection mechanism from a network's end points of a CIS security monitoring and ADS system [17–19]. Some approaches rely on securing the communication channel between end device and ADS/IDS monitoring system [18] or rely on securing the end node itself by ad-hoc, trusted computing based, mechanisms [17,20]. Without loss of generality, we can assume that the main threats on the end point security monitoring sensors can be associated with attacks on a CIS end node that disrupt the sensor logging mechanism. This disruption can be achieved by manipulation of the communication channel between the device's sensor and the remote ADS/security monitor leading to integrity or authenticity threats or can be achieved by hijacking the CIS node itself. In the second case, our threat model also considers that it is realistic for an attacker to gain access to the CIS nodes physical storage disk and not have full control of the node's memory (the attacker does not have root access to the CIS node's operating system). We also consider in our threat model threats related to physical attacks using some invasive (tampering), semi invasive (fault injection) or non invasive attack (side channel analysis), since we can assume that CIS end nodes may be left unattended in hostile environments [21,22]. We also consider failed identification and authentication and failed access attempts on the node or the sensor itself as active threats. Such failures should be logged and sent in a trusted/secure way to the anomaly detection security monitor system for analysis.

3. Introducing Trust on Software Sensors

In cases when there is a need to instill trust on a computing system, a very efficient method involves the introduction of a trusted computing base (TCB). This level of trust is usually achieved through the inclusion of a hardware component (a security token) in the device's architecture. Through this secure environment, the TCB is able to act as a point of reference for the overall system, acting as a root of trust. Apart from the security critical operations, this secure computation environment (TCB) can also be considered hard to tamper with [23]. Both industry and academia have proposed several approaches for the most efficient way this hardware root of trust mechanism should be realized. One of the most important approaches is the specifications provided by Trusted Computing Group (TCG), which aims to instill trust on a system by guarding critical data (private keys), by blocking the execution of potentially harmful code and by attesting the system's trust level to other entities. Even from boot time, the system's security status is monitored continuously in order to achieve this level of trust, usually further enhanced by hardware mechanisms included in the TCB. Since software solutions do not provide adequate protection individually, TCG specifies a Hardware Security Module called the Trusted Platform Module (TPM), capable of acting as trust anchor within a computer system [24,25].

3.1. Using Trusted Platform Modules

For a CI device to be considered trusted, the inclusion of a TPM chip is of paramount importance, coupled with the necessary software stack embedded in the OS kernel in order to support the TCG trusted computing functionality. As a result, a TCG Trusted computing enabled device is capable of creating a secure environment to execute an ADS sensor's software code [17]. Despite the broad adoption of TPMs in Personal Computer use, they are not yet utilized in the CI domain and cyberphysical systems. This issue is also present in the majority of embedded devices or CI control elements (e.g., Programmable Logic Controllers—PLCs). For these types of devices without a TPM and constrained in resources and power consumption, TCG's suggestion is the use of the Device Identifier Composition Engine (DICE) mechanism, which nonetheless is far from adopted in CI system OT and IT end nodes that are not PCs.

3.2. Using Virtual Environments

An additional approach that can be used to instill trust in a CI device is Hardware virtualization. The general concept is the creation of isolated execution environments that under certain conditions can be considered trusted. Using this technique, critical or sensitive applications and their accompanying data can be directed towards these trusted areas of virtual machines (VMs) running on virtualized hardware. By expanding this logic, an isolated OS has the ability to operate on such a VM. If access to this VM is under the control of a Trusted Computing Base (TCB) program on the CPU, the OS can be considered secure and isolated from any other untrusted VM running [23]. This process of an employed TCB running over the device hardware as a hypervisor structure has various implementation problems in practice, like hardware constraints, system real time behavior, scheduling and access control rights. Apart from TCG's TPM that offers support for virtualization, solutions based on hardware virtualization (i.e., virtualization assisted through a processor Instruction Set) have been developed for AMD and Intel based systems [26]. In the embedded system domain, though, the type of constraints imposed eliminate the option of using hardware virtualization. The most similar solution for this category of devices is the ARM TrustZone technology [27], which enables the creation of trusted and nontrusted execution environments.

4. Proposed Approach for Legacy Systems

While the newest IT based CI devices may have some mechanism of instilling trust, typical CI control devices that constitute the backbone of a CI control loop still have legacy processing units that are not created for security but rather for safety and high, real time, responsiveness. Security Monitoring loggers installed on such devices need to rely on an execution environment that is capable of supporting the ADS sensors' functionality and that is protected from malicious entities. To achieve high security in legacy devices, it has been proposed in several works to introduce external security tokens that can be considered trusted [23,28–30]. Having that in mind, extending the work in [2], we propose a Hardware Security Token (HST) that could be used as an external security element on legacy devices in order to instill a level of trust on collected ADS sensor logs and provide a series of security services to an associated host device and user. In the following subsections we extend, expand and analyze the HST architecture, functionality and services thus structuring a complete solution for CI legacy device security protection.

4.1. Hst Architecture

The HST is a synchronous System on Chip (SoC) device based on an ARM microprocessor with TrustZone support (e.g., ARM Cortex A processor class) that is connected through an AMBA AXI bus to a series of cryptographic accelerator peripheral IP cores and storage elements like RAM, ROM, and NVRAM memory modules. The cryptography accelerator peripherals act as a security element of the ARM Trustzone enabled processor and consist of an RSA signature unit, an Elliptic Curve (EC) Point Operation unit (ECPO), a SHA256 hash function unit as well as a symmetric key encryption/decryption and key generation unit (using the AES algorithm), following an architecture similar to the one presented in [29]. All the HST IP cores are protected against semi-invasive and non invasive attacks [23,31,32]. The HST, using the above cryptographic peripherals, is capable of generating and verifying digital signatures and certificates, performing key agreement schemes like Elliptic Curve Diffie Hellman Ephemeral (ECDHE) protocol or Needham–Schroeder–Lowe protocol as well as AES based encryption/decryption (AES-CBC, AES-CCM) and authenticated message integrity schemes (HMAC). In addition, the HST has a series of Input/Output Interfaces including CAN bus, USB and Ethernet. The outline of the Hardware-Software hybrid architecture is presented in Figure 1. As can be seen in the above figure, all SoC components are interconnected in the Central Interconnect bus. Apart from hardware IPs, the HST has a software stack that is capable of controlling and coordinating all hardware assisted security operations (using customized Cryptographic IP drivers and cryptography libraries) as well as all communication (through a serial console interface) with an HST host machine. Finally, in the Figure 1, special mention should be made to the non volatile RAM unit which is realized as a QSPI Flash memory. This memory acts as storage space for all cryptographic, sensitive, information like public private keys, symmetric keys, HST states, users, etc.

The AMBA AXI (Central Interconnect) bus provides access to the cryptographic accelerator peripheral IP cores using the software stack. This stack implements a software component that is executed in the ARM cortex A trusted environment. The software component also handles the communication between the HST and the host. During its operation, it polls for an input command given by the host to the HST and collects all the necessary data each specific command requires. The required drivers that enable the stable operation of each IP core are included in this software component. Thus, the input data that have been collected during the command issuance are being propagated to the corresponding cryptographic peripheral for the output result to be calculated. Once this process is completed, our custom Crypto-Library is able to correctly perform a plethora of security protocols and algorithms that inherently depend on the operations our IP Cores provide. Protocols or algorithms that the Crypto-Library features are authenticated message integrity (HMAC), certificate generation and verification, digital signature schemes (ECDSA) as well as many utility operations (key generation, key validation, communication with Flash storage, etc.). The HST outputs to the host through a secure channel the correct output data of the corresponding command.

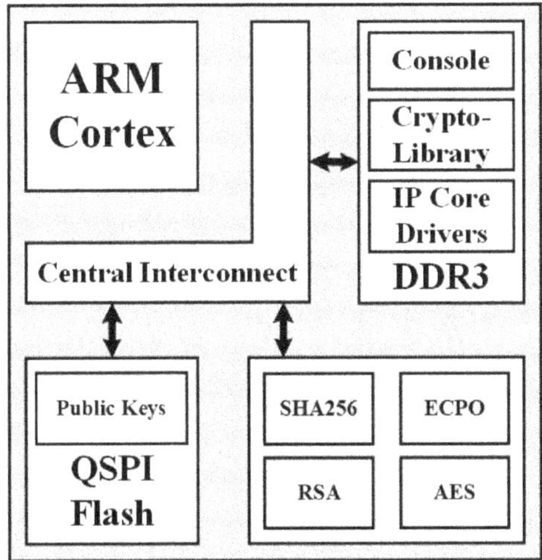

Figure 1. Hardware Security Module Architecture.

The NVRAM (flash memory) module embedded on the Zynq 7000 series FPGA board can support the validity and functionality of the HST operations in a wide range of various use cases by offering a secure, self-contained and HST controllable storage area where sensitive information can be saved. A typical configuration of an HST's flash memory contents can be viewed in Figure 2.

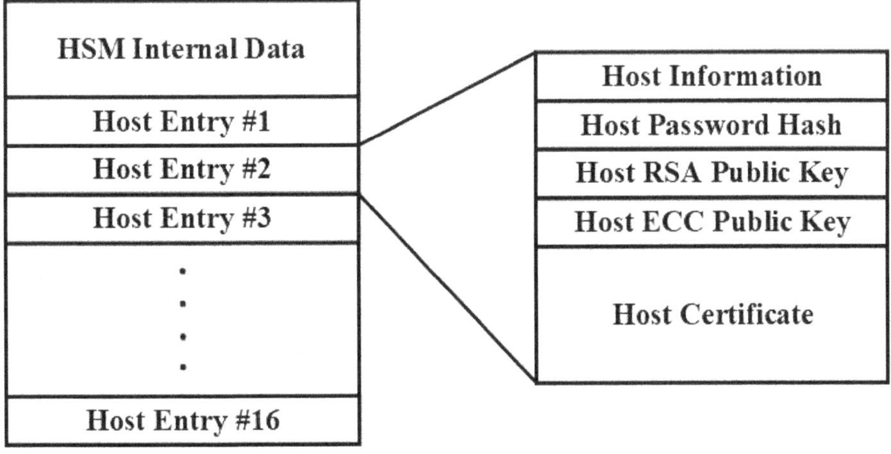

Figure 2. HSM NAND Flash memory contents.

First and foremost, stored in the flash memory are all the HST specific information, including the HST ID, status and the highly sensitive private and public RSA and ECC key pairs. The remaining available storage can be utilized in order to store multiple host entries, each containing all the necessary information of the corresponding host. More specifically, a host entry consists of basic host information, including host ID, host type, host status and a copy of the Password Hash that has been generated during the host initialization phase. Additionally, each host's RSA and ECC public keys are securely stored in the flash memory, along with a certificate that verifies the aforementioned keys (usually an ecdsa-with-sha256 based X-509 certificate due to storage limitations).

Apart from the above mentioned hardware structure, the HST has a dedicated software execution core that is retrieved from the HST flash memory and is loaded in the ARM processor RAM. This software core has dedicated components for the HST communication with the external world. More specifically, the software core enables the HST, extending the functionality described in [2,29], to connect through USB cable to a Host device and through Ethernet to an IP network. The USB serial communication channel serves as a secure means of communication between the Host device and the HST while the Ethernet based IP communication channel serves as a mean of communication between the HST and a remote ADS security monitor and analyzer. Apart from that, the HST software core is responsible for interfacing and usage of the hardware acceleration cryptography IP Cores (for the computationally demanding cryptography operations) using dedicated IP Core drivers. The HST software core also implements lightweight security operations that do not need hardware acceleration as well as security operations that during computations need some dedicated Hardware IP core output. Finally, the HST software core implements the HST API that the HST uses in order to communicate with the associated host device. The HST software component is being used in order to achieve two main operations, the host to HST associated functionality and the HST logging mechanism.

4.2. Host to Hst Functionality

The HST can be used in order to identify and authenticate a host device, to collect a series of log entries from the host devices and to transmit them through a secure channel to the ADS monitoring system. In addition, the HST is capable of providing individually, security services to the host devices like certificate generation/verification, digital signatures, key agreement and secure channel establishment. Apart from the actual HST component, the above functionality is manifested through the use of a dedicated software component (HST/host software component) on the host device that acts as a proxy between the Host and the HST. The HST/host software component is operating on an untrusted environment in host device, so it should be assumed that it should not store sensitive information on the host device in a non secure way. This component is also responsible for the authentication of both the host user and the host device to the overall ADS monitoring system. It also generates appropriate log entries using the syslog protocol and secures those entries using the HST. It does not solely rely on the Linux OS syslog mechanism but it also has a dedicated syslog client embedded in its structure in order to minimize a potential attacker's involvement in the logging approach. Finally, the HST/host software component can rely on the HST commands and messages that are issued by the host user using a dedicated HST Command Line Interface (HST CLI) as well as execute HST CLI scripts.

To achieve a secure use of the HST/host software component and its access to the HST, the approach followed in [2] is adopted and extended. Initially, it can be assumed without lose of generality that before deployment, the HST undergoes an initialization phase. The used and host device that are going to use the HST register their interest in a trusted entity (a trusted host) that uses the registration information (an initial password for the user and a device ID for the host machine) to associate these entities with the HST and the its user. The registration information will be used by the trusted entity to prepare the

HST for deployment. More specifically, the HST in its secure processing and storage core will generate a salt (a random value) for the provided password and use both these inputs in a Key Derivation Function (KDF) to create a symmetric cryptography key $Q = Q_{init}$. Apart from the above information, the HST will be used by the Trusted entity in order to generate an Asymmetric cryptography key pair (public and private key) and a associated certificate. This information will act as a secure token and will be encrypted using Q. The trusted entity stores in the HST the host/used ID, the salt, the hash outcome of the password and the host's public key. Finally, the HST generates its own Asymmetric Cryptography public/private key pair. The trusted entity concludes the initialization phase by registering the host to the HST by providing the encrypted host key pair and certificate. The host can store this information in its storage areas (a hard drive disk or flash drive). An attacker that intercepts, copies and analyzes these files will fail to retrieve the key pair since he will not know the password nor the salt. Retrieving this information is very hard for an attacker since the password should not be stored in the host machine nor in the associated HST while the salt is only stored in the HST secure area. Only by a user knowing the password and providing it to the host device connected to the HST associated with this device can he get access to the keys and use the HST services (this acts as a two factor authentication). Failure to provide the appropriate password will generate a log entry (an abnormal event) that will be transmitted through the HST Ethernet dedicated channel to the ADS (using the HST logging mechanism). Taking into account that the HST is considered trusted, the ADS security monitor can trust that the HST sensor's collected input is not tampered. To achieve that, the entry is digitally signed with the specific HST's Asymmetric Cryptography key. It can be assumed that the ADS has knowledge of all the HST sensors' Asymmetric Cryptography public keys and their associated certificates.

When the Initialization phase is finalized, the HST-host system can be deployed in a CI system and provide security services and secure logging. A CI host device and its associated HST are fully connected through USB and the HST services become available when a secure session is established between the pair.

The host–HST secure session establishment follows the key agreement scheme proposed in Figure 3. The presented protocol, extending the work in [2], supports a two factor authentication mechanism by combining information of user (user password) and device (the host device secure token). The protocol is built around the Elliptic Curve Diffie Helman Ephemeral (ECDHE) key exchange mechanism for generating a (AES) session key and establishing a secure channel between the host and the HST. The proposed protocol extends the ECDHE by providing a mechanism for securely unlocking during execution the host secure token provided during registration without revealing sensitive information in the process (based on the provided threat/attack model of Section 2.1). Initially, the host user requests to be connected to the HST and thus provides the registered password to the Host device. Note that there is not any form of the password (in clear, encrypted or hashed) stored (apart from the host's memory) in the host device (e.g., there is no password file). The host device just uses this password to generate its hash function digest. Upon receiving the password by the user (along with the associated username), the HST/host software component installed in the host device sends a request to the HST to receive the HST only stored salt (salt1) in order to decrypt the host secure token (i.e., the encrypted host's/User's Certificate and key pair) that is stored in the host disk. The HST then, internally, generates a nonce value that provides replay attack protection, retrieves the hash function digest of the password (that is securely stored) as well as the stored salt (salt1), generates a new salt (salt2) to achieve forward security, concatenates the nonce with salt1 and salt2 and digitally signs the outcome with the HST private key (KH_pr). Then, the HST concatenates the generated digital signature, the nonce (a random number), the HST stored salt (salt1) and a newly generated salt value (salt2) and using a Symmetric Key encryption algorithm ($E_K()$: AES) with key the password's hash function digests. Finally, the HST creates a digital signature $N = DS_{KH_{pr}}(K|nonce|salt1|salt2)$ and uses it in the encrypted result $E_K(N|nonce|salt1|salt2)$ that eventually sends as a reply to the user's request. Upon receipt of such message, the host generates his own version of password Hash function digest K and tries to decrypt the symmetric key encrypted message $E_K(N|nonce|salt1|salt2)$. Then he can extract the salt (salt1) and through a KDF that uses the password and salt1, recreate the key Q that is necessary to access the stored secure token (user's certificate and keys). By retrieving the nonce, salt1 and salt2 and by calculating K, the user can verify the digital signature N using the HST public key and prove that the HST has knowledge of the password digest K. Using the retrieved Host Asymmetric Encryption Keys stored only in memory, the host then sends a digital signature of the nonce to the HST, thus verifying knowledge of both the nonce and his private key ($E_{K_{pr}}$). Then, the ECDHE key agreement scheme is executed using the host retrieved key pair. The outcome of ECDHE is a common session key S that can be used for encrypting the remaining traffic between host and HST. The Certificate and Asymmetric cryptography keys are encrypted using the result of a KDF that has as inputs the password along with the new salt (salt2). This result is stored back to the host storage area.

Establishing a secure channel, where traffic is encrypted between the HST and the host, the HST/host software component can forward requests for security services as well as send log messages that will be transmitted with integrity and authenticity through the HST dedicated Ethernet IP communication to the ADS. Integrity and authenticity are achieved by digitally signing the log entries with the retrieved HST Asymmetric Cryptography keys.

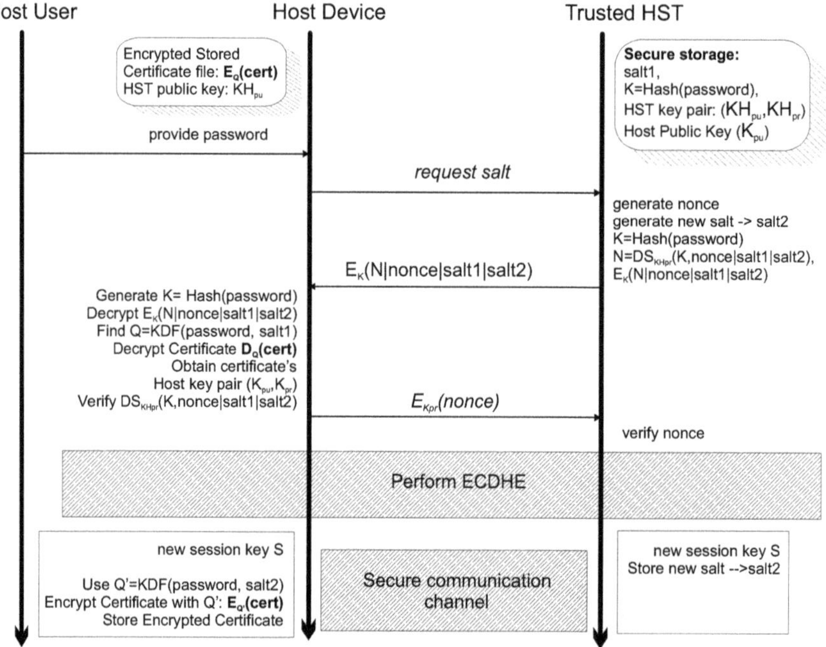

Figure 3. The proposed session key agreement protocol.

4.3. Host-Hsm Logging Mechanism

When a security related incident is taking place, it can be detected by the HST/host software component. A log entry is then generated to either be stored using the syslog protocol in the auth.log of the Linux OS as a syslog entry in the host device or to be generated internally in the HST/host software component and forwarded if confidentiality, integrity and authenticity are confirmed to a remote

ADS through the HST. Such security incidents can range between possible cryptanalytic attacks, loss of connection between the host and the HST, password authentication failure and many others. The proposed logging mechanism ought to be flexible and simple to use, while containing a sufficient amount of information that can be scaled up according to the application's needs. Through this scope, each log entry is a JSON message that includes information related to a specific event in the following format (Table 1):

Table 1. HSM Log entry JSON structure.

HSM Log Entry JSON Structure

```
{
    "HostID":<integer>,
    "HostIP":<integer>,
    "HostState":<string>,
    "HSTid":<integer>,
    "timestamp":<integer>,
    "event":{
        "type":<integer>,
        "failure":<integer>,
        "severity":<integer>
    }
    "comments": <string>
```

The above structure is always digitally signed with the keys that are stored inside the HST (the host/HST software component does not have access to them). Note that the above log format can be expanded with relative ease to include additional fields of varying type, producing a more detailed log entry that conveys a complete overview of an event with adequate information. The first five basic fields of this JSON array format are necessary for the correct identification of the specific unit that produces a log entry, as well as the exact time the log was generated. The *"HostState"* field provides characterization of the host state relative to the HST. There exist two states in which a CI host can be in, administration and user. In the administrator (*admin*) state, the host user can gain full access to the HSM services and features, including the ability to store other host entries to the HST flash memory, as was analyzed previously. In the *user* state, however, the host user lacks the authorization to add new host entries to the HST. The core of the log entry containing the most important information is the *"event"* array. Its first field, *"type"*, is an integer number indicating the type of event that has been logged. In its current HST realization, the acceptable values of this field are the following types:

- 0: Message integrity validation event.
- 1: Password based host to HST session initiation.
- 2: HST availability.
- 3: Security channel failure.

Directly linked to the above, the field *"failure"* is an integer number that indicates if the occurred event of the type specified in *"type"* has failed (*failure* = 1) or if it has concluded normally (*failure* = 0). The last field encapsulated in the *"event"* is the *"severity"* one, signifying the importance in terms of security impact in cases where event failure is detected. A value of 0 indicates low severity, while the maximum value of 3 marks the logged event with the highest severity. In the final JSON field *"comments"*, additional information related to the logged event can be provided. After generation, the log entry is sent to the ADS monitoring system, containing all the necessary information about the host device, the logged event and its severity. Accordingly, the level of trust and confidence by the ADS is enhanced, providing simultaneously valuable details that aid the appropriate management of the inflicted node or device.

5. Hst Practical Conceptualization–Realization

In order to further promote the functionality and applicability of the HST concept and highlight its importance as a flexible and scalable system that provides solutions to many different security problems, CI systems along with trustworthy anomaly detection, in this section we describe a realization of the proposed solution that was implemented during the EU project "CIPSEC:Enhancing Critical Infrastructure Protection with innovative SECurity framework" and that is being expanded in EU projects "CONCORDIA" [33] and "CPSoSaware" [34]. In this realization we specify the HST CLI environment and show its practicality in promoting and accessing the HST security services. Consistency and ease of upgrading should be essential characteristics of this CLI, focusing on scalability and adaptability to a wide range of security scenarios that need to be implemented in multiple CI Systems. These scenarios can vary from simple End-to-End secure channel establishment, to appropriate secure and trusted logging and even to a PKI-like scheme that manages host public keys. In its current realization, the HST is implemented in a Digilent Zedboard device that includes the Zynq 7000 series SoC with ARM Cortex A9 processor and an FPGA fabric on chips (used for the implementation of the HST Hardware IP cores).

5.1. *Case Study Hst Cli for Cryptographic Application Programming*

The availability of a variety of IP cores, as well as the sufficiently powerful Cortex-A9 processing unit offer the ability of implementing numerous cryptographic and security protocols available today. These operations can be accessed directly from the host machine through an in-house built serial CLI. The serial console component accepts commands for execution that adhere to a specific format. Its general structure is as follows: *"command [options] [HostID] [data]"*. For example, in order to execute the Hash-based Message Authentication Code (HMAC) protocol, the host sends to the HST through the serial secure communication channel the command *"hmac [key] [message]"*, where [key] is a secret shared key and [message] the input data of the algorithm. In cases when a command requires extra information related to a specific host like an ECC public key, an extra *[HostID]* field must be added, providing to the HST the ability to extract and use the correct host entry located in its flash memory.

The HST's hardware board cryptographic features can be accessed through the HST/Host software component. This tool is developed for Linux OS based Host machines and is currently realized for 32 bit and 64 bit x86 Linux platforms as well as ARM Linux platforms. During operation, two different modes are available. In the HST Console mode the various CLI commands are transmitted directly to the HST through a terminal console. In the HST OS mode, on the contrary, all HST commands are given as arguments upon our Linux based HST/host software component executable or as CLI scripts. As it is apparent, OS mode provides greater functionality and flexibility, enabling the development and easy deployment of different applications that want to take advantage of the HST's features. The overall CLI approach that is being used resembles the openssl library approach with the extension of hardware and dedicated trusted tokens use (hardware in the loop concept, i.e., the HST).

5.2. *Hst as a Certificate Authority*

In a plethora of industrial and CI systems, sensitive information is being exchanged continuously by a wide range of different host machines. This exchange is often exposed to attacks both at a physical and network level, as many hardware sensors and host devices operate in a variety of hostile environments. This is even more prevalent in many Legacy systems that inherently lack strong security design principles. Thus, in such cases the HST functionality can be extended to offer support of certificate management for the different host devices that a CI utilizes, promoting a unique host–HST module as a pseudo Certificate Authority (CA).

Typically, in this scenario, most of the host devices operate in *user* mode, meaning they cannot add another host entry in the HST's flash memory other than their own entry during the host initialization phase due to the lack of elevated privileges. Each host device has already generated and stored a host ECC private and public key pair. Acting as a pseudo-CA, a host–HST module is operated in *admin* mode. As already mentioned, a host in *admin* state is authorized to store additional information on the flash memory module embedded in its corresponding HST. With the functionality our Crypto-Library offers, the CA can receive Certificate Signing Requests (CSR) from any *user* host operating in the same network as the CA. The CA then checks the validity of the digitally signed CSR and upon successful validation generates an ECC Certificate bound to the specific *user* host. The Certificate is stored on the CA's corresponding host entry and sent through the network to the host that requested it. Through this process, the CA is in possession of all the user Hosts' certificates and updates this list whenever a new host is added to the network or a Host regenerates its ECC key pair. Any host from this point forward can request from the CA another host's certificate, in order to validate the authenticity of its public key and consequentially the ownership of the corresponding private one. Using this PKI-like structure, greater trust is instilled upon the different hosts' communication than utilizing a basic digital signature scheme without the existence of a CA.

5.3. Real-World Test Case Hst Validation

The functionality of the HST has been practically deployed for assessment in two specific CI testbed scenarios. For this purpose, the HST is extended to include a Raspberry Pi module connected to the Zynq 7000 series FPGA board (Zedboard) through a USB serial communication channel. This Pi module, connected to the IP network with either Ethernet or Wi-Fi connection emulates a standalone host machine. In a similar manner, we assume there exist identical host–HST pairs operating across a CI. Such a configuration can be seen in Figure 4.

Figure 4. Raspberry Pi and Zedboard host-HSM configuration.

5.3.1. Test Case A

The HST can be used to provide End-to-End encryption, integrity and authentication between CI network domains. This functionality in the proposed HST can be related to anomalous behavior since repeated failures in data integrity and authentication constitute a cyberthreat and may signal the beginning of elaborate attack schemes (e.g., advanced Distributed Denial of Service). Having this rationale in mind, the HST concept was adapted accordingly in order to offer the above described functionality. A practical evaluation of the approach was realized in the Deutsche Bahn (DB) NETZE (Germany railways interlocking mechanism provider) infrastructure testing where the host–HST pair was able to capture and redirect UDP packets (used in the RaSTA industrial control protocol used in the DB railway interlocking actuators) destined for another host device in a secure way under a Man-in-the-Middle attack (MitM) and Man-at-the-End (MATE) scenario. The overall use case test configuration is presented in Figure 5. In this mode, the host is exclusively communicating with its associated HST that is responsible for handling the network traffic related to the host machine. The source of this traffic (UDP packets) that is forwarded to the host's HST, can either be encrypted information from another host–HST pair or data that needs to be encrypted before being forwarded to a destination host–HST pair (in order to achieve End-to-End secure communication). For a mechanism like this to operate properly, the involved HSTs (one for each end point host) execute a key exchange protocol to generate a common shared session key. After this establishment, the raw data a particular host wishes to send to another host must firstly be encrypted and authenticated by the HST (using for example authenticated encryption or encryption and MAC mechanisms) and attached to a UDP packet. The UDP packet is then sent through the network accordingly to the correct destination host–HST pair, where the ciphertext is decrypted and its authenticity–integrity validated, thus revealing the original raw data to the destination host. Utilizing this design philosophy, two hosts can effectively establish a secure channel of communication even over IP, taking advantage of the encryption and decryption services only the HSTs can provide. The logging mechanism on the HST is permanently active in the above mentioned activities in order to detect any failure in the overall process (e.g., wrong key establishment, faulty session establishment, no authentic message, non authentic host user that tries to access the HST, password attacks on the HST or the host, etc.), then the ADS monitoring system is informed through an HST dedicated wireless network channel (instead of an ethernet wired one) due to the testing site policy restrictions. In the validation process of the use case, for integrity HMAC with SHA256 was used, for secure session Establishment ECDHE was used and secure communication was done using AES CCM mode. The exchanged UDP messages as well as the log entry messages to the ADS were digitally signed using ECDSA (using secp256r1 ECC) with SHA256.

To validate the above use case and measure the response time of the HST, a MiTM and MATE attack was mounted. In the MiTM attack scenario a malicious user tried to compromise the message integrity and authenticity of the messages during transmission. In the MATE attack scenario, a malicious user tries to bypass the security of the HST, by performing a dictionary attack to find the HST passwords and also maliciously alters log entries of the host device in order to hide its identity. In both the scenarios, all the attacks were captured by the HSTs and the malicious activities were reported to the anomaly detection system. In Figures 6 and 7, the ADS log entries and resulting events are presented.

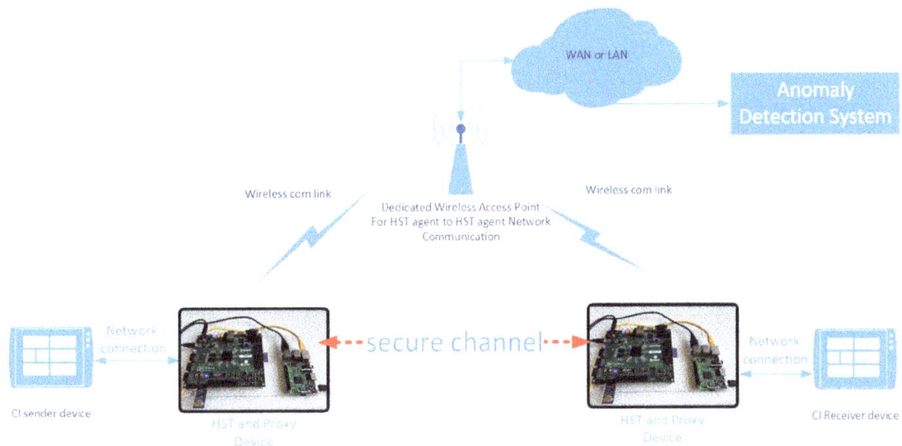

Figure 5. Use Case A: Achieving Confidentiality and Integrity under Man-in-the-Middle (MiTM) and Man-at-the-End (MATE) attack.

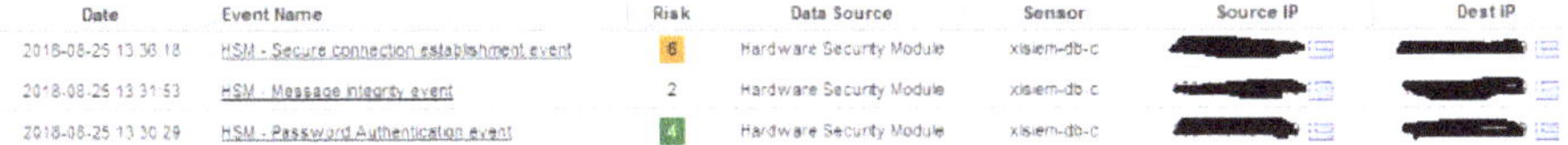

Date	Event Name	Risk	Data Source	Sensor	Source IP	Dest IP
2018-08-25 13 33 18	HSM - Secure connection establishment event	6	Hardware Security Module	xisiem-db-c	[illegible]	[illegible]
2018-08-25 13 31 53	HSM - Message integrity event	2	Hardware Security Module	xisiem-db-c	[illegible]	[illegible]
2018-08-25 13 30 29	HSM - Password Authentication event	4	Hardware Security Module	xisiem-db-c	[illegible]	[illegible]

Figure 6. Use Case A: Anomaly Detection and Security Information and Event Management (SIEM) log entries.

Event Name	Events	Risk	Duration	Source	Destination	Status
Monday 27-Aug-2018 [Delete]						
Attack, Possible MitM attack or tampered HSM module	1	3	6 secs	[illegible]	[illegible]	open
Bruteforce attack, Possible bruteforce attack against HSM module	4	6	14 secs	[illegible]	[illegible]	open
Network attack, Compromised secure communication with HSM module	4	3	9 secs	[illegible]	[illegible]	open

Figure 7. Use Case A: Anomaly Detection and SIEM extracted events.

5.3.2. Test Case B

A very common CI setup consists of a SCADA industrial monitoring system facility that collects information from in-field deployed sensors (e.g., temperature sensors). To emulate the above setup, a test case scenario was created that consists of a server (emulating the SCADA system) and a client (emulating the deployed temperature sensor). Typically, values from the sensor are transmitted through an unprotected channel (e.g., using Modbus protocols) to the SCADA system that is responsible for the fine tuning of the climate conditions in the CI facility. This configuration is prone to Man-in-the-Middle attacks from an adversary that alters the transmitted message and potentially causes severe damage to expensive CI equipment. The emulation process emulates the above scenario using two HST–host configurations as described above and seen in Figure 4 that are connected through a wireless network as a server and a client. One of the HST–host pairs has soldered on the host side (Raspberry Pi) a Pimoroni Enviro pHAT [35], which features a BMP280 temperature/pressure sensor. This sensor is continuously polling for a new temperature value, transmitting it through unprotected UDP packets to the emulated temperature control system. To mount a MitM attack, the Ettercap [36] Open Source tool for Linux was used. Under an unprotected communication channel (without HST), a successful MitM attack was executed, successfully altering the temperature value from its usual range to an extreme one and causing the temperature control system to react accordingly, with unwanted consequences. In order to demonstrate the validity of both the attack and the applied solution, a message integrity mechanism based on hash-based message authentication code (HMAC) is deployed during the communication. The message is hashed and any possible alteration to it is detected by the HST, leading to the rejection of the specific packet. A message integrity failure event is then logged appropriately by the host–HST logging mechanism and transmitted to an operating ADS or SIEM in a similar way as in the Test Case A.

5.4. Results and Discussion

Typically, the main issues when it comes to security agents/sensors is that they should be able to respond in time when an attack is taking place or is about to take place, to remain secure under the presence of attacks (hostile environment) and to be able to capture all events associated with a threat or an attack. In this subsection, the measurement results collected during the validation process of the two test cases are presented. It should be noted that the key difference of the HST in terms of response time is the hardware accelerated security primitives that are employed. Thus, the collected measurements are focused on the computation delay for each cryptography primitive operation employed in the test cases that is assisted through hardware means. In Table 2, the time delay accounted from the pointed message is inputted to the HST until it is processed and finally transmitted to the ADS (or a CI end node) is presented for the HST HMAC message integrity and the End-to-End secure communication mechanism (using AES CCM). As it can be observed, the execution time for these operations, due to hardware acceleration, is considerably small if the delays introduced by the communication channel between the HST and the host are also taken into account. When measuring the benefit of hardware implementation versus software ones individually for each security primitive operation inside the HST, as expected, our hardware approach fairs considerably better than purely software designs. For example, using hardware acceleration, an HMAC with SHA256 operation delays 42šec for small byte length inputs versus 62šec when using only software code (67% improvement). Similar improvement appears when the input byte length increases. In addition, for ECDSA digital signature scheme operations, the hardware accelerated solution needed 14.3 ms for signing and 23.7 ms for verification of signatures (using secp256r1 ECCs) versus 24.9 ms and 43.2 ms respectively using only software code (57% improvement in speed).

Table 2. Full execution time of HMAC-SHA256 and AES CCM encryption mechanisms.

Runtime Benchmark				
	16B	**64B**	**128B**	**256B**
HMAC	0.049 s	0.061 s	0.071 s	0.092 s
AES	0.052 s	0.067 s	0.089 s	0.135 s

6. Conclusions

In this paper, we propose an approach on how to collect information from CI device ADS sensors that can be trusted and are not tampered with. This approach was based on a hardware assisted dedicated security service provider, the HST, that supports a secure event log and monitoring mechanism. In our approach, the goal is to move securing of security related logs, needed by an ADS, from the Operating System of a CI host (that can be considered insecure) to the HST dedicated hardware module. The HST performs operations in a secure environment and has sole knowledge of cryptography keys that are used for providing confidentiality, integrity and authenticity of the logging mechanism. Thus, even if an attacker manages to compromise the CI host system, he still does not have knowledge of the security keys and also does not have access to the log monitoring mechanism (which in our proposal is fully manifested in the HST). In the paper, we analyzed the proposed approach based on the above described concept and detailed the HST functionality as well as the functionality of the associated HST–host security component deployed in the host device. We also described how the log mechanism could be realized (using JSON data structures) and also provided a practical realization of the HST concept. After describing a manifestation of the HST command line interface, we also described cases study scenarios where the HST can be used to provide even additional services to the secure log monitoring and reporting mechanism.

Author Contributions: The authors of the paper have contributed to the work as follows: conceptualization, A.P.F. and C.D.; methodology, A.P.F. and C.D.; software, A.P.F. and C.D.; validation, A.P.F., K.L. and C.D.; formal analysis, A.P.F. and C.D.; investigation, A.P.F. and C.D.; resources, A.P.F. and O.K.; data curation, A.P.F. and C.D.; writing—original draft preparation, A.P.F. and C.D.; writing—review and editing, K.L.; visualization, K.L.; supervision, A.P.F. and O.K.; project administration, O.K.; funding acquisition, A.P.F. and O.K. All authors have read and agreed to the published version of the manuscript.

Funding: This paper's work has received funding from the European Union's Horizon 2020 research and innovation programme CPSoSaware under grant agreement No 873718 and also the European Union's Horizon 2020 research and innovation programme CONCORDIA under grant agreement No 830927).

Conflicts of Interest: The authors declare no conflict of interest.

References

1. Ten, C.W.; Yamashita, K.; Yang, Z.; Vasilakos, A.V.; Ginter, A. Impact assessment of hypothesized cyberattacks on interconnected bulk power systems. *IEEE Trans. Smart Grid* **2017**, *9*, 4405–4425. [CrossRef]
2. Fournaris, A.P.; Lampropoulos, K.; Koufopavlou, O. Trusted hardware sensors for anomaly detection in critical infrastructure systems. In Proceedings of the 2018 7th International Conference on Modern Circuits and Systems Technologies (MOCAST), Thessaloniki, Greece, 9 May 2018; pp. 1–4.
3. Baggett, R.K.; Simpkins, B.K. *Homeland Security and Critical Infrastructure Protection*; ABC-CLIO: Santa Barbara, CA, USA, 2018.
4. Shackleford, D. Who's using Cyberthreat Intelligence and how? *SANS Inst.* **2015**, *24*, 2018.
5. Lewis, T.G. *Critical Infrastructure Protection in Homeland Security: Defending a Networked Nation*; John Wiley & Sons: Hoboken, NJ, USA, 2019.
6. Mitchell, R.; Chen, I.R. A Survey of Intrusion Detection Techniques for Cyber-physical Systems. *ACM Comput. Surv.* **2014**, *46*, 55:1–55:29. [CrossRef]
7. Rathi, M.M.; Jain, V.K.; Merchant, S.T.; Lin, V.C. Method and System for Detecting and Preventing Access Intrusion in a Network, 2014. U.S. Patent 8,707,432, 22 April 2014.
8. Manandhar, K.; Cao, X.; Hu, F.; Liu, Y. Detection of faults and attacks including false data injection attack in smart grid using Kalman filter. *IEEE Trans. Control. Netw. Syst.* **2014**, *1*, 370–379. [CrossRef]
9. Sun, Y.; Luo, H.; Das, S.K. A trust-based framework for fault-tolerant data aggregation in wireless multimedia sensor networks. *IEEE Trans. Dependable Secur. Comput.* **2012**, *9*, 785–797. [CrossRef]
10. Khan, M.T.; Serpanos, D.; Shrobe, H. A rigorous and efficient run-time security monitor for real-time critical embedded system applications. In Proceedings of the 2016 IEEE 3rd World Forum on Internet of Things (WF-IoT), Reston, VA, USA, 14 December 2016; pp. 100–105.
11. Watterson, C.; Heffernan, D. Runtime verification and monitoring of embedded systems. *IET Softw.* **2007**, *1*, 172–179. [CrossRef]
12. Fadlullah, Z.M.; Taleb, T.; Ansari, N.; Hashimoto, K.; Miyake, Y.; Nemoto, Y.; Kato, N. Combating against attacks on encrypted protocols. In Proceedings of the 2007 IEEE International Conference on Communications, Glasgow, UK, 28 June 2007; pp. 1211–1216.
13. Jing, Q.; Vasilakos, A.V.; Wan, J.; Lu, J.; Qiu, D. Security of the Internet of Things: Perspectives and challenges. *Wirel. Netw.* **2014**, *20*, 2481–2501. [CrossRef]
14. Niu, Z.; Shi, S.; Sun, J.; He, X. A survey of outlier detection methodologies and their applications. In Proceedings of the International Conference on Artificial Intelligence and Computational Intelligence, Taiyuan, China, 25 September 2011; pp. 380–387.

15. Coppolino, L.; D'Antonio, S.; Formicola, V.; Romano, L. Enhancing SIEM technology to protect critical infrastructures. In *Critical Information Infrastructures Security*; Springer: Berlin, Germany, 2013; pp. 10–21.
16. Hu, F.; Lu, Y.; Vasilakos, A.V.; Hao, Q.; Ma, R.; Patil, Y.; Zhang, T.; Lu, J.; Li, X.; Xiong, N.N. Robust cyber–physical systems: Concept, models, and implementation. *Future Gener. Comput. Syst.* **2016**, *56*, 449–475. [CrossRef]
17. Coppolino, L.; Kuntze, N.; Rieke, R. A Trusted Information Agent for Security Information and Event Management. *Secur. Anal. Syst. Behav.* **2012**, 6–12.
18. King, D.; Orlando, G.; Kohler, J. A case for trusted sensors: Encryptors with Deep Packet Inspection capabilities. In Proceedings of the MILCOM 2012 IEEE Military Communications Conference, Orlando, FL, USA, 1 November 2012; pp. 1–6. [CrossRef]
19. Zuech, R.; Khoshgoftaar, T.M.; Wald, R. Intrusion detection and Big Heterogeneous Data: A Survey. *J. Big Data* **2015**, *2*, 3. [CrossRef]
20. Wazid, M.; Das, A.K.; Bhat K, V.; Vasilakos, A.V. LAM-CIoT: Lightweight authentication mechanism in cloud-based IoT environment. *J. Netw. Comput. Appl.* **2020**, *150*, 102496. [CrossRef]
21. Fraile, L.P.; Fournaris, A.P.; Koufopavlou, O. Revisiting Rowhammer Attacks in Embedded Systems. In Proceedings of the 2019 14th International Conference on Design Technology of Integrated Systems in Nanoscale Era (DTIS), Mykonos, Greece, 16 April 2019; pp. 1–6. [CrossRef]
22. Fournaris, A.P.; Pocero Fraile, L.; Koufopavlou, O. Exploiting hardware vulnerabilities to attack embedded system devices: A survey of potent microarchitectural attacks. *Electronics* **2017**, *6*, 52. [CrossRef]
23. Fournaris, A.P.; Keramidas, G. From hardware security tokens to trusted computing and trusted systems. In *System-Level Design Methodologies for Telecommunication*; Springer: Cham, Switzerland, 2014; Volume 9783319006, pp. 99–117. [CrossRef]
24. Trusted Computing Group. TCG TPM Specification Version 1.2. 2007. Available online: https:/www.trustedcomputinggroup.org/specs/TPM/ (accessed on 29 May 2020).
25. Challener, D.; Yoder, K.; Catherman, R.; Safford, D.; Van Doorn, L. *A Practical Guide to Trusted Computing*; IBM Press: Armonk, NY, USA, 2007.
26. Intel. Intel® Trusted Execution Technology (Intel® TXT). 2011. Available online: https://www.intel.com/content/www/us/en/architecture-and-technology/trusted-execution-technology/trusted-execution-technology-security-paper.html (accessed on 29 May 2020).
27. ARM. ARMTrustZone. Available online: http://www.arm.com/products/processors/technologies/trustzone.php (accessed on 29 May 2020)
28. Hein, D.; Toegl, R. An Autonomous Attestation Token to Secure Mobile Agents in Disaster Response. In Proceedings of the First International ICST Conference on Security and Privacy in Mobile Information and Communication Systems (MobiSec 2009), Torino, Italy, 3–5 June 2009.
29. Fournaris, A.; Lampropoulos, K.; Koufopavlou, O. Hardware Security for Critical Infrastructures—The CIPSEC Project Approach. In Proceedings of the 2017 IEEE Computer Society annual Symposium on VLSI (ISVLSI), Bochum, Germany, 3 July 2017. [CrossRef]
30. Heiser, G.; Andronick, J.; Elphinstone, K.; Klein, G.; Kuz, I.; Ryzhyk, L. The road to trustworthy systems. In Proceedings of the Fifth ACM Workshop on Scalable Trusted Computing, Chicago, IL, USA, 18 July 2010; pp. 3–10.
31. Fournaris, A.P.; Koufopavlou, O. Protecting CRT RSA against Fault and Power Side Channel Attacks. In Proceedings of the 2012 IEEE Computer Society Annual Symposium on VLSI, Amherst, MA, USA, 19 August 2012; pp. 159–164. [CrossRef]
32. Fournaris, A.P.; Zafeirakis, I.; Koulamas, C.; Sklavos, N.; Koufopavlou, O. Designing efficient elliptic Curve Diffie-Hellman accelerators for embedded systems. In Proceedings of the 2015 IEEE International Symposium on Circuits and Systems (ISCAS), Lisabon, Portugal, 27 May 2015; pp. 2025–2028. [CrossRef]
33. EU H2020 Project CONCORDIA: ACybersecurity Competence Network with Leading Research, Technology, Industrial and Public Competences. Available online: https://www.concordia-h2020.eu/ (accessed on 29 May 2020).
34. EU H2020 Project CPSoSaware: Cross-Layer Cognitive Optimization Tools & Methods for the Lifecycle Support of Dependable CPSoS. Available online: http://cpsosaware.eu/ (accessed on 29 May 2020).

35. Pimoroni. Enviro pHAT. Available online: https://github.com/pimoroni/enviro-phat (accessed on 29 May 2020).
36. Ornaghi, A.; Valleri, M. Ettercap Project. Available online: https://www.ettercap-project.org/index.html (accessed on 29 May 2020).

Article

Protection of Superconducting Industrial Machinery Using RNN-Based Anomaly Detection for Implementation in Smart Sensor†

Maciej Wielgosz [1,2,*], Andrzej Skoczeń [3] and Ernesto De Matteis [4]

1 Faculty of Computer Science, Electronics and Telecommunications, AGH University of Science and Technology, al. Adama Mickiewicza 30, 30-059 Cracow, Poland
2 Academic Computer Centre CYFRONET AGH, ul. Nawojki 11, 30-072 Cracow, Poland
3 Faculty of Physics and Applied Computer Science, AGH University of Science and Technology, al. Adama Mickiewicza 30, 30-059 Cracow, Poland; skoczen@agh.edu.pl
4 CERN European Organization for Nuclear Research, CH-1211 Geneva 23, Switzerland; ernesto.de.matteis@cern.ch
* Correspondence: wielgosz@agh.edu.pl; Tel.: +48-12-617-27-92

† This paper is an extended version of "Looking for a Correct Solution of Anomaly Detection in the LHC Machine Protection System" published in the Proceedings of the 2018 International Conference on Signals and Electronic Systems (ICSES), Kraków, Poland, 10–12 September 2018.

Received: 27 October 2018; Accepted: 11 November 2018; Published: 14 November 2018

Abstract: Sensing the voltage developed over a superconducting object is very important in order to make superconducting installation safe. An increase in the resistive part of this voltage (quench) can lead to significant deterioration or even to the destruction of the superconducting device. Therefore, detection of anomalies in time series of this voltage is mandatory for reliable operation of superconducting machines. The largest superconducting installation in the world is the main subsystem of the Large Hadron Collider (LHC) accelerator. Therefore a protection system was built around superconducting magnets. Currently, the solutions used in protection equipment at the LHC are based on a set of hand-crafted custom rules. They were proved to work effectively in a range of applications such as quench detection. However, these approaches lack scalability and require laborious manual adjustment of working parameters. The presented work explores the possibility of using the embedded Recurrent Neural Network as a part of a protection device. Such an approach can scale with the number of devices and signals in the system, and potentially can be automatically configured to given superconducting magnet working conditions and available data. In the course of the experiments, it was shown that the model using Gated Recurrent Units (GRU) comprising of two layers with 64 and 32 cells achieves 0.93 accuracy for anomaly/non-anomaly classification, when employing custom data compression scheme. Furthermore, the compression of proposed module was tested, and showed that the memory footprint can be reduced four times with almost no performance loss, making it suitable for hardware implementation.

Keywords: anomaly detection; recurrent neural networks; neural networks compression; LHC

1. Introduction

The benefit of using superconductivity in industrial applications is well understood. However actual application of superconducting devices is still limited by difficulties in the maintenance of cryogenic stability of superconducting cables and coils. In many cases, it is hard to design safe superconducting circuit. A small mechanical rearrangement releases enough energy to initiate local avalanche process (quench) leading to loss of superconducting state and next to overheating of the machine and then even to melting. Therefore superconducting machines require a special protection system.

In general, such a system consists of data producers and processing servers. The producers are electronic devices located close to individual superconducting machines which require protection. This device (producer) is capable of collecting and processing data and generating the activation signal for local actuators. The servers are capable of storing and analyzing data delivered by producers through the network.

The overarching research goal is to improve data processing within the producers and to move a part of the analysis task into them in order to reduce network load. The preliminary results of this research were presented in Ref. [1], which focused on testing the suitability of Recurrent Neural Networks (RNNs) for this application. In this work, the additional aspects for hardware implementation, especially model compression, are explored.

The presented research main contributions are as follows:

- development of a neural algorithm dedicated to detecting anomaly occurring in the voltage time series acquired on the terminals of superconducting machines in electrical circuits,
- design and verification of the complete processing flow,
- introduction of the RNN-based solution for edge computing which paves the way for low-latency and low-throughput hardware implementation of the presented solution,
- development of a system level model suited for future experiments with the adaptive grid-based approach; the software is available online (see Supplementary Material section).

1.1. Protection System for Superconducting Machinery

One of the biggest superconducting systems is installed at the Large Hadron Collider (LHC) accelerator at the European Organization for Nuclear Research (CERN). Despite its scientific purpose, the LHC should be considered as a huge industrial system. The final product of this factory is the total number of particle collisions. During the steady operation, every second, 40,000,000 collisions are performed in three different interaction points. The LHC consists of a chain of superconducting magnets. The chain is located in an underground circular tunnel, 100 m under the Earth's surface. The Figure 1 presents the view on the magnet's chain. The perimeter of the tunnel is about 27 km long. Superconducting magnets, responsible for shaping the beam trajectory, are crucial elements of the accelerator that require permanent monitoring. The details of the design of the LHC accelerator are described in Ref. [2].

When the LHC collides particles, even a tiny fraction of energy stored in each proton beam can cause a magnet to leave the superconducting state. Such an occurrence is named a quench, and it can severely damage the magnets in case of machine protection procedures failure. These procedures mainly relay on triggering the power down of the whole accelerator when resistive part of voltage on one superconducting element exceeds a predefined threshold. This study concerns only a protection system of superconducting elements inside the LHC.

A protection system known as Quench Protection System (QPS) has been installed at the LHC since the beginning and it successfully works since ten years of the LHC operation. The detailed description of the existing system can be found in Refs. [3,4]. The protection unit, visible in Figure 1, installed under the magnet performs acquisition, processing and buffering of samples of voltage existing between terminals of the superconducting coil. These units are end-points of a massive distributed system covering the whole length of the LHC. The acquisition is performed using Analog-to-Digital Converter (ADC). The processing relies on filtering and compensation of inductive voltage. The pure resistive voltage is compared with the threshold, and the output of comparator activates actuators and generates trigger signals for other subsystems of the LHC. The actuators are also included to the yellow rack installed under each magnet in the tunnel (see Figure 1). The task of actuators is to inject energy to the objective coil in order to heat it homogeneously.

Figure 1. The LHC tunnel. The blue cryostat contains superconducting main dipole magnets. The protection unit is visible on the floor under the magnet (yellow rack). The photo taken by A.S. in 2007.

The data is continuously buffered in a circular buffer. Some of the samples (both total and resistive voltage) are directly transmitted to the CERN Accelerator Logging Service (CALS) database in the cloud. The CALS serves permanent monitoring of almost every device in the CERN's accelerator's complex. The sampling rate for this service is very low, in the best case it is 10 Hz (100 ms). The circular buffer consists of two parts. The first part is filled with samples all the time in a circular manner. The second part is filled only in case of triggering or in case of a request sent by an operator. A trigger (or a request) freezes first part of the buffer. Then the whole buffer is transmitted to a cloud and stored in a dedicated database Post Mortem (PM) System. Examples of voltage time series taken from this database are presented in Figure 2. The sampling rate of the PM data is much higher, and in our case, it is 500 Hz (2 ms).

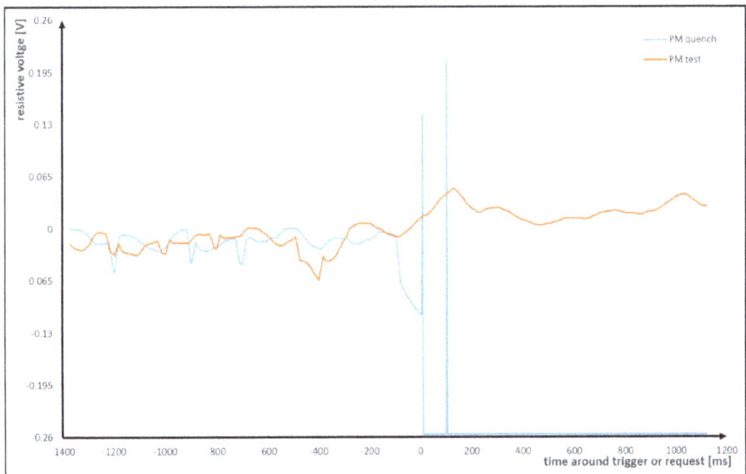

Figure 2. The presentation of a contents of U_{RES} field of two PM data files for one of the superconducting magnets (with electrical current 600 A). The voltage range of the ADC is from −256 mV to 256 mV. Time 0 ms refers to trigger (request) time stored in the field QUENCHTIME in the PM data files.

The presented protection system underwent many upgrades introduced during breaks of the LHC operation. However, the emergence of new superconducting materials opens a question concerning detection algorithms for application in the future protection system again. Such experiments are conducted in SM18 CERN's facility (see Figure 3), currently testing the magnets for the High Luminosity LHC phase.

Figure 3. Test facilities of SM18 for testing MQXFS inner triplet quadrupole magnet, including the rack used for the data acquisition and tests (photos provided by E.M.).

1.2. State of the Art

Anomaly and novelty detection methods have been researched over many years [5–7] which resulted in the development of many successful algorithms. They may be in general divided into three different categories of density-based, distance-based and parametric methods. In addition to the standard procedures, neural algorithms in a majority of cases employing RNN-based architectures [8–11] slowly pave the way to the basic set of the anomaly detection procedures.

Training dataset is different for novelty and outlier detection. In novelty detection, it is not contaminated by anomalies—all outliers need to be removed beforehand. On the contrary, the training procedure of the outlier detection model involves incorporating anomalous data into the training dataset. Both flows employ an unsupervised approach as a training procedure, although some of the procedures may be boosted using some external hyper-parameters such as contamination factor, thresholds or max features that are taken into account in the training process.

In this work three different algorithms were used as a baseline for the RNN-based approach proposed by the authors: Elliptic Envelope, Isolation Forest (IF), and One-Class Support Vector Machine (OC-SVM) [12–18]. Elliptic Envelope belongs to a set of methods with an underlying assumption of known distribution (usually Gaussian) for normal data and all the points distant from the center of the ellipse are considered outliers. The Mahalanobis distance [19] is used as a measure of distance and an indicator that a given data point may be considered as an outlier.

Another useful method, invented in 2008, is the Isolation Forest [20] which performs anomaly detection using the random forest. The underlying concept of this approach is based on an idea of a random selection of features and a random selection of split within the tree nodes between maximum and minimum values of the selected features. A concept of the decision function in IF algorithm defines deviation of an averaged path over a forest of random trees. Random partitioning creates significantly shorter paths for anomalies which results from the fact that outliers are concentrated close to extreme values of features which in turn locates them on the border of the trees.

OC-SVM is usually considered to be a novelty-detection method, and the training data should not contain outliers. It performs well in high-dimensional space where there is no assumption regarding the distribution of the underlying data. However, if the data is well distributed (e.g., Gaussian) the Isolation Forest or Elliptic Envelope may perform better.

The presented methods assume the spatial structure of the data with no temporal relationships. There is a set of methods such as ESD, ARIMA, Holt-Winters [21,22] which take time component into account and have proven to be effective. However, due to their complexity, it is challenging to implement them in very low-latency systems in hardware.

When detecting anomalies in time series, RNNs are both scalable and adaptable which is critical when it comes to design and implementation of complex anomaly detection systems [23–25]. RNNs were introduced long ago [26] and have been successfully applied in many domains [8], according to the authors' knowledge, there are no studies on the performance of compressed RNNs in anomaly detection. Nevertheless, several works on effective quantization and compression of RNNs are available [27,28]. Consequently, decision was made to explore the feasibility and performance of several compression techniques of RNNs in low-latency anomaly detection domain of LHC machinery monitoring. Furthermore, the adopted approach addresses both data and coefficients quantization with in-depth analysis of correlation of employing different techniques.

2. Materials and Methods

2.1. Quantization Algorithm

2.1.1. Previous Work

In the authors' previous works concerning superconducting magnets monitoring Root-Mean-Square Error (RMSE) [24] and both *static* [23] and *adaptive* data quantization [1,25] approaches were used. Based on experiments conducted and described therein, a conclusion can be drawn that RNNs can be used to model magnets behavior and detect anomalous occurrences.

The initially introduced RMSE approach had several drawbacks, the main of which was a necessity to select an arbitrary detection threshold. In Ref. [23], the *static* quantization was used, mapping the input data into a set of m equal-width bins. This method, however, resulted in sub-par results, stemming from the uneven distribution of the samples in the bins, up to the point where nearly all samples occupied only one or two bins (see *static* samples counts in Figure 4).

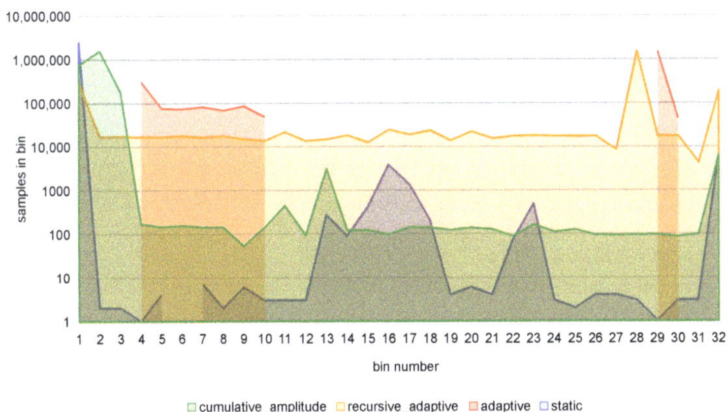

Figure 4. Samples per bin for PM dataset U_{RES} channel ($m = 32$). Note the logarithmic scale.

In Ref. [25], an approach based on *adaptive* data quantization and automatic thresholds selection was introduced. *Adaptive* data quantization resulted in much better use of bins and consequently significantly improved the accuracy results. Its principle of operation is mapping the input space to a fixed number of categories (bins) in such a way, that all categories have (ideally) the same samples cardinality (see Appendix A.1 for more details). Resulting bins widths are uneven, explicitly adjusted to each of the input signal channels (see *adaptive* samples counts in Figures 4 and 5 to compare bin edges generated with various approaches).

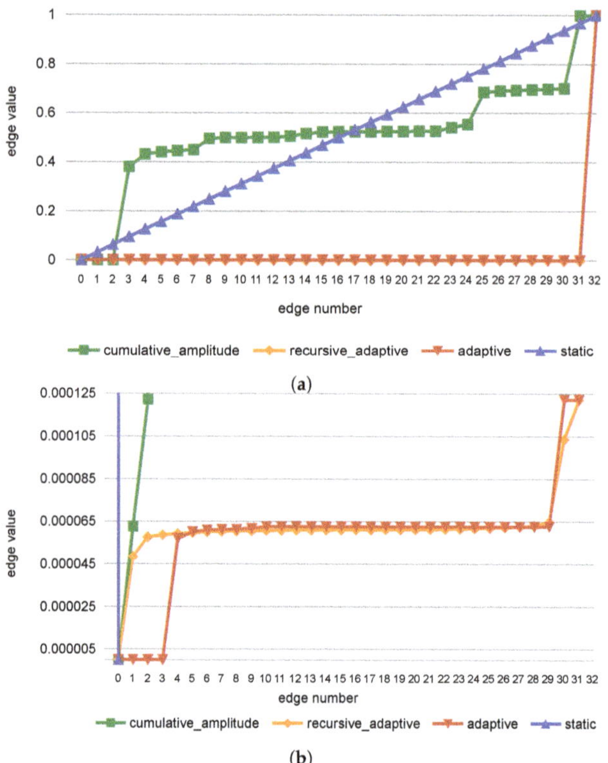

(**b**)

Figure 5. Full (**a**) and zoomed-in (**b**) bin edges for PM dataset U_{RES} channel ($m = 32$). Please note that *adaptive* quantization algorithm effectively yields only 10 bins, since some edges values occur multiple times.

2.1.2. Other Quantization Approaches

The drawback of *adaptive* algorithm is that it can effectively generate fewer bins than requested when some values occur in the dataset in significant numbers (see Figure 5). To mitigate this effect, which became apparent when working with PM data, a modification of *adaptive* algorithm, called *recursive_adaptive*, was introduced. The $m + 1$ initially found edges are treated as candidates, and if duplicates are detected, the recursive process is started. At first, the duplicated edges are added to the final edges list, and all repeating values are removed from the dataset. Then, the remaining data is used as an entry point to find $m + 1$–*number of final edges* new edge candidates. The process is repeated until there are no duplicates in candidate edges or there is no more data left in the dataset. As a result of this process, the bins are more evenly used.

An alternative edge-finding algorithm, called *cumulative_amplitude*, is based on the idea of equalizing the sum of the samples amplitudes in each bin. As in *adaptive* algorithm, before edges selection, the samples are normalized and sorted. Then, the threshold value is computed as a sum of amplitudes of samples left in the dataset divided by the required edges number (see Appendix A.2 for equations). Contrary to the *adaptive* approach of determining the edges based on the samples count, in *cumulative_amplitude* the edge value is chosen when the sum of samples' amplitudes crosses this threshold. As a result, the maximal values are not grouped with smaller ones. It may, however, level the differences between smaller values, that may contain crucial information. In the implementation, the concept described above was modified to also use recursive duplication removal, with the threshold value determined anew for each recursion level.

2.2. Implementation Overview

The presented anomaly detection system was created in Python, using Keras [29] library, with both Theano [30] and Tensorflow [31] backends depending on availability. The reference methods were implemented using scikit-learn [32] library. It is prepared to work with normalized data, with all the available data (both training and testing) used during the normalization process. The focal system modules and data flow can be seen in Figure 6.

The number of input categories (`in_grid`), the bins' edges calculation algorithm (`in_algorithm`), the history window length (`look_back`), the model and its hyper-parameters used during the anomaly detection process and other options are specified in the configuration file. The particular setup is also saved while results are reported, ensuring the particular test environment can be recreated even if configuration included several possible values. Each of the input channels is quantized using the same grid/algorithm combination.

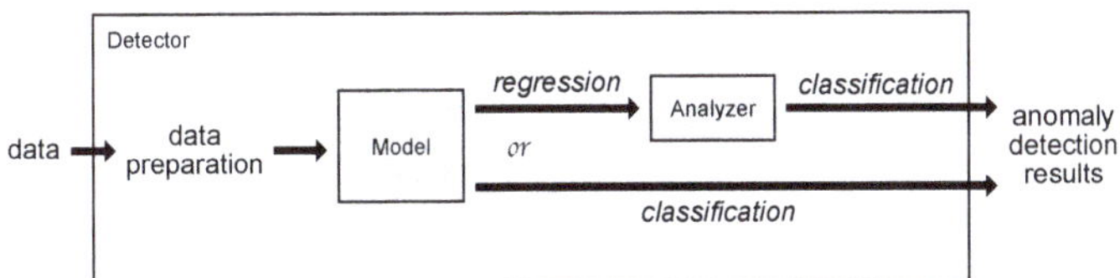

Figure 6. High-level system architecture. © 2018 IEEE. Reprinted, with permission, from Wielgosz, M.; Skoczeń, A.; Wiatr, K. Looking for a Correct Solution of Anomaly Detection in the LHC Machine Protection System. 2018 International Conference on Signals and Electronic Systems (ICSES), 2018, pp. 257–262 [1].

The model is an abstraction layer over the actual classifier. Currently implemented models include Random (for baseline testing), Elliptic Envelope, Isolation Forest, OC-SVM, Long Short-Term Memory (LSTM) and Gated Recurrent Unit (GRU).

Depending on the system configuration (Figure 6), some models can be used for either classification or regression. In the regression mode, the model is trained on data without anomalies and yields the output that needs to be further processed by the analyzer to obtain anomaly detection results.

In the classification mode, used in the experiments presented in this paper, the model is trained using data containing anomalies (except for models belonging to the novelty detection category). Instead of trying to predict the next (quantized) value, it directly classifies the sample as either anomaly or not.

The models can be roughly divided into the ones working with either spatial (Elliptic Envelope, Isolation Forest, OC-SVM) or temporal data (LSTM, GRU). Relevant data preprocessing and structuring, as well as model training and testing, is coordinated by one of the possible detectors. The model/detector combination used in particular setup is defined in configuration file. This option ensures the system extensibility since the detector does not need to know about all possible models in advance.

Currently, the experiments are carried out using the software implementation of the system. The target system, however, will need to be implemented in hardware to ensure it complies with latency requirements. To fit the Neural Network (NN) model onto the Field-Programmable Gate Array (FPGA) or Application-Specific Integrated Circuit (ASIC) board, it needs to be compressed while retaining the high accuracy (Figure 7). The ready module can potentially be used as a stand-alone detector or in conjunction with the currently used system (Figure 8).

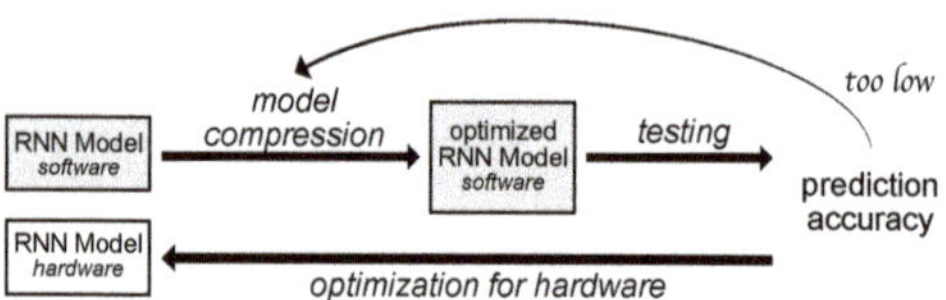

Figure 7. Design flow for hardware implementation. © 2018 IEEE. Reprinted, with permission, from Wielgosz, M.; Skoczeń, A.; Wiatr, K. Looking for a Correct Solution of Anomaly Detection in the LHC Machine Protection System. 2018 International Conference on Signals and Electronic Systems (ICSES), 2018, pp. 257–262 [1].

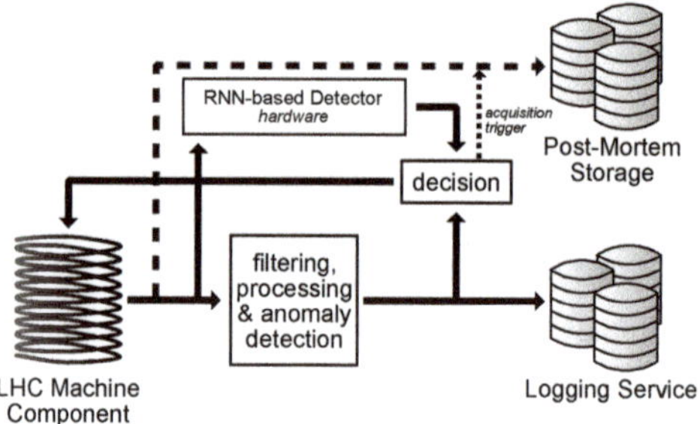

Figure 8. Proposed system. © 2018 IEEE. Reprinted, with permission, from Wielgosz, M.; Skoczeń, A.; Wiatr, K. Looking for a Correct Solution of Anomaly Detection in the LHC Machine Protection System. 2018 International Conference on Signals and Electronic Systems (ICSES), 2018, pp. 257–262 [1].

2.3. Model Complexity Reduction

Deep Learning models have a range of features which render them superior to other similar Machine Learning models. However, they usually have high memory footprint as well as require substantial processing power [33]. Computing requirements are especially crucial when it comes to embedded implementation of Deep Learning models in edge processing nodes like in the case of the system described in this paper. Fortunately, there are multiple ways to mitigate these issues and preserve all the benefits of the models, for example by using techniques such as pruning and quantization [28].

During a quantization, a floating-point number x from a quasi-continuous space of IEEE-754 notation is mapped to fixed-point value q, represented using **total** bits. The **total** is conventionally equal to 8 or 16, which are bit-widths supported by GPUs and the latest embedded processors. For FPGA and ASIC it is, however, possible to use an arbitrary number of bits. The quantization is done separately for each layer's weights $\mathcal{W}$.

2.3.1. Linear Quantization

The *linear* quantization used during experiments can be described by the following equation:

$$q = s \cdot \text{clip}\left(\left\lfloor \frac{x}{s} + \frac{1}{2} \right\rfloor, 1 - 2^{\textbf{total}-1}, 2^{\textbf{total}-1} - 1\right), \tag{1}$$

where scaling factor:

$$s = \frac{1}{2^{\textbf{total}-1-\lceil \log_2 \max(\mathcal{W}) \rceil}} \tag{2}$$

and clipping function:

$$\text{clip}(a, min, max) = \begin{cases} min & \text{if } a < min, \\ a & \text{if } min \leq a \leq max, \\ max & \text{otherwise.} \end{cases} \tag{3}$$

2.3.2. MinMax Quantization

The *minmax* quantization used during experiments can be described by the following equation:

$$q = s \cdot \left\lfloor \frac{x - \min(\mathcal{W})}{s} + \frac{1}{2} \right\rfloor + \min(\mathcal{W}), \tag{4}$$

where scaling factor:

$$s = \frac{\max(\mathcal{W}) - \min(\mathcal{W})}{2^{\text{total}} - 1}. \tag{5}$$

Also tested was *log_minmax* quantization, where:

$$q = \operatorname{sign}(x) \cdot e^{minmax(\ln|x|)} \tag{6}$$

2.3.3. Hyperbolic Tangent Quantization

The *tanh* quantization used during experiments can be described by the following equation:

$$q = \operatorname{arctanh}\left(s \cdot \left\lfloor \frac{\tanh(x) + 1}{s} + \frac{1}{2} \right\rfloor - 1\right), \tag{7}$$

where scaling factor:

$$s = \frac{2}{2^{\text{total}} - 1}. \tag{8}$$

The main idea behind coefficients quantization is using a dynamic range of the available number representation to its fullest, and meet the requirements of the hardware platform to be used for deploying the system at the same time. In our implementation, the so-called dynamic fixed-point notation was used. It allows emulating fixed-point number representation using floating point container. It is worth noting that most of edge computing platforms require linear quantization due to the fixed, a priori defined size of internal registers and arithmetic processing elements. FPGA and custom-designed ASIC which this work is targeting have no such limitation.

3. Results

A series of experiments were conducted to practically examine performance of the proposed methods. Different configurations of the module setup were used in order to expose impact of different parameters of the proposed algorithm on the overall performance of the anomaly detection system. Furthermore, the performance of the proposed solution was compared with a range of state-of-the-art algorithms.

3.1. Dataset

The dataset used in the experiments contained 2500 series retrieved from PM database, with 64-16-20 training-validation-testing split. 2415 of those series had a length of 1248 samples, while the remaining 85 series had a length of 1368 samples. For each of the series, the four input channels were available:

U_{DIFF}—total voltage measured between terminals of superconducting coil,
U_{RES}—resistive voltage extracted from the total voltage U_{DIFF} using the electric current I_{DCCT},
I_{DCCT}—current flowing through superconducting coil measured using Hall sensor, and
I_{DIDT}—time derivative of the electric current I_{DIDT} calculated numerically.

Anomalies were marked based on the value of QUENCHTIME field found in PM data, with each anomaly starting at the indicated point and continuing until the end of a series. As such, the data can be considered to be weakly labelled. 874 training and 225 testing series contained anomalies and over 26% of samples in the dataset were marked. Over 84% of the anomalies had a length of 750 samples, over 7%—566, and over 4%—1320. The length of the remaining anomalies varied between 214 and 908 samples.

Before the start of experiments, the data was normalized. The example data series (and results) visualizations can be seen in Figures 9 and 10. Even in just those two figures, it can be seen that the quenches vary in shape and it is really difficult to find apparent similarities just by visual examination. This makes tasks of data labeling and manual verification of the detection results not feasible without heavy experts involvement.

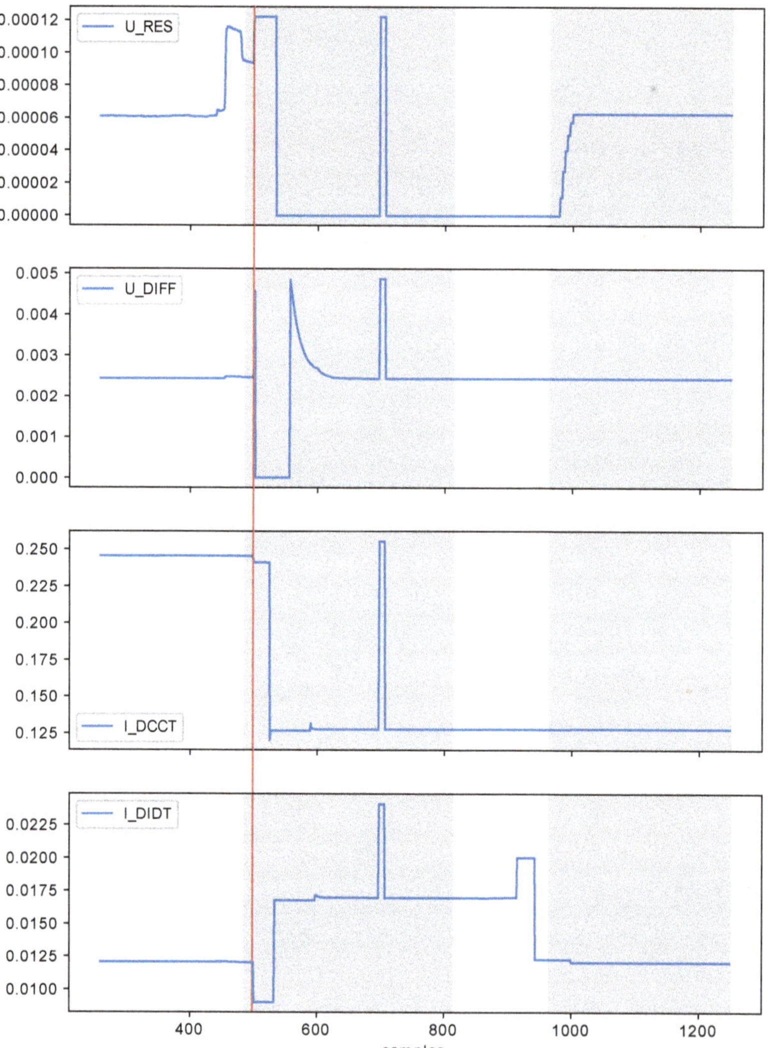

Figure 9. Example single series results visualization (`in_grid` = 32, `in_algorithm` = *adaptive*, `look_back` = 256). Red line across all subplots marks the QUENCHTIME and gray spans indicate the anomalies found by the system.

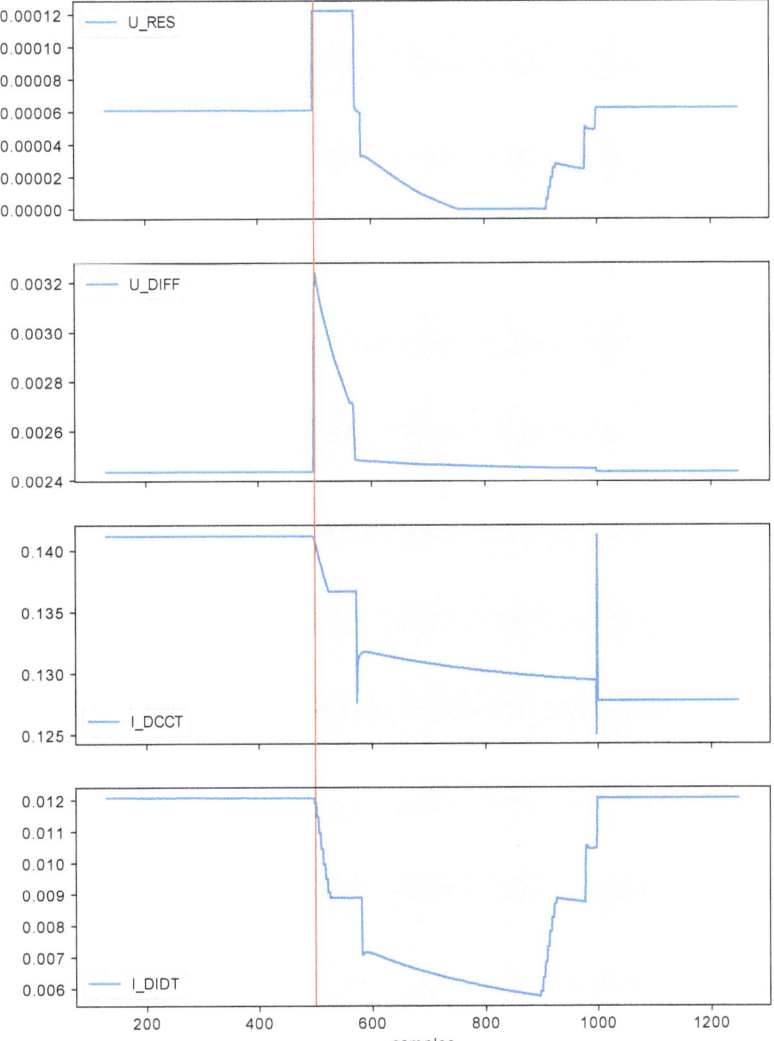

Figure 10. Example single series results visualization (`in_grid` = 32, `in_algorithm` = *recursive_adaptive*, `look_back` = 128). Red line across all subplots marks the QUENCHTIME and gray spans indicate the anomalies found by the system.

3.2. Quality Measures

During the experiments two main quality measures were used: F-measure and accuracy. While F-measure is better suited to evaluate the results of anomaly detection, in case of PM data the relative lack of imbalance between anomalous and normal samples (especially factoring in the required history length) makes the accuracy also a viable metric.

Additionally, the NN models quantization results are usually measured in terms of accuracy, so its usage allows to relate our results with others found in literature. For example, the drop in accuracy resulting from quantization should be no higher than one percentage point [34].

An accuracy can be defined as:

$$\text{accuracy} = \frac{tp + tn}{tp + tn + fp + fn}, \tag{9}$$

where:

- tp—true positive—item correctly classified as an anomaly,
- tn—true negative—item correctly classified as a part of normal operation,
- fp—false positive—item incorrectly classified as an anomaly,
- fn—false negative—item incorrectly classified as a part of normal operation.

An F-measure is calculated using two helper metrics, a recall (10), also called sensitivity, and a precision, also called specificity (11):

$$\text{recall} = \frac{tp}{tp + fn}, \tag{10}$$

$$\text{precision} = \frac{tp}{tp + fp}. \tag{11}$$

The β parameter controls the recall importance in relevance to the precision when calculating an F-measure:

$$\text{F}_\beta = (1 + \beta^2) \cdot \frac{\text{recall} \cdot \text{precision}}{\text{recall} + \beta^2 \cdot \text{precision}}. \tag{12}$$

During the experiments two β values were used, 1 and 2, to show the impact of the recall on the final score. Recall as a quality assessment measure reflect an ability of an algorithm to find all entities. On the other hand precision, describes several found entities were correctly classifier. Those measures have to some extent opposite effect on each other. This means that raise of precision usually leads to a drop of recall and vice-versa.

The Receiver Operating Characteristic (ROC) curve is a graph used to analyze the operation of the binary classifiers as the one presented in this work. It shows the performance of the model taking into account all classification thresholds. True Positive Rate (recall) is plotted as a function of False Positive Rate ($1 -$ precision). When a classification threshold is lowered, the classifier tends to classify more input data items as positive, which leads to an increase of both fp and tp.

To derive quantitative conclusions from ROC curve Area Under Curve (AUC) may be employed. It measures the whole two-dimensional area under the ROC curve. It may be considered as an integral operation performed from points (0,0) to (1,1) on a ROC graph. AUC values fall into a range between 0 and 1. A model whose predictions are 100% wrong has an AUC of 0.0, the one which works perfectly hasAUC of 1.0.

3.3. *History Length and Data Quantization*

The initial experiments attempted to determine the impact of history length (`look_back`) and quantization levels (`in_grid`) on RNN models performance. Models were trained on full dataset and four channels for 7 epochs, with batch size equal to 16,384.

Use of dynamic range of the data representation is one of the most important indicators of the quantization algorithm effectiveness since it affects the potential information loss due to the lack of a proper representation capacity, i.e., 'wasting' resources on empty bins, while other bins contain 'too many' of the values or the number of bins could be reduced altogether. Figure 11 shows that recursive and cumulative adaptive approaches provide full grid use in contrast to adaptive quantization method which exhibits significant grid underuse.

Experiments with different sizes of `in_grid` were conducted and it turned out that this parameter has a very little impact on the performance of the model (see Figure 12). Aside from `in_grid` $= 8$, which universally performs the worst regardless of the used algorithm, other values yield similar results starting with `look_back` $= 128$. This allows reducing the size of the input (`in_grid`) to 32 which can be encoded using 5 bits. The biggest impact on the performance has `look_back` which was presented in Figures 13–15 and Table 1. The models with `look_back` of 512 reach AUC close to 0.98 and significantly outperform the models with `look_back` of 16. It also can be seen that in current tests only setups with `look_back` = 256 and `look_back` = 512 were capable of reaching the *recall* = 1, ensuring all anomalies were found, while retaining high precision. The avoidance of false negatives is crucial in this use case, since quench after-effects, resulting in the equipment destruction, can be both dangerous and extremely costly.

It is also worth to keep in mind that the developed model is supposed to work in highly demanding environment where response latency is a critical factor which decides how much time other sections of the global protection system have to execute their procedures. Thus the work needs to be done towards reducing discrepancy in AUC value between setups of different `look_back`. For instance, AUC for `look_back` = 64 is 0.87 and for `look_back` = 256 equals to 0.97 (see Figure 13).

Table 1. The parameters of NN built with GRU cells for three different algorithms (two layers, 64 and 32 cells + Dense, `in_grid` = 32).

In_Algorithm	Look_Back	Accuracy	F_1 Score	F_2 Score
adaptive	16	0.8462	0.6722	0.6167
	32	0.8506	0.7031	0.6687
	64	0.8611	0.7376	0.7124
	128	0.8838	0.7973	0.7835
	256	0.9162	0.8743	0.8796
	512	0.9543	0.9474	0.9522
recursive_adaptive	16	0.8507	0.6920	0.6481
	32	0.8543	0.7022	0.6561
	64	0.8652	0.7350	0.6928
	128	0.8868	0.8040	0.7939
	256	0.9172	0.8746	0.8749
	512	0.9571	0.9506	0.9560
cumulative_amplitude	16	0.8436	0.6609	0.5999
	32	0.8473	0.6664	0.5968
	64	0.8562	0.7115	0.6620
	128	0.8853	0.7927	0.7622
	256	0.9231	0.8830	0.8805
	512	0.9669	0.9625	0.9779

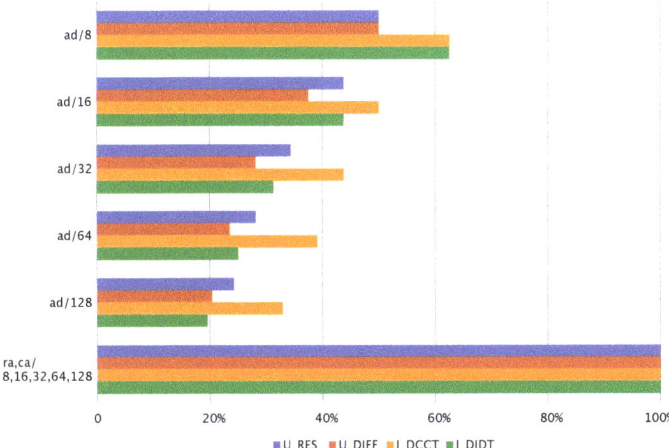

Figure 11. Grid use for various `in_grid` values and `in_algorithm`. ad—*adaptive*, ra—*recursive_adaptive*, ca—*cumulative_amplitude*.

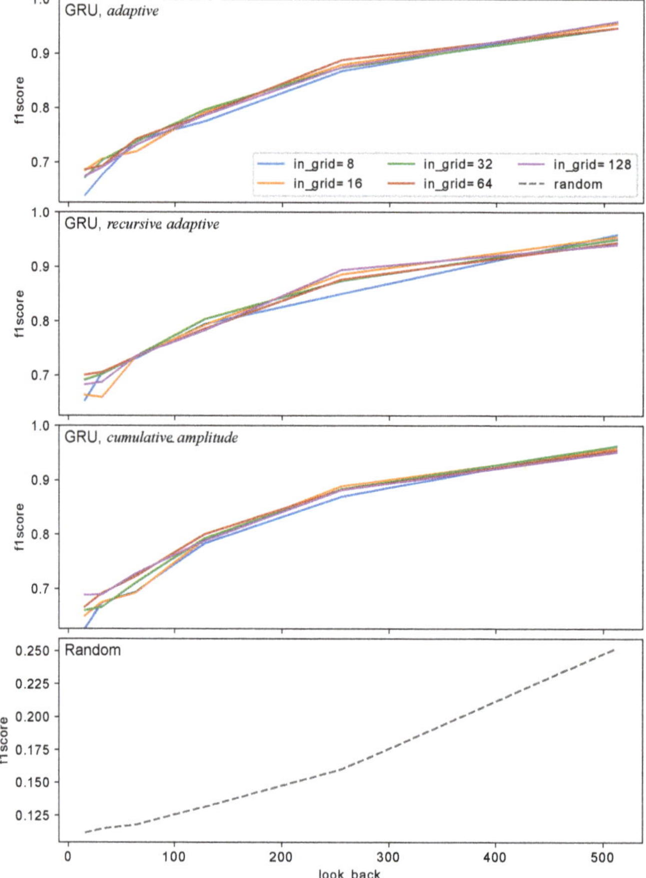

Figure 12. F_1 score as a function of `look_back` for several `in_grid` and `in_algorithm` values. Dashed line shows Random baseline model performance for the same `look_back`.

The comparative tests were run using full training set (`samples_percentage` = 1). The goal was to study the behavior of the RNN-based methods and other classic anomaly detection methods with respect to the quantization algorithm. Most of the tests were run using single U_{RES} input channel (Table 2), with additional experiments using four input channels run for RNN-based methods (Table 3).

As a baseline, the Random model was used. It generates predictions by respecting the training set's class distribution ("stratified" strategy). Since it ignores the input data entirely, its results do not depend on the `in_algorithm` choice. This model performance turned out to be at the level of 0.6334.

For the other models' tests, based on the previous experiments, the values of `look_back` = 256 and `in_grid` = 32 were selected as providing a good tradeoff between resources consumption and the results quality. For the spatial methods, the history vector was flattened. The amount of contamination (the proportion of outliers in the data set) was calculated based on the whole training set and passed to the methods. For the Elliptic Envelope, all points were included in the support of the raw Minimum Covariance Determinant (MCD) estimate. The OC-SVM was tested with both 332 Radial Basis Function (RBF) and linear kernel, with RBF kernel coefficient (γ) equal to 0.1 and an upper bound on the fraction of training errors and a lower bound of the fraction of support vectors (ν) equal to $0.95 * contamination + 0.05$. Each of the RNNs had the same testing architecture (two layers, with 64 cells in the first one and 32 cells in second, followed by fully connected layer) and was trained for seven epochs.

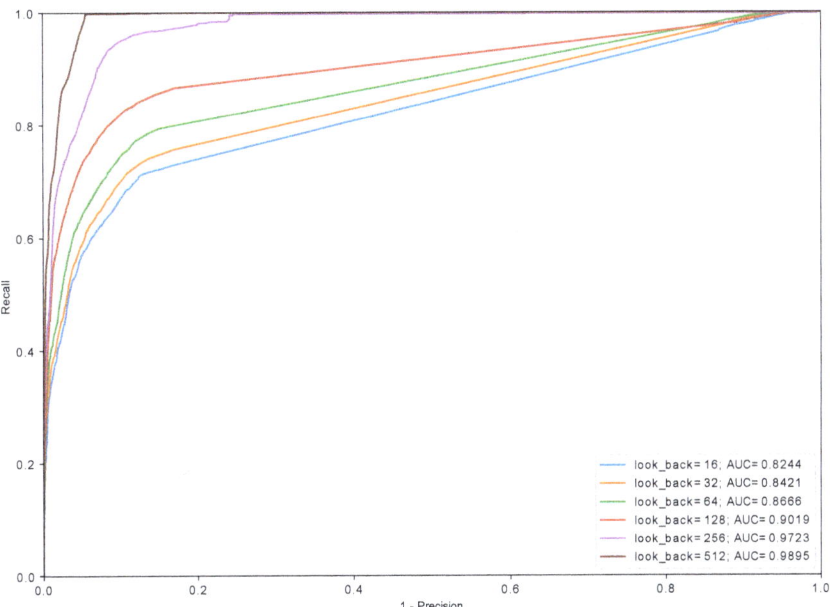

Figure 13. The ROC curve for algorithm *adaptive* (`in_algorithm` = *adaptive*, `in_grid` = 32, GRU (two layers, 64 and 32 cells) + Dense).

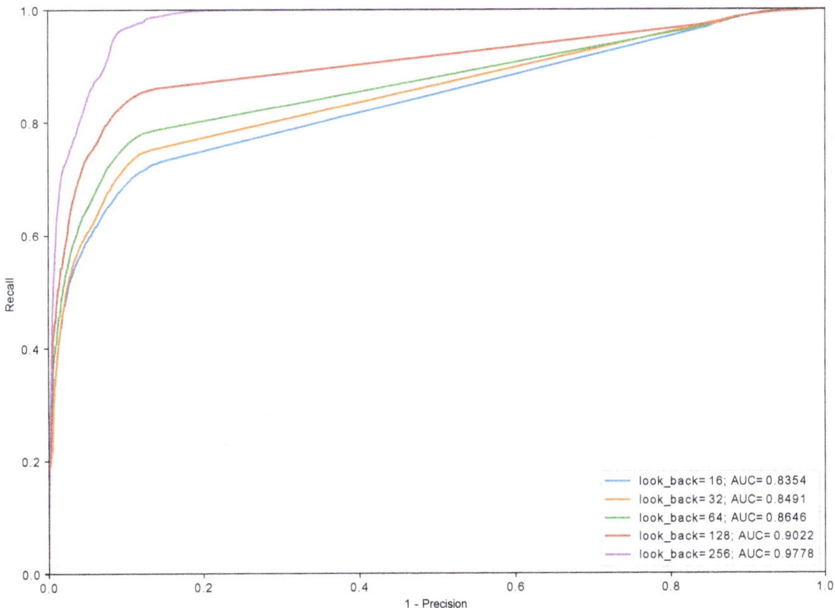

Figure 14. The ROC curve for algorithm *recursive_adaptive* (`in_algorithm` = *recursive_adaptive*, `in_grid` = 32, GRU (two layers, 64 and 32 cells) + Dense).

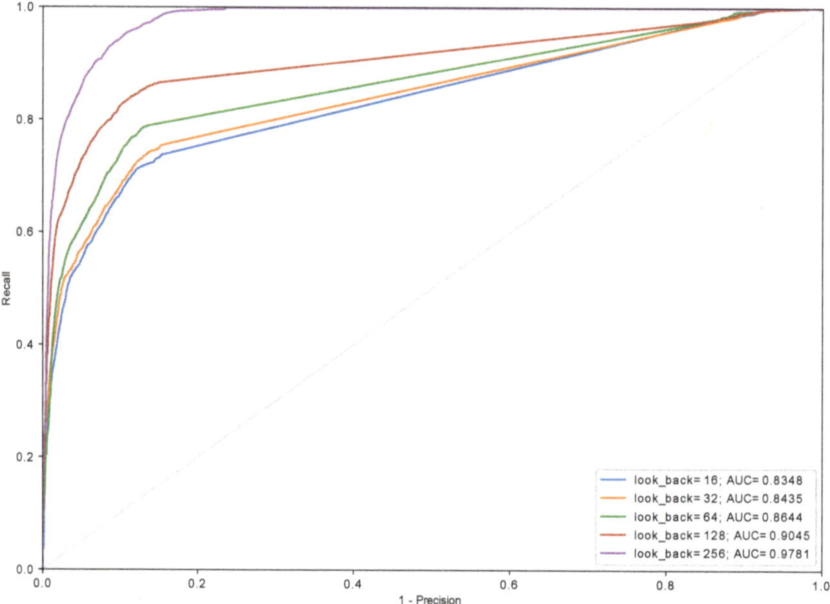

Figure 15. The ROC curve for algorithm *cumulative_amplitude* (`in_algorithm` = *cumulative_amplitude*, `in_grid` = 32, GRU (two layers, 64 and 32 cells) + Dense).

Table 2. Testing accuracy (20% of dataset). All models were run with `in_grid` = 32 and `look_back` = 256, using single input channel (U_{RES}), NNs were trained for 7 epochs. The best result is marked in bold.

Model	In_Algorithm		
	Adaptive	*Recursive_Adaptive*	*Cumulative_Amplitude*
Random (stratified)	0.6334	0.6334	0.6334
Elliptic Envelope	0.6700	0.7775	0.6700
Isolation Forest	0.7947	0.7596	0.8094
OC-SVM (RBF kernel)	0.3300	0.8232	0.3300
OC-SVM (linear kernel)	0.2959	0.7881	0.2528
GRU (two layers, 64 and 32 cells)	0.8928	**0.9005**	0.8842
LSTM (two layers, 64 and 32 cells)	0.8271	0.8552	0.7402

Table 3. Testing accuracy (20% of dataset). Models were run with `in_grid` = 32 and `look_back` = 256, using four input channels (U_{RES}, U_{DIFF}, I_{DCCT}, I_{DIDT}), NNs were trained for 7 epochs. The best result is marked in bold.

Model	In_Algorithm		
	Adaptive	*Recursive_Adaptive*	*Cumulative_Amplitude*
GRU (two layers, 64 and 32 cells)	0.9235	**0.9300**	0.8842
LSTM (two layers, 64 and 32 cells)	0.9194	0.9092	0.9023

It is worth noting that the presented methods such as OC-SVM, Isolation Forest, or Elliptic Envelope are more sensitive to the data distribution and perform well for specific kind of underlying data (i.e., specific distribution or feature extraction which can bring the original data to this distribution before the methods are applied). In the case of the presented CERN data, the distribution is not always stable (varies across setups and magnets). It is also worth noting that the distribution as such does not tell much about temporal aspects of the analyzed data. It may happen that the signals of the same distribution have different temporal shape. This is even more pronounced for more temporarily complex signals (see Figures 9 and 10).

3.4. Coefficients Quantization

Table 4 shows results of coefficients quantization for the neural models from Tables 2 and 3 using several different methods. For all methods, the quantization above the ten bits yields results nearly identical to original. A significant drop in accuracy is observed below 8 bits of representation. It may be noted (Table 3) that for lower number of bits the performance oscillates between ≈0.7 and ≈0.3 which means classifying all the features as one category.

Table 4. Coefficients Quantization Results for GRU (two layers, 64 and 32 cells) + Dense, trained on four input channels. Accuracy as a function of bit-width.

Bits	Method	In_Algorithm		
		Adaptive	*Recursive_Adaptive*	*Cumulative_Amplitude*
Original Model		**0.9235**	**0.9300**	**0.8842**
10	*linear*	0.9236	0.9287	0.8841
	minmax	0.9233	0.9300	0.8841
	log_minmax	0.9235	0.9298	0.8842
	tanh	0.9232	0.9283	0.9232
9	*linear*	0.9236	0.9279	0.8838
	minmax	0.9237	0.9295	0.8842
	log_minmax	0.9231	0.9293	0.8843
	tanh	0.9219	0.9260	0.8842
8	*linear*	0.9206	0.9257	0.8830
	minmax	0.9238	0.9311	0.8838
	log_minmax	0.9207	0.9283	0.8844
	tanh	0.9161	0.9143	0.8836
7	*linear*	0.9177	0.3989	0.8850
	minmax	0.9194	0.9250	0.8841
	log_minmax	0.9218	0.9236	0.8833
	tanh	0.9131	0.9033	0.8851
6	*linear*	0.8952	0.9008	0.8871
	minmax	0.9144	0.8839	0.8842
	log_minmax	0.9111	0.9076	0.8844
	tanh	0.8702	0.8782	0.8788
5	*linear*	0.3722	0.8442	0.8802
	minmax	0.9031	0.9058	0.8810
	log_minmax	0.3948	0.8878	0.8812
	tanh	0.8247	0.3306	0.8670
4	*linear*	0.8500	0.2745	0.8587
	minmax	0.8678	0.8702	0.8775
	log_minmax	0.8649	0.3848	0.8734
	tanh	0.7491	0.8464	0.3017
3	*linear*	0.7928	0.8135	0.8190
	minmax	0.3391	0.7900	0.8530
	log_minmax	0.7664	0.8023	0.8564
	tanh	0.6922	0.2833	0.7985
2	*linear*	0.3006	0.6700	0.7065
	minmax	0.7371	0.3391	0.3466
	log_minmax	0.7908	0.7369	0.3110
	tanh	0.7216	0.7549	0.2309
1	*linear*	0.6700	0.3300	0.3300
	minmax	0.6706	0.7003	0.6717
	log_minmax	0.7171	0.7459	0.2121
	tanh	0.7171	0.7459	0.2121

4. Discussion and Conclusions

The protection system for superconducting machinery existing at the LHC is a vast distributed system installed around the whole circular tunnel. It consists of many individual units connected with a dedicated network. The approach used to design protection units is hard to scale and requires laborious manual adjustment of working parameters. The presented work explores the possibility of using the RNN to build a protection device of a new generation. The idea is to perform on-line analysis inside local protection unit using data acquired with a much higher sampling rate without sending such a massive amount of data to the cloud.

One of the main advantages of the proposed methodology is the simplicity of the parameters setting and adjustment through a complete workflow. Only the model architecture and data quantization levels need to be selected, and even those can be automatically optimized. Additionally, since the solution is based on NNs, it can be extended (scaled) to use more sensors (or data streams) or even incorporate text tags (present in many cases in historical CERN superconducting magnets data), while keeping the overall architectural design the same, even for various types of magnets. It is also possible to fine-tune or retrain RNN-based modules when data has changed (e.g., underlying architecture was modified or aged) when in the traditional approach it would require reconsideration and restructuring of the existing solution. This architectural uniformity makes it a good candidate for implementation in a distributed edge-computing cluster of sensors, trained in an end-to-end fashion. Such a holistic approach can significantly reduce overall resources consumption, latency, and throughput.

The conducted experiments showed that using large `look_back` significantly boost the performance of the model, while the number of quantization levels (`in_grid`) as low as 32 is sufficient for the task. The framework demonstrated to be capable of achieving 93% of testing accuracy for GRU (two layers, 64 and 32 cells). The accuracy results are affected by the weak labeling of the data, e.g., sometimes the system labels as anomalous samples occurring for a bit before QUENCHTIME marker. Such results lower the accuracy, while in fact being the desired outcome. The proposed system also often creates a 'gap' in the anomaly, at which point the system shutdown signal would already be sent (such example can be seen in Figure 9). Considering that, the availability of the data manually labeled by experts could improve the system performance.

The coefficients quantization level should also be considered a meta-parameter of the model optimization. Experiments showed that the selection of any value equal to or above 8 bits does not lead to noticeable performance degradation. Careful choice of the quantization level may allow reducing memory footprint even more; however, it must be noted that below 8 bits the accuracy of the model oscillates.

Overall, due to the relatively small size of the neural models and the possibility of significantly reducing their memory footprint (4×) with a minimal performance loss, the presented model is a good candidate for hardware implementation in FPGA or ASIC.

Supplementary Materials: The software of presented anomaly detection system is available online at https://bitbucket.org/maciekwielgosz/anomaly_detection.

Author Contributions: Conceptualization, M.W., A.S. and E.D.M.; methodology, M.W.; software, M.W.; writing—original draft preparation, M.W. and A.S.; writing—review and editing, M.W. and A.S.; resources, M.W.; visualization, M.W., data curation, A.S. and E.D.M.; supervision, M.W.; funding acquisition, E.D.M.

Funding: This work was partially financed (supported) by the statutory tasks of Faculty of Physics & Applied Computer Science and Faculty of Computer Science, Electronics & Telecommunications of AGH University of Science and Technology in Cracow (Poland), within the subsidy of Polish Ministry of Science and Higher Education. The publication fee was funding by CERN.

Acknowledgments: We give special thanks to Andrzej Siemko the Group Leader of Machine Protection and Electrical Integrity Group (MPE) at Technology Department (TE) at CERN and Reiner Denz the Section Leader of Electronics for Protection Section (EP) at MPE-TE at CERN. They spend their time on long discussions and revealed a great interest in our investigations.

Conflicts of Interest: The authors declare no conflict of interest.

Abbreviations

The following abbreviations are used in this manuscript:

ADC	Analog-to-Digital Converter
ASIC	Application-Specific Integrated Circuit
AUC	Area Under Curve
CALS	CERN Accelerator Logging Service
CERN	European Organization for Nuclear Research
EP	Electronics for Protection Section
FPGA	Field-Programmable Gate Array
GRU	Gated Recurrent Unit
IF	Isolation Forest
LHC	Large Hadron Collider
LSTM	Long Short-Term Memory
MCD	Minimum Covariance Determinant
MPE	Machine Protection and Electrical Integrity Group
NN	Neural Network
OC-SVM	One-Class Support Vector Machine
PM	Post Mortem
QPS	Quench Protection System
RBF	Radial Basis Function
RMSE	Root-Mean-Square Error
RNN	Recurrent Neural Network
ROC	Receiver Operating Characteristic
TE	Technology Department

Appendix A. Data Quantization

This section presents the equations that can help to better understand the data quantization performed during experiments.

Appendix A.1. Adaptive Data Quantization

Adaptive data quantization principle of operation is mapping the input space to a fixed number of categories (bins) in such a way, that all categories have (ideally) the same samples cardinality as described by Equations (A1) and (A3) (see Table A1 for notation).

$$S_{norm} \xrightarrow{\Pi_{qa}(m)} S_{qa} : \{0 \ldots m-1\}^{1\times n}, \tag{A1}$$

$$\Pi_{qa}(m) : \bigwedge_{x\in S_{norm}} \bigvee_{y\in S_{qa}} y = \begin{cases} \mathbf{edges}_y \leqslant x \cdot m < \mathbf{edges}_{y+1} \\ \qquad\qquad\qquad\qquad \text{if } x < 1 \\ y = m-1 \text{ if } x = 1, \end{cases} \tag{A2}$$

$$\mathbf{edges} : \bigwedge_{0\leqslant y\leqslant m} \mathbf{edges}_y = \begin{cases} 0 \text{ if } y = 0 \\ \mathbf{srt\_samples}_{y \cdot \lceil \frac{n}{m} \rceil} \\ \qquad \text{if } 0 < y < m \\ 1 \text{ if } y = m. \end{cases} \tag{A3}$$

Appendix A.2. Cumulative Amplitude Data Quantization

Cumulative_amplitude method is based on the idea of equalizing the sum of the samples amplitudes in each bin. As in *adaptive* algorithm, before edges selection, the samples are normalized and sorted. Then, the threshold value Θ_m is computed as a sum of amplitudes of samples left in the dataset divided by the required edges number (A4).

$$\Theta_m = \frac{\sum \textbf{srt\_samples}}{m}, \tag{A4}$$

$$\textbf{edges} : \bigwedge_{0 \leqslant y \leqslant m} \textbf{edges}_y = \begin{cases} 0 \text{ if } y = 0 \\ \textbf{srt\_samples}_{\text{idx}(y)} \\ \qquad \text{if } 0 < y < m \\ 1 \text{ if } y = m, \end{cases} \tag{A5}$$

$$\text{idx}(y) = \begin{cases} -1 \text{ if } y = 0 \\ \min(k) : \left(\sum_{i=\text{idx}(y-1)+1}^{k} |\textbf{srt\_samples}_i| \right) \geqslant \Theta_m \\ \qquad \text{if } 0 < y < m. \end{cases} \tag{A6}$$

Contrary to the *adaptive* approach of determining the edges based on the samples count, in *cumulative_amplitude* the edge value is chosen when the sum of samples' amplitudes crosses this threshold (A5)–(A6).

Table A1. Notation used in quantizaton Equations (A1)–(A6).

Symbol	Meaning
n	number of samples
m	number of classes (categories, bins); $m \in \mathbb{N}_{>0}$
S_{norm}	normalized input space
S_{qa}	signal space after *adaptive* quantization
$\textbf{edges}_i$	i-th quantization edge, see (A3)
$\textbf{srt\_samples}_i$	i-th sample in the ascending sorted array of all available signal samples

References

1. Wielgosz, M.; Skoczeń, A.; Wiatr, K. Looking for a Correct Solution of Anomaly Detection in the LHC Machine Protection System. In Proceedings of the 2018 International Conference on Signals and Electronic Systems (ICSES), Kraków, Poland, 10–12 September 2018; pp. 257–262.
2. Evans, L.; Bryant, P. LHC Machine. *J. Instrum.* **2008**, *3*, S08001. [CrossRef]
3. Denz, R. Electronic Systems for the Protection of Superconducting Elements in the LHC. *IEEE Trans. Appl. Supercond.* **2006**, *16*, 1725–1728. [CrossRef]
4. Steckert, J.; Skoczen, A. Design of FPGA-based Radiation Tolerant Quench Detectors for LHC. *J. Instrum.* **2017**, *12*, T04005. [CrossRef]
5. Chandola, V.; Mithal, V.; Kumar, V. Comparative Evaluation of Anomaly Detection Techniques for Sequence Data. In Proceedings of the 2008 Eighth IEEE International Conference on Data Mining, Pisa, Italy, 15–19 December 2008; pp. 743–748.
6. Chandola, V.; Banerjee, A.; Kumar, V. Anomaly Detection: A Survey. *ACM Comput. Surv.* **2009**, *41*, 15. [CrossRef]
7. Pimentel, M.A.; Clifton, D.A.; Clifton, L.; Tarassenko, L. A Review of Novelty Detection. *Signal Process.* **2014**, *99*, 215–249. [CrossRef]
8. Graves, A. *Supervised Sequence Labelling with Recurrent Neural Networks*; Springer: Berlin/Heidelberg, Germany, 2012.
9. Morton, J.; Wheeler, T.A.; Kochenderfer, M.J. Analysis of Recurrent Neural Networks for Probabilistic Modelling of Driver Behaviour. *IEEE Trans. Intell. Transp. Syst.* **2016**, *18*, 1289–1298. [CrossRef]
10. Pouladi, F.; Salehinejad, H.; Gilani, A.M. Recurrent Neural Networks for Sequential Phenotype Prediction in Genomics. In Proceedings of the 2015 International Conference on Developments of E-Systems Engineering (DeSE), Duai, UAE, 13–14 December 2015; pp. 225–230.

11. Chen, X.; Liu, X.; Wang, Y.; Gales, M.J.F.; Woodland, P.C. Efficient Training and Evaluation of Recurrent Neural Network Language Models for Automatic Speech Recognition. *IEEE Trans. Audio Speech Lang. Process.* **2016**, *24*, 2146–2157. [CrossRef]
12. Ma, J.; Perkins, S. Time-series Novelty Detection Using One-Class Support Vector Machines. In Proceedings of the International Joint Conference on Neural Networks, Portland, OR, USA, 20–24 July 2003; Volume 3, pp. 1741–1745.
13. Zhang, R.; Zhang, S.; Muthuraman, S.; Jiang, J. One Class Support Vector Machine for Anomaly Detection in the Communication Network Performance Data. In Proceedings of the 5th Conference on Applied Electromagnetics, Wireless and Optical Communications; World Scientific and Engineering Academy and Society (WSEAS) ELECTROSCIENCE'07, Stevens Point, WI, USA, 14–16 December 2007; pp. 31–37.
14. Su, J.; Long, Y.; Qiu, X.; Li, S.; Liu, D. Anomaly Detection of Single Sensors Using OCSVM_KNN. In Proceedings of the Big Data Computing and Communications: First International Conference, BigCom 2015, Taiyuan, China, 1–3 August 2015; Springer International Publishing: Cham, Switzerland, 2015; pp. 217–230.
15. Ruiz-Gonzalez, R.; Gomez-Gil, J.; Gomez-Gil, F.J.; Martínez-Martínez, V. An SVM-Based Classifier for Estimating the State of Various Rotating Components in Agro-Industrial Machinery with a Vibration Signal Acquired from a Single Point on the Machine Chassis. *Sensors* **2014**, *14*, 20713–20735. [CrossRef] [PubMed]
16. Hornero, R.; Escudero, J.; Fernández, A.; Poza, J.; Gómez, C. Spectral and Nonlinear Analyses of MEG Background Activity in Patients With Alzheimer's Disease. *IEEE Trans. Biomed. Eng.* **2008**, *55*, 1658–1665. [CrossRef] [PubMed]
17. Schölkopf, B.; Williamson, R.C.; Smola, A.J.; Shawe-Taylor, J.; Platt, J.C. Support Vector Method for Novelty Detection. In Proceedings of the 12th International Conference on Neural Information Processing Systems, Denver, CO, USA, 29 November–4 December 2000; pp. 582–588.
18. Schölkopf, B.; Platt, J.C.; Shawe-Taylor, J.C.; Smola, A.J.; Williamson, R.C. Estimating the Support of a High-Dimensional Distribution. *Neural Comput.* **2001**, *13*, 1443–1471. [CrossRef] [PubMed]
19. Masnan, M.J.; Mahat, N.I.; Shakaff, A.Y.M.; Abdullah, A.H.; Zakaria, N.Z.I.; Yusuf, N.; Subari, N.; Zakaria, A.; Aziz, A.H.A. Understanding Mahalanobis distance criterion for feature selection. *AIP Conf. Proc.* **2015**, *1660*, 050075. [CrossRef]
20. Liu, F.T.; Ting, K.M.; Zhou, Z.H. Isolation Forest. In Proceedings of the 2008 Eighth IEEE International Conference on Data Mining, Pisa, Italy, 15–19 December 2008; pp. 413–422.
21. Wang, C.; Viswanathan, K.; Choudur, L.; Talwar, V.; Satterfield, W.; Schwan, K. Statistical Techniques for Online Anomaly Detection in Data Centers. In Proceedings of the 12th IFIP/IEEE International Symposium on Integrated Network Management (IM 2011) and Workshops, Dublin, Ireland, 23–27 May 2011; pp. 385–392.
22. Ekberg, J.; Ylinen, J.; Loula, P. Network behaviour anomaly detection using Holt-Winters algorithm. In Proceedings of the 2011 International Conference for Internet Technology and Secured Transactions, Abu Dhabi, UAE, 11–14 December 2011; pp. 627–631.
23. Wielgosz, M.; Skoczeń, A. Recurrent Neural Networks with Grid Data Quantization for Modeling LHC Superconducting Magnets Behaviour. In *Contemporary Computational Science*; Kulczycki, P., Kowalski, P.A., Łukasik, S., Eds.; AGH University of Science and Technology: Kraków, Poland, 2018.
24. Wielgosz, M.; Skoczeń, A.; Mertik, M. Using LSTM recurrent neural networks for detecting anomalous behavior of LHC superconducting magnets. *Nuclear Inst. Methods Phys. Res. A* **2017**, *867*, 40–50. [CrossRef]
25. Wielgosz, M.; Mertik, M.; Skoczeń, A.; Matteis, E.D. The model of an anomaly detector for HiLumi LHC magnets based on Recurrent Neural Networks and adaptive quantization. *Eng. Appl. Artif. Intell.* **2018**, *74*, 166–185. [CrossRef]
26. Hochreiter, S.; Schmidhuber, J. Long Short-Term Memory. *Neural Comput.* **1997**, *9*, 1735–1780. [CrossRef] [PubMed]
27. He, Q.; Wen, H.; Zhou, S.; Wu, Y.; Yao, C.; Zhou, X.; Zou, Y. Effective Quantization Methods for Recurrent Neural Networks. *arXiv* **2016**, arXiv:1611.10176.
28. Shin, S.; Hwang, K.; Sung, W. Fixed-Point Performance Analysis of Recurrent Neural Networks. In Proceedings of the 2016 IEEE International Conference on Acoustics, Speech and Signal Processing (ICASSP), Shanghai, China, 20–25 March 2016; pp. 976–980.
29. Chollet, F. Keras. 2015. Available online: https://keras.io (accessed on 10 November 2018).

30. Theano Development Team. Theano: A Python Framework for Fast Computation of Mathematical Expressions. 2016. Available online: http://deeplearning.net/software/theano/ (accessed on 10 November 2018).
31. Abadi, M. TensorFlow: Large-Scale Machine Learning on Heterogeneous Systems. 2015. Available online: https://tensorflow.org (accessed on 10 November 2018).
32. Pedregosa, F.; Varoquaux, G.; Gramfort, A.; Michel, V.; Thirion, B.; Grisel, O.; Blondel, M.; Prettenhofer, P.; Weiss, R.; Dubourg, V.; et al. Scikit-Learn: Machine Learning in Python. *J. Mach. Learn. Res.* **2011**, *12*, 2825–2830.
33. Sze, V.; Chen, Y.H.; Emer, J.; Suleiman, A.; Zhang, Z. Hardware for Machine Learning: Challenges and Opportunities. In Proceedings of the 2017 IEEE Custom Integrated Circuits Conference (CICC), Austin, TX, USA, 30 April–3 May 2017.
34. Wu, J.; Leng, C.; Wang, Y.; Hu, Q.; Cheng, J. Quantized Convolutional Neural Networks for Mobile Devices. In Proceedings of the IEEE Conference on Computer Vision and Pattern Recognition (CVPR), Las Vegas, NV, USA, 27–30 June 2016; pp. 4820–4828.

Article

Deep CNN Sparse Coding for Real Time Inhaler Sounds Classification

Vaggelis Ntalianis [1,*], Nikos Dimitris Fakotakis [1], Stavros Nousias [1,2,*], Aris S. Lalos [2], Michael Birbas [1], Evangelia I. Zacharaki [1] and Konstantinos Moustakas [1]

1. Department of Electrical & Computer Engineering, University of Patras, 26504 Patras, Greece
2. Industrial Systems Institute, Athena Research Center, 26504 Patras, Greece

* Correspondence: up1020072@upnet.gr (V.N.); snousias@upatras.gr (S.N.); Tel.: +30-2610996170 (S.N.)

Received: 31 January 2020; Accepted: 16 April 2020; Published: 21 April 2020

Abstract: Effective management of chronic constrictive pulmonary conditions lies in proper and timely administration of medication. As a series of studies indicates, medication adherence can effectively be monitored by successfully identifying actions performed by patients during inhaler usage. This study focuses on the recognition of inhaler audio events during usage of pressurized metered dose inhalers (pMDI). Aiming at real-time performance, we investigate deep sparse coding techniques including convolutional filter pruning, scalar pruning and vector quantization, for different convolutional neural network (CNN) architectures. The recognition performance has been assessed on three healthy subjects following both within and across subjects modeling strategies. The selected CNN architecture classified drug actuation, inhalation and exhalation events, with 100%, 92.6% and 97.9% accuracy, respectively, when assessed in a leave-one-subject-out cross-validation setting. Moreover, sparse coding of the same architecture with an increasing compression rate from 1 to 7 resulted in only a small decrease in classification accuracy (from 95.7% to 94.5%), obtained by random (subject-agnostic) cross-validation. A more thorough assessment on a larger dataset, including recordings of subjects with multiple respiratory disease manifestations, is still required in order to better evaluate the method's generalization ability and robustness.

Keywords: deep sparse coding; convolutional neural networks; signal analysis; respiratory diseases; medication adherence

1. Introduction

The respiratory system is a vital structure vulnerable to airborne infection and injury. Respiratory diseases are leading causes of death and disability across all ages in the world. Specifically, nearly 65 million people suffer from chronic obstructive pulmonary disease (COPD) and 3 million die from it each year. About 334 million people suffer from asthma, the most common chronic disease of childhood, affecting 14% of all children globally [1]. The effective management of chronic constrictive pulmonary conditions lies, mainly, in the proper and timely administration of medication. However, as recently reported [2], a large proportion of patients use their inhalers incorrectly. Studies have shown that possible technique errors can have an adverse impact on clinical outcome for users of inhaler medication [3,4]. Incorrect inhaler usage and poor adherence were found to be associated with high COPD assessment test scores [5], short durations of COPD, high durations of hospitalization and high numbers of exacerbations.

Several methods have been introduced to monitor a patient's adherence to medication. As a series of studies indicate, effective medication adherence monitoring can be defined by successfully identifying actions performed by the patient during inhaler usage. Several inhaler types are available in the market, among which the pressurized metered dose inhalers and dry powder inhalers are the most common. In any case, the development of a smart inhaler setup, that allows better monitoring and direct feedback to the user independently of the drug type, is expected to lead to more efficient drug delivery, thereby becoming the main product used by patients.

The pMDI usage technique is characterized as successful, if a certain sequence of actions is followed [6]. Appropriate audio based monitoring could help patients synchronize their breath with drug activation and remind them to keep their breath after inhalation, for a sufficient amount of time. Several methodologies that engage electronic monitoring of medication adherence, have been introduced in the past two decades [7], aiming to alter patient behavioural patterns [8,9]. In the field of inhaler based health monitoring devices, a recent comprehensive review by Kikidis et al. [10] provides a comparative analysis of several research and commercial attempts in this direction. Cloud based self-management platforms and sensor networks constitute the next step towards effective medication adherence and self-management of respiratory conditions [11,12].

In all cases, it is crucial to successfully identify audio events related to medication adherence. In this direction, several approaches have been proposed in the literature, presenting mainly decision trees or other state of the art classifiers, applied on a series of extracted features. However, the aforementioned methodologies come with high computational cost, limiting the applicability of monitoring medication adherence to offline processing or online complex distributed cloud-based architectures, that are able to handle the need for resources. Therefore, the demand for computationally fast, yet highly accurate, classification techniques still remains.

Motivated by the aforementioned open issues, this study lies on the same track as several data-driven approaches [13–15], presenting a method that recognizes the respiration and drug delivery phases on acoustic signals derived from pMDI usage. The main focus of this work is the investigation of acceleration aspects, namely filter pruning, scalar pruning and vector quantization, applied on convolutional neural networks (CNNs). The adaptation of such strategies allows to reduce computational complexity and improve performance and energy efficiency. The CNNs are trained to differentiate four audio events, namely, drug actuation, inhalation, exhalation and other sounds. Five different CNN architectures are investigated and the classification accuracy is examined as a function of compression rate. More specifically, the benefits of this work can be summarized in the following points:

- The presented methods are applied directly on the time-domain avoiding computationally expensive feature extraction techniques.
- The overall classification accuracy for the proposed CNN architecture is high (95.7%–98%), for both within and across subjects cross-validation schemes.
- A compression rate by a factor of 7.0 can be achieved with accuracy dropping only by 1%.
- The investigated deep sparse coding oriented strategies (Implementation of this work and a part of the dataset used to validate it, is available online at: https://github.com/vntalianis/Deep-sparse-coding-for-real-time-sensing-of-medication-adherence), namely filter pruning and vector quantization, allow compliance with real-time requirements and open the path for adaptation of the inhaler device into Internet of Things (IoT).

The rest of the paper is organized as follows: Section 2 presents an extensive overview on relevant literature, Section 3 describes the CNN architectures and our methodology to enforce sparsity, Section 4 presents the experimental setup and the evaluation study and, finally, Section 5 provides future directions on the analysis of inhaler sounds.

2. Related Work

This section examines classical and data-driven approaches on classification of inhaler sounds. Early methodologies encompass electronic or mechanical meters integrated into the device, activated with the drug delivery button. Howard et al. [16] reported the existence of several such devices, able to record the time of each drug actuation, or the total number of them. The use of audio analysis came up later as a method, which can characterize the quality of inhaler usage, while, also, monitoring the timings of each audio event. The classical audio analysis involves transformation of the time-domain into a set of features, mainly, in the frequency domain, including Spectrogram, Mel-Frequency Ceptral Coefficients (MFCCs), Cepstrogram, Zero-Crossing Rate (ZCR), Power Spectral Density (PSD) and Continuous Wavelet Transform (CWT). Subsequently, audio-based evaluation employs the extracted features via classification approaches to locate and identify medication-related audio events.

Holmes et al. [17–19] designed decision trees in the scientific sub-field of blister detection and respiratory sound classification. This study includes detection of drug activation, breath detection and inhalation-exhalation differentiation and provides feedback, regarding to patient adherence. As a first step, the audio signal is segmented into frames of specific length, with overlaps. The mean power spectral density is calculated for defined frequencies and is used as a threshold to differentiate between blister and non-blister sounds. Also, the maximum normalized amplitude and the time duration are used to remove false positive sounds. The algorithm, then, examines the mean PSD, in specific frequency band, as the last threshold for blister sounds categorization. At the second stage, this algorithmic approach detects breath sounds. In this case, the audio signal is first filtered to remove high frequency components above a threshold, using a low-pass type I 6th order Chebyshev filter. Many window techniques exist for the design of Finite Impulse Response (FIR) and Infinite Impulse Response (IIR) Filters [20–22], such as Hamming, Hanning and Blackman for FIR Filter design, and Butterworth and Chebyshev for IIR Filters.

After signal segmentation, one set of 12 MFCCs is calculated for each frame, forming a short-time Cepstrogram of the signal. Ruinskiy et al. [23] perform singular value decomposition to capture the most important characteristics of breath sounds obtained from MFCC calculations. They set an adaptive threshold, according to the lowest singular vector in the inhaler recording, and mark the singular vectors above this threshold as potential breath events. For the last threshold, at this stage, the ZCR is extracted for each frame. Finally, the algorithms find the differentiation between inhalations and exhalations. The mean PSD of identified breaths is calculated for a determined frequency band and is used as a threshold for classification. Then, the standard deviation of the ZCR was found to be higher for inhalations in comparison to exhalations and a value is set, from empirical observations.

Taylor et al. [24,25] used the CWT to identify pMDI actuations, in order to quantitatively assess the inhaler technique, focusing only on the detection of inhaler actuation sounds. As a step forward data-driven approaches learn by example from features and distributions found in the data. Taylor et al. [26] compared Quadratic Discriminant Analysis (QDA) and Artificial Neural Network (ANN) based classifiers using MFCC, Linear Predictive Coding, ZCR and CWT features.

Nousias et al. [13,27] compared feature selection and classification strategies using Spectrogram, Cepstrogram and MFCC with supervised classifiers, such as Random Forest (RF), ADABoost and Support Vector Machines (SVMs), demonstrating high classification accuracy.

Pettas et al. [15] employed a deep learning based approach using the Spectrogram as a tool to develop a classifier of inhaler sounds. The Spectrogram is swept across the temporal dimension with a sliding window with length $w = 15$ moving at a step size equal to a single window. The features of each sliding window are inserted into a recurrent neural network with long-short memory units (LSTM), demonstrating high performance in transitional states where mixture of classes appear.

Ntalianis et al. [14] employed five convolutional networks, applied directly in the time-domain, and showed that CNNs can automatically perform feature extraction and classification of audio samples with much lower computational complexity, at similar or higher classification accuracy than classical approaches. Each model uses a vector of $n = 4000$ samples reshaped in a two-dimensional array (250×16), that is introduced in the deep CNN. Evaluation was limited to five-fold cross-validation in a subject-agnostic way, in which different samples from the same subject might be part of the training and test set, respectively.

This study aims to build upon previous CNN-based approaches for the identification of inhaler events, by investigating also acceleration strategies. CNNs have been established as a reliable state-of-the-art, data-driven approach for biosignal classification [28–32]. The adaptation of acceleration approaches, including filter pruning, scalar pruning and vector quantization, aims to lead to lower computational complexity and higher energy efficiency, facilitating IoT targeted implementations. In Section 4 we present the classification accuracy of the aforementioned studies, aiming to compare results of previous studies with our current approach.

3. Monitoring Medication Adherence through Deep Sparse Convolutional Coding

This section provides a comprehensive analysis of the deep architecture employed to perform the inhaler audio classification. Based on a main convolutional neural network architecture, five different variations are being investigated. Furthermore, compression and acceleration strategies, namely filter and scalar values pruning and vector quantization, are also being analyzed.

3.1. Convolutional Neural Network Architecture

The CNN architecture, presented in Figure 1, consists of three convolutional layers with a max-pooling layer, a dropout function [33] and four fully connected layers. Using this structure, five different CNNs were developed as presented in Table 1. For the convolutional kernels the stride is set equal to one with zero padding, in order to keep the shape of the output of each filter constant and equal to its input's dimensionality. Every model utilizes the same sequence of layers, but with a different number of filters in the convolutional layers, or a different number of neurons in the fully connected layers, or different activation functions. Specifically, Table 1 presents the stacked layers for each model, the values of dropout layers, the number of filters in each convolutional layer, the number of neurons in fully connected layers and the activation function. In the fifth model, we select Exponential Linear Unit (ELU), due to the fact that the recordings contain both negative and positive values and ELU, in contrast to ReLu, does not zero out negative values. As far as the training parameters is concerned, the learning rate is set to 0.001, the batch size is equal to 100 and the categorical cross entropy loss function is employed. Training is executed through 5-fold cross validation with 20 epochs and Adam optimizer. In order to train the five CNN architectures, raw recordings in the time domain are directly used as input. The initial audio files contain multiple events, namely inhalation, exhalation, drug delivery and environmental noise. The final stage of preprocessing contains the formation of sound samples of 0.5 s duration (i.e., 4000 samples) collected with a sliding step of 500 samples. Only samples with unique classes are retained in the dataset. For a given convolutional layer, the previous layer's feature maps are convolved with learnable kernels and passed through the activation function to form the output feature map described by Equation (1).

$$\mathbf{x}_j^{\ell} = f\left(\sum_{i \in M_j} \left(\mathbf{x}_i^{\ell-1} * \mathbf{k}_{ij}^{\ell}\right) + b_j^{\ell}\right), \tag{1}$$

where M_j represents a selection of input feature maps. The output is fed to a set of four dense layers. The aforementioned architectures were chosen experimentally to keep the classification accuracy high and, simultaneously, the computational complexity as low as possible. We also experimented with both shallower and deeper architectures, but did not observe any further improvement.

Table 1. Convolutional neural network (CNN) architecture variations for tested models.

Layers	Layer Parameters	Model 1	Model 2	Model 3	Model 4	Model 5
Convolutional Layer	Filters	16	16	8	8	16
	Kernel Size	4 × 4	4 × 4	4 × 4	4 × 4	4 × 4
	Activation Function	ReLu	ReLu	ReLu	ReLu	ELU
Max Pooling	Kernel Size	2 × 2	2 × 2	2 × 2	2 × 2	2 × 2
Dropout		0.2	0.2	0.2	0.2	0.2
Convolutional Layer	Filters	16	16	8	8	16
	Kernel Size	5 × 5	5 × 5	5 × 5	5 × 5	5 × 5
	Activation Function	ReLu	ReLu	ReLu	ReLu	ELU
Max Pooling	Kernel Size	2 × 2	2 × 2	2 × 2	2 × 2	2 × 2
Dropout		0.1	0.1	0.1	0.1	0.1
Convolutional Layer	Filters	16	16	8	8	16
	Kernel Size	6 × 6	6 × 6	6 × 6	6 × 6	6 × 6
	Activation Function	ReLu	ReLu	ReLu	ReLu	ELU
Max Pooling	Kernel Size	2 × 2	2 × 2	2 × 2	2 × 2	2 × 2
Dense	Neurons	64	64	64	64	64
	Activation Function	ReLu	ReLu	ReLu	ReLu	ELU
Dense	Neurons	128	32	32	128	128
	Activation Function	ReLu	ReLu	ReLu	ReLu	ELU
Dense	Neurons	64	16	16	64	64
	Activation Function	ReLu	ReLu	ReLu	ReLu	ELU
Dense	Neurons	4	4	4	4	4
	Activation Function	ReLu	ReLu	ReLu	ReLu	ELU
Test Loss		0.2413	0.2459	0.1891	0.2040	0.2145
Test Accuracy		0.9440	0.9397	0.9483	0.9586	0.9570

For the implementation we used NumPy and SciPy, mainly for data mining and numerical computation tasks, as they are the fundamental packages to define, optimize and evaluate mathematical expressions, for scientific computing. These libraries also optimize the utilization of GPU and CPU, making the performance of data-intensive computation even faster. We developed this approach using Scikit-learn, which is built on top of the two aforementioned libraries and, also, using Tensorflow, which focuses on building a system of multi-layered nodes (multi-layered nodes system with high-level data structures). That allowed us to train and run the convolutional networks on either CPU or GPU. We, furthermore, used Pandas, which focuses on data manipulation and analysis (grouping, combining, filtering, etc.) and Keras, which is a high-level neural networks API, running on top of Tensorflow. Lastly, we used Matplotlib, as a standard Python library for data visualization (2D plots and graphs).

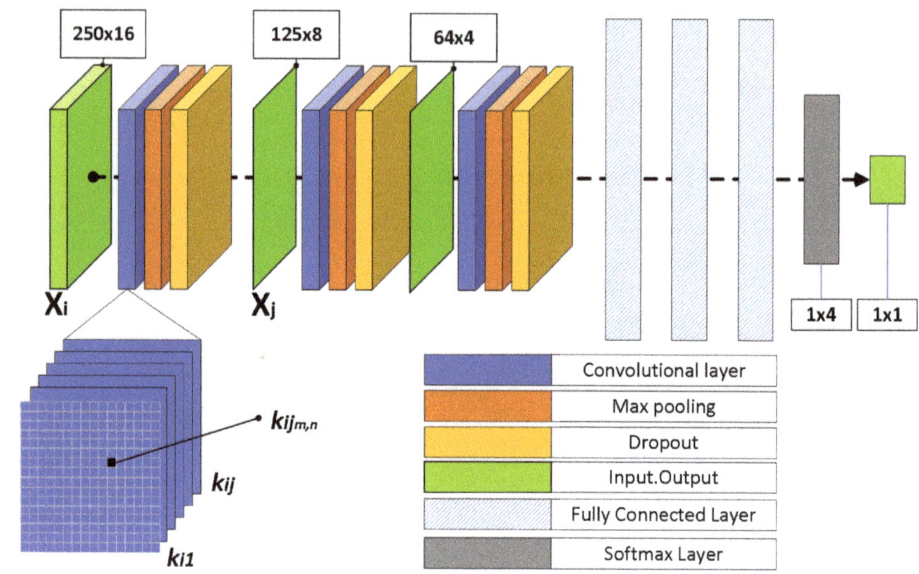

Figure 1. CNN Architecture.

3.2. Filter Pruning

In order to reduce the computational requirements of the developed CNN architectures, we performed filter pruning as described in Reference [34], aiming to remove the less significant kernels in the convolutional layers, without deteriorating performance. In particular, we evaluate the contribution of each kernel, at the output of the layer, by calculating the sum of its absolute weights. For a convolutional layer with an input feature map of

$$x_i \in \Re^{h_i \times w_i \times n_i}, \tag{2}$$

where h_i, w_i, n_i are the height, the width and the number of channels of the input respectively, the output feature response has a shape of

$$x_{i+1} \in \Re^{h_{i+1} \times w_{i+1} \times j_{i+1}}, \tag{3}$$

after applying j_{i+1} filters with a kernel matrix $k \in \Re^{k_1 \times k_2 \times n_i}$. The filter pruning process can be summarized in the following steps:

1. Compute the sum $s_j = \sum_{l=1}^{n_i} \sum k$ of the absolute values of the weights in each filter.
2. Sort s_j and remove n filters with the lowest sum s_j.
3. The rest of the weights remain unchanged.

For the filter pruning operation, two different approaches can be employed. Each layer can be pruned independently from others, referred to as independent pruning, or by ignoring the removed filters, also referred to as greedy pruning. The removal of a filter in the i-th layer leads to the removal of the corresponding feature map, which in turn leads to the removal of the kernels that belong to the $i+1$th layer and are applied on the aforementioned feature map. So, with independent pruning we sort the filters by taking into consideration the sum of the weights in these kernels. On the other hand, greedy pruning does not include them in the computation of the sum. Note that both approaches produce a weight matrix with the same dimensions and differ only on the filters chosen to be pruned. Additionally, in order to affect the accuracy of the prediction model as less as possible, two training strategies can be followed: (1) Prune once and retrain. This approach executes the pruning procedure first and only after all layers are processed the classifier is retrained so that its classification accuracy reaches its initial value. (2) Iterative pruning and retraining. In contrast to the first approach, with this method, when a layer is pruned, the rest of the network is immediately retrained, before the following layer is pruned. In this way, we let the weights of the model adjust to the changes occurred in previous layers, thus retaining its classification accuracy.

3.3. Pruning Scalar Values

The parameters k_{k_1,k_2,n_i} of the different filters within each layer of a CNN are distributed in a range of values, with standard deviation σ. Weights very close to zero have an almost negligible contribution to a neuron's activation. To this end, a threshold $\ell \in [\ell_{min}, \ell_{max}]$ is defined so that

$$k_{k_1,k_2,n_i} = 0 \quad if \quad |k_{k_1,k_2,n_i}| < \ell \cdot \sigma, \tag{4}$$

ℓ_{min} and ℓ_{max} define the range of values during hyperparameter optimization and were set experimentally to $\ell_{min} = 0.1$ and $\ell_{max} = 1.0$. It is important to clarify that we employ the standard deviation in order to control the maximum number of parameters to be pruned. Finally, this method was evaluated by directly carrying out pruning on every layer of the classifier and by retraining the layers that follow the pruned one.

3.4. Vector Quantization

3.4.1. Scalar Quantization

One way to decrease the number of parameters in a layer is to perform scalar quantization to them [35]. For example, for a fully connected layer with weight matrix $W \in \Re^{m \times n}$, we can unfold the matrix so that $W \in \Re^{1 \times m \cdot n}$ and perform k-means clustering as described by the following formula:

$$\min \sum_{i}^{m \cdot n} \sum_{j}^{N_{cl}} \|\mathbf{w}_i - \mathbf{c}_j\|_2^2. \tag{5}$$

The codebook can be extracted from the N_{cl} cluster centers $\mathbf{c}_j$ produced by the k-means algorithm. The initial parameters are then assigned with cluster indexes to map them to the closest center. Consequently, we can reconstruct the initial weight matrix W as:

$$\hat{W}_{ij} = c_z \tag{6}$$

where

$$\min_z \|\mathbf{W}_{ij} - \mathbf{c}_z\|_2^2. \tag{7}$$

In respect to the convolutional layers, first, we have to decide on which dimension the k-means algorithm is going to be applied [36]. In the $i + 1$ convolutional layer, the weight matrix is a 4-dimensional tensor $W \in \Re^{k_1 \times k_2 \times n_i \times n_{i+1}}$, where n_i is the number of channels of the input feature map and n_{i+1} the number of channels of the output feature map. It is preferable to perform k-means along the channels of the output feature maps in order to reduce the computational requirements by reusing pre-computed inner products.

3.4.2. Product Quantization

The main concept of product quantization is to split the vector space into multiple sub-spaces and perform quantization in each subspace separately. In this way, we are able to better exploit the redundancy in each subspace [35]. In particular, let a weight matrix of a fully connected layer $W \in \Re^{m \times n}$. We partition it column-wise so that:

$$W = [W^1, W^2, ..., W^s], \tag{8}$$

where $W^i \in \Re^{m\times(n/s)}$. We apply k-means clustering in each submatrix W^i:

$$\min \sum_{z}^{n/s} \sum_{j}^{N_{cl}} \left\| \mathbf{w}_z^i - \mathbf{c}_j^i \right\|_2^2, \tag{9}$$

where w_z^i represents the z-th column of the sub-matrix W^i and c_j^i the column of the sub-codebook $C_i \in \Re^{m\times N_{cl}}$. The reconstruction process is performed based on the assigned cluster and the codebook for every sub-vector w_z^i. So, the reconstructed matrix is:

$$\hat{W} = [\hat{W}^1, \hat{W}^2, ..., \hat{W}^s], \quad \text{where} \tag{10}$$

$$\hat{w}_j^i = c_j^i, \quad \text{where} \quad \min_j \left\| \mathbf{w}_z^i - \mathbf{c}_j^i \right\|_2^2. \tag{11}$$

It is important to highlight, that product quantization can be performed to either the x-axis or the y-axis of W, but most importantly the partitioning parameter s must divide n.

Regarding the convolutional layers, following the same idea as in scalar quantization, we split the initial space vector into sub-spaces along the channel axis and we perform k-means at the same dimension. Let a convolutional layer have n channels. Then, its weight matrix has a shape of $W \in \Re^{k\times k\times c\times n}$. Introducing the splitting parameter s, each sub-space has a shape of $W^l \in \Re^{k_1\times k_2\times n_i\times(n_{i+1}/s)}$.

3.5. Compression Rate and Computational Complexity

The objectives of the aforementioned algorithms are to reduce the storage requirements as well as the computational complexity of the classifier. This section provides an insight of the computation of compression and theoretical speed up that pruning and quantization achieve. As mentioned above, pruning causes sparsity by forcing weights to become zero. Consequently, these weights are not stored in memory achieving a compression rate, described by the following expression. This equation is used to compute the extent of the compression for a model when filter pruning is applied as well.

$$compression\ rate = \frac{number\ of\ parameters\ in\ not\ pruned\ model}{number\ of\ non\ zero\ parameters\ after\ pruning} \tag{12}$$

On the other hand, quantizing with k-means clustering results in a more complex expression for the compression rate, because it depends on the magnitude of both codebook and the index matrix. In particular, for scalar quantization in fully connected layer, the codebook contains N_{cl} values and only $log_2 N_{cl}$ bits are necessary to encode the centers. Thus, if 32-bit (single-precision) representation is used for the calculations, the total amount of bits required to store is $32 \cdot k + m \cdot n \cdot log_2(N_{cl})$ and the compression rate that this method achieves, per layer, is:

$$\frac{32 \times m \times n}{32 \times k + m \times n \times log_2(N_{cl})}. \tag{13}$$

However, the burden of memory requirements due to the codebook is considered negligible, comparing to the requirements of index matrix. Therefore the compression rate can be approximated with the simpler formula [35]:

$$32/log_2 N_{cl} \tag{14}$$

In case of the convolutional layers the k-means algorithm is performed on the channels axis. Therefore the number of the weights required to be stored is:

$$\texttt{new\_weights} = k_1 \times k_2 \times n_i \times N_{cl}, \tag{15}$$

instead of the initial, which is:

$$\texttt{old\_weights} = k_1 \times k_2 \times n_i \times n_{i+1} \tag{16}$$

and the compression rate can be computed by the following formula:

$$\frac{32 \times \texttt{old\_weights}}{32 \times \texttt{new\_weights} + ch \times log_2(N_{cl})}. \tag{17}$$

Note that the index matrix contains only as many positions as the number of channels in the convolutional layer. This lies on the fact that by performing clustering along the channels axis, we produce filters with the same weight values, thus we only need to map the initial filters to the new ones. Finally, the compression rate that product quantization approach presented in Section 3.4.2 can achieve, per layer, in the fully connected layer, is calculated by:

$$\frac{32 \times m \times n}{32 \times m \times N_{cl} \times s + n \times log_2(N_{cl} \times s)}. \tag{18}$$

For this method both the cluster indexes and the codebook for each sub-vector should be stored. Regarding the convolutional layers, performing the k-means algorithm with N_{cl} clusters along the channels the compression rate is calculated by:

$$\frac{32 \times k_1 \times k_2 \times n_i \times n_{i+1}}{32 \times k_1 \times k_2 \times n_i \times N_{cl} \times s + n_{i+1} \times log_2 N_{cl}}. \tag{19}$$

Next, we present the gain in floating point operations required to perform the classification task, when the aforementioned techniques are employed. Among these methods, pruning scalar values and scalar quantization on fully connected layer do not offer any computational benefit. The zeroed out weights are scattered inside the weight matrix of each layer and, therefore, they do not form a structural pattern. For example, this method does not guarantee that all zeroed out weights belong to a certain kernel or to a certain neuron. On the other hand, with filter pruning, we remove whole filters, reducing efficiently the computational burden. In a convolutional layer the amount of necessary operations depends on the dimensionality of the input feature map and the number of the weights in the layer. The total number of floating point operations in a convolutional layer is:

$$n_{i+1} \times n_i \times k_1 \times k_2 \times h_{i+1} \times w_{i+1}, \tag{20}$$

where n_i is the number of channels of the input feature map, k the dimensionality of the kernel, n_{i+1} the channels of the output feature map and h_{i+1}, w_{i+1} the height and the width of the layer's output, respectively. The product of $n_{i+1} \cdot n_i \cdot k_1 \cdot k_2$ determines the amount of the weights that the layer contains.

The pruning of a single filter saves $n_i \cdot k_1 \cdot k_2 \cdot h_{i+1} \cdot w_{i+1}$ operations from the current layer and $n_{i+2} \cdot k_1 \cdot k_2 \cdot h_{i+2} \cdot w_{i+2}$ from the next layer. These additional operations can be avoided, because the certain kernels of the next convolutional layer are also removed. Specifically, the aforementioned kernels are applied on pruned feature maps.

Regarding the fully connected layer, the amount of floating point operations (flops) can be directly calculated from the dimensions of the weight matrix of a layer. For the weights matrix $W \in \Re^{m \times n}$, the number of flops is calculated as $m \cdot n$. The value of m corresponds to the dimension of the layer's input and n to the dimension of its output, which is equal to the number of neurons in each layer. In order to reduce the number of flops in a fully connected layer, we should remove neurons, decreasing the dimensionality of the layer's output.

Performing scalar quantization in convolutional layers with N_{cl} clusters along the channel axis, we need to execute $N_{cl} \cdot n_i \cdot k_1 \cdot k_2 \cdot h_{i+1} \cdot w_{i+1}$ operations per layer which results in a ratio of:

$$\frac{N_{cl} \times n_i \times k_1 \times k_2 \times h_{i+1} \times w_{i+1}}{n_{i+1} \times n_i \times k_1 \times k_2 \times h_{i+1} \times w_{i+1}} \tag{21}$$

and by introducing the splitting parameter s to perform product quantization, the previous ratio becomes:

$$\frac{N_{cl} \times s \times n_i \times k_1 \times k_2 \times h_{i+1} \times w_{i+1}}{n_{i+1} \times n_i \times k_1 \times k_2 \times h_{i+1} \times w_{i+1}}. \tag{22}$$

Finally, performing product quantization on fully connected layers with N_{cl} clusters and s sub-spaces, the ratio of the flops required after quantization, to the flops required before quantization, is calculated as

$$\frac{m \times N_{cl} \times s}{m \times n} = \frac{N_{cl} \times s}{n}. \tag{23}$$

4. Experimental Procedure And Evaluation

4.1. Data Acquisition

Audio recordings from pMDI use were received, using a standard checklist of steps, recommended by National Institute of Health (NIH) guidelines, as it was essential to ensure that the actuation sounds were accurately recorded. The data were acquired from three subjects, between 28 and 34 years old, who all used the same inhaler device loaded with placebo canisters. The recordings were performed in an acoustically controlled indoor environment, free of ambient noise, at the University of Patras, to reflect possible use in real-life conditions and to ensure accurate data acquisition. The study supervisors were responsible for inhaler actuation sounds and respiratory sounds and followed a protocol, that defined all the essential steps of pMDI inhalation technique. Prior training of the participants, on this procedure, allowed to reduce the experimental variability and increase the consistency of action sequences. Each participant annotated in written form the onset and duration of each respiratory phase, during the whole experiment. Also, the annotation of the different actions was subsequently verified and completed by a trained researcher and based on visual inspection of the acquired temporal signal. In total, 360 audio files were recorded with a duration of twelve seconds each [13–15].

The acoustics of inhaler use were recorded as mono WAV files, at a sampling rate of 8000 Hz. After quantization, the signal had a resolution (bit depth) of 16 bits/sample. Throughout the processing of the audio data, no further quantization on the data took place, except the quantization of the CNN weights into clusters of similar values of the convolutional and fully connected layer. The device for the recordings is presented in Reference [13]. Figure 2 depicts an overview of the processing pipeline. The sensor's characteristics are 105 dB-SPL sensitivity and 20 Hz–20 kHz bandwidth. Each recording contains a full inhaler usage case. The first person (male) submitted 240 audio files, the second person (male) 70 audio files and the third subject (female) 50 audio files. Each subject, at first, breathes out and after bringing the device to their mouth he/she starts to inhale. Simultaneously, the subject presses the top of the inhaler to release the medication and continues to inhale until having taken a full breath. Then, breath holding follows for about 10 s and, finally, exhaling.

In order to train and test the proposed classifier, the audio recordings were segmented into inhaler activation, exhalation, inhalation, and noise (referring to environmental or other sounds) by a human expert using a graphical annotation tool [13]. A user interface visualizes the audio samples while the user selects parts of the audio files and assigns a class. The annotated part is stored in a separate audio file. A full audio recording timeseries example is presented in Figure 3, colored according to the annotated events. Any signal part that has not been annotated, was considered as noise, during the stage of validation. This dataset has the potential to allow in-depth analysis of patterns on sound classification and data analysis of inhaler use in clinical trial settings.

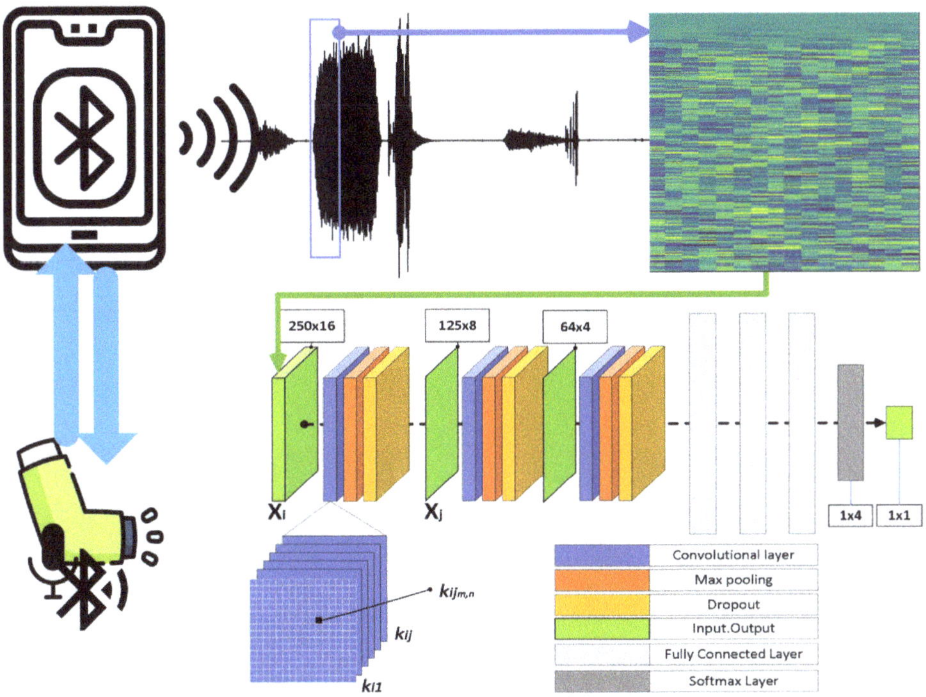

Figure 2. Overview of the processing pipeline.

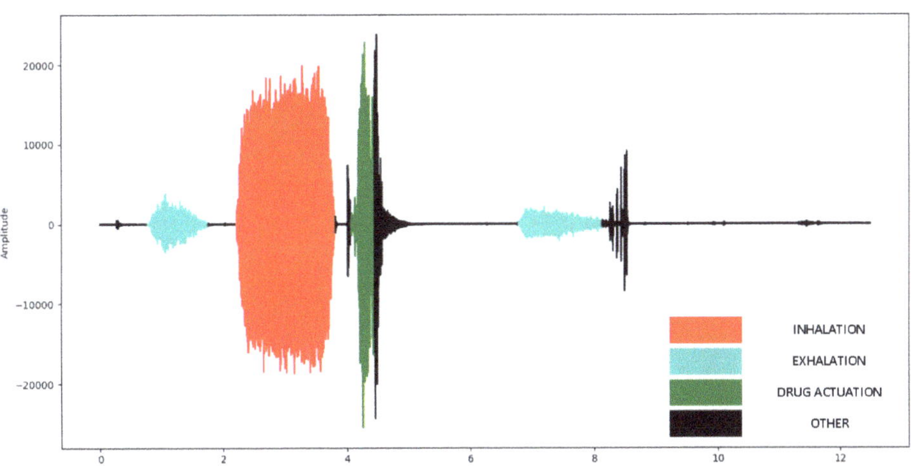

Figure 3. Annotated audio file of 12 s. Red color corresponds to inhalation, cyan to exhalation, green to drug activation and black to other sounds.

Each sound sample has a total duration of 0.5 s, sampled with 8 kHz sampling rate and 16-bit depth. The audio files used for training and testing were loaded through appropriate libraries in a vector of 4000×1 dimension and, then, reshaping is performed in order to employ two-dimensional convolutions. In particular, the first 16 samples are placed in the first row of the matrix, the next 16 samples in the second, and so on, until a 250×16 matrix is constructed. An example of the reshaping procedure is given in Figure 4, while Figure 5 visualizes examples of sounds per class after this reshaping procedure.

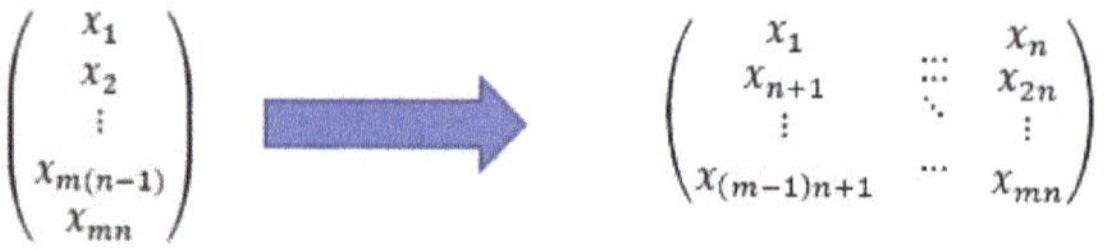

$$\begin{pmatrix} x_1 \\ x_2 \\ \vdots \\ x_{m(n-1)} \\ x_{mn} \end{pmatrix} \Rightarrow \begin{pmatrix} x_1 & \cdots & x_n \\ x_{n+1} & \cdots & x_{2n} \\ \vdots & \ddots & \vdots \\ x_{(m-1)n+1} & \cdots & x_{mn} \end{pmatrix}$$

Figure 4. Illustration of reshaping of a vector into a two-dimensional matrix.

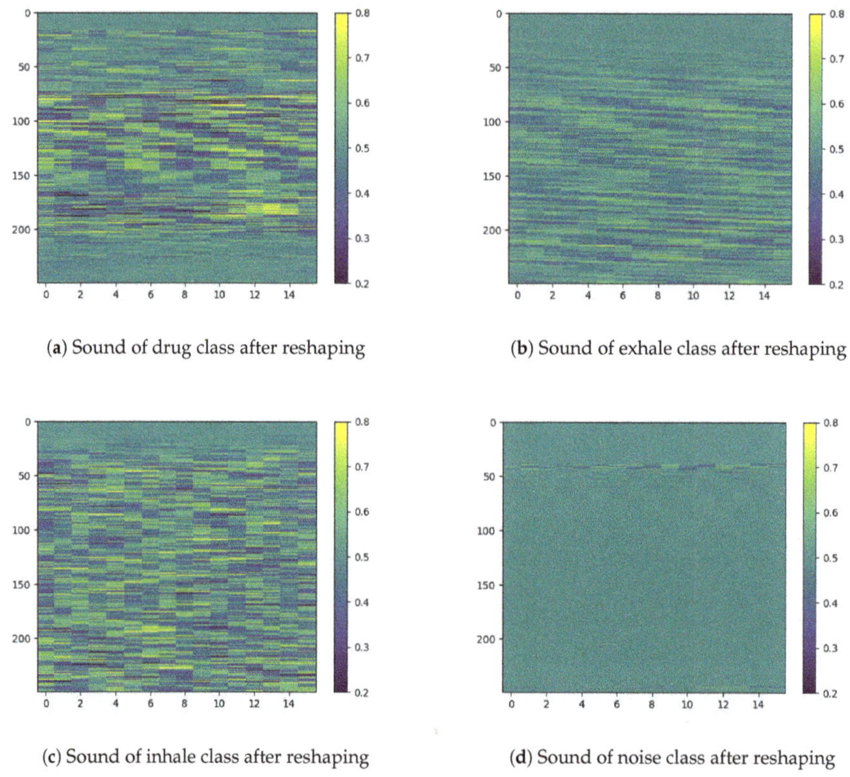

(**a**) Sound of drug class after reshaping
(**b**) Sound of exhale class after reshaping
(**c**) Sound of inhale class after reshaping
(**d**) Sound of noise class after reshaping

Figure 5. Visualization of the segmented audio files for each respiratory phase after the reshaping procedure.

4.2. Evaluation Schemes

The training and assessment of the five CNN models is performed in three different cross-validation settings. Firstly, we consider the *Multi Subject* modeling approach. In this case, the recordings of all three subjects are used to form a large dataset, which is divided in five equal parts used to perform five-fold cross-validation, thereby allowing different samples from the same subject to be used in training and test set, respectively. This validation scheme was followed in previous work [14] and thus performed, also, here for comparison purposes.

The second case includes the *Single Subject* setting, in which the performance of the classifier is validated through training and testing, within each subject's recordings. Specifically, the recordings of each subject are split in five equal parts, to perform cross-validation. The accuracy is assessed for each subject separately and, then, the overall performance of the classifier is calculated by averaging the three individual results.

Finally, *Leave-One-Subject-Out (LOSO)* method is employed. With this approach we use the recordings of two subjects for training and the recordings of the third subject for testing. This procedure is completed, when all subjects have been used for testing, and the accuracy is averaged to obtain the overall performance of the classifiers.

4.3. Results

4.3.1. Comparison with Relevant Previous Work

In order to better assess the contribution of the proposed approach, we first summarize in Table 2 the classification performance of previous state of the art algorithms that were presented in Section 2. In more details, Holmes et al. [17] presented, in 2012, a method that differentiates blister and non-blister events with an accuracy of 89.0%. A year later, Holmes et al. [18,19], also, developed an algorithm that recognizes blister events and breath events (with an accuracy of 92.1%) and separates inhalations from exhalations (with an accuracy of more than 90%). Later, Taylor et al. developed two main algorithms for blister detection [26,37] based on Quadratic Discriminant Analysis and ANN, and achieved an accuracy of 88.2% and 65.6%, respectively. Nousias et al. in Reference [13] presented a comparative study between Random Forest, ADABoost, Support Vector Machines and Gaussian Mixture Models, reaching the conclusion that RF and GMM yield a 97% to 98% classification accuracy on the examined dataset, when utilizing MFCC, Spectrogram and Cepstrogram features.

Pettas et. al [15] developed a recurrent neural network with long short term memory (LSTM), which was tested on the same dataset with this study and using the same modeling schemes, that is, *SingleSubj*, *MultiSubj* and *LOSO*. For the subject-specific modeling case the overall prediction accuracy was 94.75%, with higher accuracy in the prediction of breathing sounds (98%). Lower accuracy is demonstrated in drug administration and environmental sounds. Much higher accuracy is reported for *MultiSubj* modeling, where the training samples are obtained from all subjects and shuffled across time. It yielded a drug administration prediction accuracy of 93%, but a lower prediction accuracy of environmental sounds (79%), demonstrating a total of 92.76% accuracy over all cases. Furthermore, the *LOSO* validation demonstrated similar results, with the *SingleSubj* case. The high classification accuracy obtained by LSTM-based deep neural networks, is also in agreement with other studies [13,19]. Specifically, the recognition of breathing sounds is more accurate than the drug administration phase, which reaches a value of 88%, while the overall accuracy is 93.75%. In order to compare our approach with previous studies, we followed the same validation strategies for each different convolutional neural network architecture and summarize the comparative results in Table 3.

From Tables 2 and 3, it is apparent that the classification accuracy achieved by our approach does not exceed the performance of the relevant state of the art approaches. In fact, our approach performs, similarly, with the methods developed by Holmes et al. [17,18], Taylor et al. [24] and Pettas et al. [15], but the approach of Nousias et al. [13] outperforms our algorithm, mainly, for the drug and environmental noise classes. However, the utilization of a CNN architecture in the time domain allows for an implicit signal representation, that circumvents the need of additional feature extraction (e.g., in the spectral domain) and, thereby, results in significantly lower execution times. We compare the computational cost of Model 5 of our method with the Random Forest algorithm presented in Reference [13], both executed in the same machine (Intel(R) Core(TM) i5-5250U CPU @ 2.7 GHz). The results are summarized in Figure 6.

Table 2. State of the Art with multi-subject validation setting.

			Accuracy per Class (%)				Overall Accuracy (%)
			Drug	Inhale	Exhale	Noise	
Holmes et al. (2012)			89.0	-	-	-	89.0
Holmes et al. (2013-14)			92.1	91.7	93.7	-	92.5
Taylor et al. (2017)	QDA		88.2	-	-	-	88.2
	ANN		65.6	-	-	-	65.6
Nousias et al. (2018)	SVM	MFCC	97.5	97.7	96.1	96.7	97.0
		SPECT	97.5	94.9	58.9	95.4	86.6
		CEPST	99.4	98.6	98.2	98.8	98.7
	RF	MFCC	97.1	96.7	95.9	95.1	96.2
		SPECT	97.7	98.0	97.0	96.5	97.3
		CEPST	99.0	98.2	97.4	96.5	97.7
	ADA	MFCC	97.5	96.9	96.8	93.6	96.2
		SPECT	98.8	98.4	97.0	97.9	98.0
		CEPST	99.2	97.5	97.4	97.9	98.0
	GMM	MFCC	96.7	97.7	96.1	96.3	96.7
		SPECT	99.2	98.2	93.3	88.4	94.8
		CEPST	99.4	98.6	99.2	96.9	98.5
Proposed Approach	Model 1		88.4	99.4	92.2	85.7	94.4
	Model 2		83.9	99.1	97.5	81.9	94.0
	Model 3		86.4	98.9	94.5	80.2	94.8
	Model 4		83.6	98.7	96.2	83.4	95.9
	Model 5		86.7	97.9	98.3	85.5	95.7

Table 3. State of the Art with all validation settings.

			Accuracy per Class (%)				Overall Accuracy (%)
			Drug	Inhale	Exhale	Noise	
Pettas et al. (2019)	Single Subject		83.0	98.0	98.0	87.0	94.8
	Multi Subject		93.0	96.0	98.0	79.0	92.8
	LOSO		88.0	98.0	96.0	86.0	93.8
Proposed Approach	Model 1	Single Subject	71.5	99.3	98.1	93.1	97.4
		Multi Subject	88.4	99.4	92.2	85.7	94.4
		LOSO	100.0	96.3	98.8	-	93.2
	Model 2	Single Subject	76.7	99.7	96.6	80.9	97.6
		Multi Subject	83.9	99.1	97.5	81.9	94.0
		LOSO	100.0	93.6	95.4	-	83.4
	Model 3	Single Subject	65.3	99.6	98.9	84.2	97.5
		Multi Subject	86.4	98.9	94.5	80.2	94.8
		LOSO	100.0	92.2	89.0	-	98.0
	Model 4	Single Subject	68.4	99.6	99.0	84.9	98.2
		Multi Subject	83.6	98.7	96.2	83.4	95.9
		LOSO	85.7	82.6	99.2	-	86.0
	Model 5	Single Subject	85.0	99.5	99.5	95.0	98.0
		Multi Subject	86.7	97.9	98.3	85.5	95.7
		LOSO	100.0	92.6	97.9	-	96.2

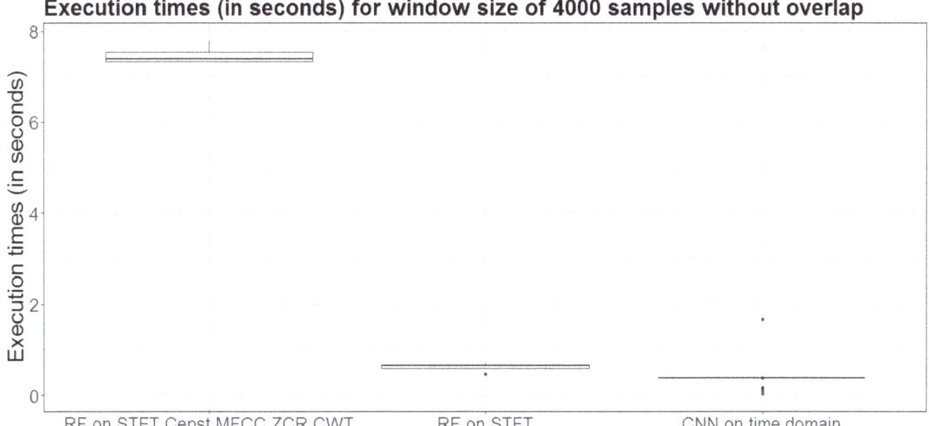

Figure 6. Comparison of the computational cost of our approach and other studies. Boxplots from left to right: RF with multiple features (mean time: 7.5 s), RF with only STFT (mean time: 0.6 s) and Model 5 of our CNN (mean time: 0.4 s).

This figure highlights the gain in computational speed up of our approach, compared to the time consuming Random Forest algorithm with feature extraction. Specifically, Figure 6 shows that classification by RF, using multiple features, requires more than 7 s, whereas the CNN Model 5 requires less than half a second. Finally, it is important to note that our approach is faster even when only STFT is extracted and used as input to the Random Forest.

4.3.2. Pruning Scalar Weights

In order to evaluate the performance of this algorithm, we present the classification accuracy as well as the compression rate, when no retraining is applied, in Table 4. The parameter l, which determines the threshold for pruning, varies from 0.1 to 1.0 with a step of 0.1. It is clear that when increasing the parameter l and consequently the threshold for pruning, the accuracy of the classifier decays. Among the five models, more robust to changes appears to be Models 1, 4 and 5, because they retain their performance above 90%, even when the parameter l is set to 0.8. On the other hand, model 3 and 4 show the worst performance dropping below 90% for intermediate values of l.

The results, presented in Table 5, corresponding to the approach that employs the retraining technique, show that the classifiers are able to adapt to the changes made in the previous layers, retaining their high accuracy, independently of the threshold defined by l and σ. It is worth mentioning that with this approach the lowest classification accuracy is 93% achieved by model 3, whereas model 5 reaches up to 96%, improving its initial performance. Additionally, we are able to compress the architectures two times more than the previous approach, where retraining process is not included. This occurs because retraining the network results in larger standard deviation of the weights in each layer, but with their mean value almost equal to zero. Thus, more weights will be zeroed out.

It should be highlighted that pruning scalar values can only reduce the memory requirements since there are fewer non zero weights. However, it does not perform structural pruning, meaning that it is uncertain if the pruned parameters belong to a particular filter or a neuron and therefore it does not improve the computational time.

Table 4. Evaluation of the performance for the developed architectures with the method of pruning scalar weights without retraining. Factor *l* corresponds to the percentage of the standard deviation used to determine the threshold for pruning.

	Factor *l*	**0.1**	**0.2**	**0.3**	**0.4**	**0.5**	**0.6**	**0.7**	**0.8**	**0.9**	**1.0**
	Loss	0.24	0.25	0.24	0.24	0.24	0.23	0.26	0.31	0.38	0.51
Model 1	**Accuracy (%)**	94.83	94.14	93.97	93.97	93.80	93.97	93.45	91.91	90.01	87.60
	Compression Rate	1.08	1.18	1.29	1.42	1.58	1.79	1.99	2.25	2.55	2.91
	Loss	0.24	0.26	0.28	0.33	0.60	0.69	1.54	0.98	0.97	1.34
Model 2	**Accuracy (%)**	93.97	93.45	93.28	91.56	84.16	82.09	58.86	65.95	69.53	55.93
	Compression Rate	1.08	1.19	1.31	1.45	1.63	1.84	2.10	2.40	2.77	3.19
	Loss	0.19	0.19	0.19	0.23	0.24	0.38	0.74	1.44	1.67	1.56
Model 3	**Accuracy (%)**	94.83	94.83	94.32	93.11	92.95	87.77	74.69	56.45	51.80	43.02
	Compression Rate	1.08	1.19	1.31	1.45	1.64	1.84	2.08	2.39	2.74	3.1712
	Loss	0.20	0.20	0.19	0.18	0.19	0.19	0.19	0.22	0.48	0.82
Model 4	**Accuracy (%)**	95.69	95.69	95.69	95.52	95.18	95.18	95.00	93.97	85.71	75.21
	Compression Rate	1.08	1.17	1.28	1.41	1.56	1.74	1.94	2.17	2.45	2.78
	Loss	0.21	0.21	0.21	0.20	0.20	0.21	**0.21**	0.39	0.31	0.64
Model 5	**Accuracy (%)**	95.70	95.87	95.87	95.87	95.53	95.01	**94.50**	88.83	90.20	79.89
	Compression Rate	1.08	1.17	1.29	1.42	1.57	1.76	**1.98**	2.24	2.54	2.89

Table 5. Evaluation of the performance for the developed architectures with the method of pruning scalar weights with retraining. Factor *l* corresponds to the portion of standard deviation used to determine the threshold for pruning.

	Factor *l*	**0.1**	**0.2**	**0.3**	**0.4**	**0.5**	**0.6**	**0.7**	**0.8**	**0.9**	**1.0**
	Loss	0.54	0.53	0.66	0.60	0.65	0.53	0.58	0.48	0.50	0.56
Model 1	**Accuracy (%)**	94.32	95.35	94.32	95.00	94.83	94.66	95.18	94.83	95.52	95.00
	Compression Rate	1.09	1.20	1.35	1.55	1.81	1.90	2.48	2.84	3.26	3.72
	Loss	0.51	0.57	0.53	0.52	0.48	0.64	0.61	0.54	0.50	0.51
Model 2	**Accuracy (%)**	93.80	94.32	94.14	94.66	94.49	94.32	94.66	95.00	95.18	94.49
	Compression Rate	1.09	1.22	1.38	1.62	1.74	2.35	2.81	3.28	3.74	4.28
	Loss	0.39	0.53	0.55	0.56	0.49	0.40	0.45	0.44	0.43	0.41
Model 3	**Accuracy (%)**	94.66	94.83	94.14	93.63	94.83	93.80	94.14	93.45	93.45	93.97
	Compression Rate	1.09	1.22	1.39	1.61	1.90	2.00	2.59	3.00	3.47	3.92
	Loss	0.44	0.52	0.49	0.60	0.43	0.52	0.50	0.50	0.47	0.50
Model 4	**Accuracy (%)**	95.52	95.18	95.35	95.18	94.83	94.83	95.00	95.00	95.18	94.83
	Compression Rate	1.09	1.20	1.35	1.55	1.63	2.11	2.44	2.82	3.21	3.63
	Loss	0.48	0.55	0.54	0.49	0.51	0.24	0.34	0.40	0.38	**0.39**
Model 5	**Accuracy (%)**	95.70	95.70	95.87	95.53	96.04	95.01	95.01	94.55	94.50	**94.50**
	Compression Rate	1.09	1.20	1.32	1.52	1.78	5.24	5.56	6.03	6.60	**7.14**

4.3.3. Pruning Filters in Convolutional Layers

In this section, we present the results for the evaluation of all five developed models, after applying the filter pruning method according to which the filters with the smallest magnitude are removed. We tested the classification accuracy of the pre-trained models for multiple combinations of pruned filters and, additionally, we investigated the effect of iterative pruning and retraining. For every model we chose to leave at least one filter at each convolutional layer. Thus, for Models 1, 2, 5 the number of the removed filters varies from 1 to 15, whereas for Models 3, 4 it is between 1 and 7. Tables 6 and 7 present the performance of the models in terms of test loss and test accuracy, as well as results for the compression and the theoretical speed up of each architecture.

Table 6. Results for filter pruning with no retraining.

	Pruned Filters	1	2	3	4	5	6	7	8	9	10	11	12	13	14	15
Model 1	Loss	0.22	0.20	0.23	**0.24**	0.71	1.22	0.85	1.59	1.31	1.48	2.32	2.10	2.09	1.79	1.77
	Accuracy (%)	93.97	93.80	93.97	**93.97**	78.66	59.38	66.95	44.92	49.05	49.40	37.52	46.99	46.30	24.44	16.87
	Compression Rate	1.06	1.14	1.22	**1.31**	1.41	1.53	1.67	1.83	2.02	2.25	2.52	2.86	3.30	3.88	4.67
	Flops	0.89	0.78	0.68	**0.59**	0.50	0.42	0.35	0.28	0.22	0.17	0.12	0.09	0.06	0.03	0.01
Model 2	Loss	0.24	0.25	0.67	2.58	1.35	0.85	1.08	0.66	0.80	0.88	1.13	1.24	1.28	1.46	1.47
	Accuracy (%)	93.80	94.15	76.07	51.63	62.48	74.18	69.88	71.25	63.51	70.74	58.00	34.42	27.37	16.87	16.87
	Compression Rate	1.08	1.16	1.26	1.38	1.52	1.68	1.89	2.13	2.44	2.84	3.39	4.17	5.38	7.51	12.17
	Flops	0.89	0.78	0.68	0.59	0.50	0.42	0.35	0.28	0.22	0.17	0.12	0.08	0.05	0.03	0.01
Model 3	Loss	1.16	4.90	2.25	1.35	1.40	1.47	1.66	-	-	-	-	-	-	-	-
	Accuracy (%)	71.43	27.37	25.13	37.18	29.60	35.28	35.28	-	-	-	-	-	-	-	-
	Compression Rate	1.14	1.33	1.59	1.96	2.53	3.53	5.72	-	-	-	-	-	-	-	-
	Flops	0.79	0.61	0.44	0.31	0.19	0.10	0.04	-	-	-	-	-	-	-	-
Model 4	Loss	0.22	0.61	2.78	9.03	1.64	8.08	1.68	-	-	-	-	-	-	-	-
	Accuracy (%)	94.15	80.38	57.38	16.87	21.69	16.87	16.87	-	-	-	-	-	-	-	-
	Compression Rate	1.10	1.23	1.38	1.56	1.80	2.12	2.55	-	-	-	-	-	-	-	-
	Flops	0.79	0.61	0.45	0.31	0.20	0.11	0.05	-	-	-	-	-	-	-	-
Model 5	Loss	0.33	2.44	1.56	0.52	0.48	0.53	0.72	1.14	1.26	1.21	1.90	1.67	1.79	1.61	1.64
	Accuracy (%)	93.64	66.95	67.99	78.83	79.69	78.83	71.60	47.50	42.68	40.79	35.97	35.46	34.08	35.46	35.46
	Compression Rate	1.06	1.14	1.22	1.31	1.41	1.53	1.67	1.83	2.02	2.25	2.52	2.86	3.30	3.88	4.67
	Flops	0.89	0.78	0.68	0.59	0.50	0.42	0.35	0.28	0.22	0.17	0.13	0.09	0.0568	0.03	0.01

Table 7. Results for filter pruning with iterative retraining.

	Pruned Filters	1	2	3	4	5	6	7	8	9	10	11	12	13	14	15
Model 1	Loss	0.40	0.38	0.47	0.42	0.40	0.39	0.39	0.36	0.32	0.25	0.26	0.2704	0.29	0.20	0.25
	Accuracy (%)	94.32	95.18	94.32	94.49	95.18	95.18	94.32	94.66	95.01	94.84	94.84	93.63	92.08	94.32	92.94
	Compression Rate	1.06	1.14	1.22	1.31	1.41	1.53	1.67	1.83	2.02	2.25	2.52	2.86	3.30	3.88	4.67
	Flops	0.89	0.78	0.68	0.59	0.50	0.42	0.35	0.28	0.22	0.17	0.13	0.09	0.06	0.03	0.01
Model 2	Loss	0.40	0.50	0.37	0.36	0.34	0.25	0.32	0.30	0.29	0.36	0.27	0.21	0.21	0.26	1.32
	Accuracy (%)	93.97	93.46	95.18	93.80	94.15	95.52	94.15	94.66	93.97	93.46	94.66	94.66	93.97	93.11	35.46
	Compression Rate	1.08	1.16	1.26	1.38	1.52	1.68	1.88	2.12	2.43	2.84	3.39	4.17	5.38	7.51	12.17
	Flops	0.89	0.78	0.68	0.59	0.50	0.42	0.36	0.28	0.22	0.17	0.12	0.08	0.05	0.03	0.01
Model 3	Loss	0.33	0.32	0.25	0.26	0.20	0.18	0.27	-	-	-	-	-	-	-	-
	Accuracy (%)	93.63	94.15	94.84	93.29	93.80	94.15	91.05	-	-	-	-	-	-	-	-
	Compression Rate	1.14	1.33	1.59	1.96	2.53	3.53	5.72	-	-	-	-	-	-	-	-
	Flops	0.79	0.61	0.44	0.31	0.19	0.10	0.04	-	-	-	-	-	-	-	-
Model 4	Loss	0.37	0.33	0.36	0.25	0.28	0.23	0.35	-	-	-	-	-	-	-	-
	Accuracy (%)	95.87	94.49	95.15	94.84	93.80	92.94	87.78	-	-	-	-	-	-	-	-
	Compression Rate	1.10	1.23	1.38	1.56	1.80	2.12	2.55	-	-	-	-	-	-	-	-
	Flops	0.79	0.61	0.45	0.31	0.20	0.11	0.05	-	-	-	-	-	-	-	-
Model 5	loss	0.28	0.32	0.30	0.31	0.2854	0.28	0.23	0.32	0.29	0.31	0.26	0.2545	**0.25**	0.22	0.22
	Accuracy (%)	94.66	95.52	95.52	94.49	95.52	95.00	94.84	94.84	95.00	94.49	93.46	94.15	**95.18**	93.46	93.11
	Compression Rate	1.06	1.14	1.22	1.31	1.41	1.53	1.67	1.83	2.02	2.25	2.52	2.86	**3.30**	3.88	4.67
	Flops	0.89	0.78	0.68	0.59	0.50	0.42	0.35	0.28	0.22	0.17	0.13	0.09	**0.06**	0.03	0.01

In Table 6 we observe that the classification accuracy of every model is significantly deteriorating, even at low compression rates. The reason for this is that filter pruning is employed on pre-trained models and therefore the values of their weights are not the optimal for the new, shallower architectures. In addition, Model 2 can be compressed at a larger scale than the others, due to its architecture. It has the most filters in the convolutional layers and, at the same time, the smallest number of neurons in the fully connected layers, as shown in Table 1, with the amount of parameters belonging to convolutional layers being approximately 19% of the total number of parameters, whereas for the other models it is 16% or lower. Note that even with half of the filters removed, the compression rate is low, indicating that the majority of the weights belongs to the fully connected layer. On the other hand, the removal of a filter reduces the computational requirements. For example, when we prune 2 out of 16 filters from model 1, the new, more shallow, architecture requires the 78% of the initial floating point operations to perform the classification task, providing a reduction of over 20%. Approaching the maximum number of the pruned filters (leaving only one filter), the required operations are, as expected, considerably reduced to only 1% of the operations required by un-pruned models.

A countermeasure against the drop of classification accuracy, due to filter removal, is the utilization of retraining technique, as described in Section 3.2. The results of filter pruning method with iterative retraining are shown in Table 7. It can be observed that the classification accuracy for all models except from Models 2, 4 remains over 90%, whereas for Models 2, 4 it drops to 35% and 87% when $15/16$ and $7/8$ filters are pruned in each layer, respectively. Thus, by applying this method we can significantly reduce the computational time without sacrificing efficiency. A characteristic example is Model 5, which reaches up to 95% classification accuracy, even with 13 filters pruned. For the same model, the respective performance achieved without retraining is 34%, while for both cases, the pruned models require 5% of the operations needed by the initial un-pruned architecture.

4.3.4. Quantizing Only the Convolutional Layers

To evaluate the performance of the vector quantization method, we applied both scalar and product quantization to convolutional layers, as well as to fully connected layers of the network. This paragraph shows the classification accuracy of the developed models with respect to the compression rate and the number of required floating point operations, when the quantization methods are applied only on convolutional layers. As mentioned earlier, both scalar and product quantization are performed along the channel's dimension. We tested different combinations regarding the number of clusters and the value of the splitting parameter s.

In particular, for scalar quantization the number of clusters varies between 1 and 8, whereas for product quantization we tried $s = 1, 2, 4$ and the maximum number of clusters was set to 8, 4, 2 respectively. Note that for $s = 1$ we essentially perform scalar quantization. Table 8 shows the classification accuracy and the achieved compression, as well as the speed up in terms of flops. It is clear, that by increasing the number of clusters and therefore the number of filters that contain different kernels, the accuracy of the classifiers increases as well. This originates from the fact that with more different filters more features of the input can be extracted in convolutional layers. It should be also mentioned that the compression rate achieved by this method, is lower than the rate achieved by filter pruning. This happens because an index matrix is required, to map the filters in the codebook to the filters in the original structure, which increases the memory requirements.

Concerning the amount of required operations in convolutional layers, as described before, it can be reduced with this approach by reusing pre-computed inner products. In particular, for similar convolutional kernels we only need to convolve one of them with the input feature map and then the result is shared. Then, the biases are added and the result passes the activation and pooling function, to produce the input feature map of the next layer. It is worth mentioning that the percentage of the required operations is directly proportional to the percentage of the filters needed to be stored. For example, clustering of 16 filters to 4 clusters causes a 25% reduction in required floating point operations. Again, comparing the flops for filter pruning and scalar quantization, the first is more efficient. This is because the removal of a filter reduces the dimensions of the next layer's input feature map which is not the case for the scalar quantization.

Next, we evaluate the effect of product quantization on the performance of the models. Similarly, to scalar quantization on convolutional layer, we examine the fluctuation of the accuracy with respect to compression rate and the ratio of the required floating operations of the quantized architectures to the amount of flops for the initial structure, as shown in Table 9. The splitting parameter takes the values 1, 2, 4. As s increases, the number of clusters in each subspace decreases, since there are fewer filters. Because both the separation of the weight matrix of each layer and the k-means algorithm are performed on the channel axis, when $s = 1$ the results are identical to those with scalar quantization. Additionally, the increase in the value of s result in a slight decrease in the classification accuracy of the model. For example, Model 1 with $s = 1$ and $clusters = 4$ reaches an accuracy of 93%, whereas with $s = 2$ and $clusters = 2$, a combination that produces 4 distinct filters in each layer, leads to 91%. This decrease indicates that apart from how many filters we group together, it is also crucial which filters are grouped. By splitting the original space in smaller sub-spaces, we narrow the available combinations of filters and, thus, filters that differ a lot from each other could be combined forming one cluster.

Table 8. Results for scalar quantization on convolutional layers only.

	Number of Clusters	1	2	3	4	5	6	7	8
Model 1	**Loss**	4.61	0.38	0.33	0.26	0.25	0.23	0.24	0.24
	Accuracy (%)	16.87	90.36	92.94	94.32	94.84	94.84	94.84	94.66
	Compression Rate	1.18	1.16	1.15	1.14	1.12	1.11	1.10	1.09
	Flops	0.07	0.13	0.19	0.26	0.32	0.38	0.44	0.50
Model 2	**Loss**	4.35	0.28	0.25	0.22	0.23	0.23	0.24	0.25
	Accuracy (%)	16.87	92.08	93.80	94.32	94.32	94.32	94.66	94.32
	Compression Rate	1.18	1.17	1.15	1.13	1.12	1.10	1.10	1.08
	Flops	0.07	0.13	0.19	0.25	0.32	0.38	0.44	0.50
Model 3	**Loss**	10.57	0.40	0.25	0.20	0.20	0.18	0.19	0.19
	Accuracy (%)	16.87	89.16	93.80	94.49	94.84	95.01	94.84	94.84
	Compression Rate	1.10	1.08	1.07	1.05	1.04	1.03	1.01	1.00
	Flops	0.14	0.26	0.38	0.51	0.63	0.75	0.88	1.00
Model 4	**Loss**	10.88	1.08	0.24	0.22	0.22	0.22	0.21	0.20
	Accuracy	16.87	72.80	95.18	95.87	95.52	95.52	95.70	95.87
	Compression Rate	1.07	1.06	1.05	1.04	1.03	1.01	1.01	1.00
	Flops	0.14	0.26	0.39	0.50	0.63	0.75	0.88	1.00
Model 5	**Loss**	1.78	0.65	**0.24**	0.22	0.21	0.21	0.21	0.21
	Accuracy (%)	35.49	83.47	**94.49**	95.18	95.35	95.52	95.35	95.70
	Compression Rate	1.18	1.16	**1.15**	1.14	1.12	1.11	1.10	1.09
	Flops	0.07	0.13	**0.19**	0.26	0.32	0.38	0.44	0.50

It is also important to note that the increase of the s parameter leads to a slight decrease of the compression rate. This is because with higher values of the splitting parameter, the lowest number of clusters in the weight matrix is increased as well. For example, for $s = 1$ and $clusters = 4$, the amount of different filters is 4, but if we set $s = 2$ the respective amount would be 8, since we form 4 clusters in each subspace. Therefore, for the minimum number of clusters (1 cluster) and for $s = 1$, one filter will be created. For $s = 2$, two distinct filters will be formed and finally for $s = 4$, four filters will have different weight values.

Concerning the performance of the architectures, with respect to the computational complexity, we observe in Table 9 that 75% of the initial flops can be avoided for Models 1, 2, 5 without any drop in classification accuracy. On the other hand, for the remaining models we save 50% of the initial required operations, with no drop in classification accuracy. For $s = 2$, we are able to cut the majority for the operations with Models 1, 2, 5 reaching up to 94% accuracy with 38% of the initial amount of floating point operations. However, in order to achieve a classification accuracy higher than 90% we can reduce the amount of the operations by half at most. At 0.5 of the initial number of flops, model 3 reaches up to 95% and model 4 to 93%.

Table 9. Product quantization for all combinations of s and *clusters* on convolutional layers only.

	Splitting Parameter	s = 1								s = 2				s = 4	
	Clusters	1	2	3	4	5	6	7	8	1	2	3	4	1	2
Model 1	Loss	4.61	0.35	0.33	0.26	0.23	0.23	0.24	0.24	2.89	0.35	0.28	0.27	2.65	**0.22**
	Accuracy (%)	16.87	91.74	92.94	93.80	94.66	95.18	94.66	94.66	16.87	91.91	94.15	93.46	39.76	**94.66**
	compression rate	1.18	1.16	1.15	1.14	1.12	1.11	1.10	1.09	1.16	1.14	1.11	1.09	1.14	**1.09**
	Flops	0.07	0.13	0.19	0.26	0.32	0.38	0.44	0.50	0.13	0.26	0.38	0.50	0.25	**0.50**
Model 2	Loss	4.35	0.27	0.25	0.22	0.22	0.23	0.23	0.24	1.69	0.28	0.23	0.24	1.24	0.26
	Accuracy (%)	16.87	91.91	93.63	94.32	94.49	94.49	94.66	94.66	37.00	92.94	93.97	94.32	53.87	93.11
	Compression Rate	1.21	1.20	1.18	1.16	1.15	1.13	1.12	1.10	1.20	1.16	1.13	1.10	1.16	1.10
	Flops	0.0690	0.13	0.19	0.25	0.32	0.38	0.44	0.50	0.13	0.25	0.38	0.50	0.25	0.50
Model 3	Loss	10.57	0.45	0.27	0.20	0.20	0.19	0.19	0.19	10.30	0.28	0.20	0.1892	2.21	0.19
	Accuracy (%)	16.87	45.13	27.00	20.17	19.91	18.84	18.62	18.92	17.04	95.01	95.01	94.83	38.21	94.84
	Compression Rate	1.10	1.08	1.07	1.05	1.04	1.03	1.01	1.00	1.08	1.05	1.03	1.00	1.05	1.00
	Flops	0.14	0.26	0.38	0.51	0.63	0.75	0.88	1.00	0.26	0.51	0.75	1.00	0.50	1.00
Model 4	Loss	10.88	1.00	0.22	0.22	0.23	0.22	0.21	0.20	11.93	0.28	0.25	0.20	4.41	0.20
	Accuracy (%)	16.87	74.35	95.70	96.04	95.52	95.52	95.70	95.87	16.87	93.46	95.35	95.87	37.52	95.87
	Compression Rate	1.07	1.06	1.05	1.04	1.03	1.02	1.01	1.00	1.06	1.04	1.02	1.00	1.04	1.00
	Flops	0.14	0.26	0.39	0.51	0.63	0.75	0.88	1.00	0.26	0.51	0.75	1.00	0.51	1.00
Model 5	Loss	1.78	0.65	**0.23**	0.21	0.21	0.21	0.21	0.21	2.01	0.64	**0.22**	0.21	1.69	0.29
	Accuracy (%)	35.59	83.99	**94.32**	95.01	95.70	95.35	95.70	96.04	35.46	83.47	**94.66**	95.35	47.85	91.91
	Compression Rate	1.18	1.16	**1.15**	1.14	1.12	1.11	1.10	1.09	1.16	1.14	**1.11**	1.09	1.14	1.09
	Flops	0.07	0.13	**0.19**	0.26	0.32	0.38	0.44	0.50	0.13	0.26	**0.38**	0.50	0.26	0.50

To sum up, quantizing convolutional layers using k-means algorithm, either with the scalar or the product method, we can compress the structure and at the same time we can speed up the production of the output feature map and, consequently, the prediction of the classifier. Between these two benefits, the computation gain is greater, since we efficiently can remove up to 75% of the operations required initially, whereas the maximum compression rate achieved reaches up to 1.2. This result is consistent with the theory suggesting that convolutional layers are computationally expensive and they do not add excessive memory requirements. Finally, for product quantization, increasing the value of the parameter s, the performance of the classifier is deteriorated.

4.3.5. Quantizing Only Fully Connected Layer

Similarly, in this paragraph we present the results for scalar and product quantization, but in this case they are performed on the fully connected layers. For this approach we selected to perform quantization with k-means at the y axis of the weights matrix. In this way, we force the neurons to have the same output response and, therefore, we are able to reduce the computational requirements of the layers. Subsequently, we compare the requirements in storage and computation between convolutional and fully connected layer and validate that convolutions are time consuming, whereas fully connected layers significantly increase memory requirements. For Models 1, 2, 5 we perform scalar quantization with number of clusters up to 128 and for Models 3, 4 up to 52. We also executed tests for $s = 1, 2, 4$ and clusters up to 32, 16, 8 respectively. It is important to mention that because we force some neurons to have the same output, we do not perform quantization at the output layer of the classifier i.e the last fully connected layer.

From the results shown in Table 10 it is clear that Models 1, 5, which share the same structure, have the same behaviour retaining their initial performance until a compression rate of 4.6. Furthermore, we are able to achieve a larger compression for Models 3, 4 because both of them have the shallowest convolutional structure, with three convolutional layers of 8 filters in each layer.

This means that for these models the weights of fully connected layers occupy a greater portion of the total amount of the trainable parameters. It is important to highlight that we can compress model 4 more than six times and yet achieve a high classification accuracy up to 93%. Finally, for Models 2, 3 the maximum number of clusters is 52 because the last layer contains $16 \times 4 = 64$ weights and therefore there is no reason to increase it further. Also, as described above, scalar quantization does not contribute to the speed up of the classification task and this is why Table 10 does not contain the flops that the quantized models require.

Table 10. Results for scalar quantization on fully connected layers only.

	Number of Clusters	1	4	8	16	24	32	40	52	64	72	96	112	128
Model 1	Loss	1.38	0.18	0.23	0.23	0.24	0.24	0.24	0.24	0.24	0.24	0.24	0.24	0.24
	Accuracy (%)	35.46	93.97	94.49	95.00	94.15	94.49	94.32	94.32	94.32	94.66	94.49	94.49	94.94
	Compression Rate	6.08	4.61	4.12	3.71	3.51	3.38	3.28	3.17	3.09	3.05	2.94	2.89	2.84
Model 2	Loss	1.38	0.44	0.18	0.26	0.25	0.26	0.27	0.26	-	-	-	-	-
	Accuracy (%)	16.87	86.75	94.66	93.97	93.80	93.63	93.46	93.63	-	-	-	-	-
	Compression Rate	5.26	4.15	3.76	3.43	3.26	3.15	3.07	2.98	-	-	-	-	-
Model 3	Loss	1.39	0.26	0.18	0.18	0.18	0.19	0.19	0.19	-	-	-	-	-
	Accuracy (%)	16.87	91.91	95.00	95.35	94.84	95.00	95.00	94.49	-	-	-	-	-
	Compression Rate	9.56	6.23	5.30	4.60	4.27	4.06	3.91	3.74	-	-	-	-	-
Model 4	Loss	0.35	0.23	**0.19**	0.20	0.21	0.20	0.20	0.20	0.20	0.20	0.20	0.20	0.20
	Accuracy	35.46	93.29	**95.52**	95.87	95.70	96.04	95.87	95.87	95.70	96.04	95.87	95.70	95.70
	Compression Rate	12.44	7.26	**5.10**	5.10	4.69	4.43	4.25	4.04	3.90	3.81	3.65	3.53	3.45
Model 5	Loss	1.37	0.19	0.21	0.21	0.21	0.21	0.21	0.21	0.21	0.21	0.21	0.21	0.21
	Accuracy (%)	35.46	94.32	95.70	95.52	95.00	95.87	95.52	95.70	95.87	95.70	95.87	95.87	95.70
	Compression Rate	6.08	4.61	4.12	3.71	3.51	3.38	3.28	3.17	3.09	3.05	2.94	2.89	2.84

The next approach to compress and accelerate the fully connected layers is product quantization through k-means algorithm. Tables 11 and 12 present the classification accuracy, compression rate as well as the extent of the reduction of floating point operations for different number of clusters and different values of the splitting parameter s. In Table 11 Models 2, 3 we stop at 12 clusters, because they have a fully connected layer with 16 neurons and therefore there is no point in increasing the number of clusters beyond 12. Recall that quantization is performed on the columns of the weight matrix, that is, on the output response of a layer. It should be noted that the achieved compression rate is higher as the number of clusters is increasing than the respective rate with scalar quantization. This lies on the fact that the index matrix for this approach is smaller than the index matrix for scalar quantization, containing as many slots as the neurons in each layer are. Furthermore, the difference in the compression rate between the models is due to the difference between their structure. For example, Model 4 has smaller convolutional layers from Models 1, 2, 5 and larger fully connected layer from Model 3 resulting in a higher compression rate. Moreover, it is clear that by quantizing fully connected layers we do not have any gain in computational cost, since the smallest ratio with an acceptable performance is 0.977, which means that the quantized model needs to execute 97.7% of initial amount of floating point operations. Finally, Table 12 shows the performance of the classifiers for $s = 2$ and 4. In this case, it should be highlighted that increasing the value of s the classification accuracy of the models decreases, despite the same compression rate. For example, model 1 achieves a classification accuracy of 94% with $s = 2$ and 4 clusters but for $s = 4$ and 2 clusters its accuracy drops to 90%.

Table 11. Product quantization for all combinations of *clusters* and $s = 1$ on fully connected layer only.

	Splitting Parameter	**s = 1**									
	Clusters	**1**	**2**	**4**	**8**	**12**	**16**	**20**	**24**	**28**	**32**
Model 1	**Loss**	1.39	0.59	0.22	0.24	0.24	0.24	0.24	0.24	0.24	0.24
	Accuracy (%)	35.46	80.03	94.49	93.97	94.84	94.14	94.32	94.15	94.49	94.32
	Compression Rate	6.16	6.14	6.11	6.05	5.99	5.93	5.88	5.82	5.77	5.72
	Flops	0.99	0.99	0.99	0.99	0.99	0.99	0.99	0.99	0.99	0.99
Model 2	**Loss**	1.40	0.47	0.41	0.31	0.25	-	-	-	-	-
	Accuracy	16.87	83.30	88.12	92.25	93.80	-	-	-	-	-
	Compression Rate	5.29	5.27	5.25	5.20	5.16	-	-	-	-	-
	Flops	0.99	0.99	0.99	0.99	0.99	-	-	-	-	-
Model 3	**Loss**	1.40	0.76	0.19	0.19	0.20	-	-	-	-	-
	Accuracy (%)	35.46	63.85	94.32	95.18	94.66	-	-	-	-	-
	Compression Rate	9.74	9.69	9.59	9.41	9.24	-	-	-	-	-
	Flops	0.99	0.99	0.99	0.99	0.99	-	-	-	-	-
Model 4	**Loss**	1.37	0.57	**0.17**	0.20	0.20	0.20	0.20	0.20	0.20	0.20
	Accuracy (%)	35.97	84.68	**94.84**	95.69	95.52	95.69	95.52	95.52	95.87	95.87
	Compression Rate	13.12	13.03	**12.86**	12.57	12.31	12.05	11.81	11.59	11.37	11.15
	Flops	0.98	0.98	**0.98**	0.98	0.98	0.99	0.99	0.99	0.99	0.99
Model 5	**Loss**	1.36	0.41	0.21	0.20	0.21	0.21	0.21	0.21	0.21	0.21
	Accuracy (%)	35.46	88.81	95.35	95.52	95.87	95.87	96.04	96.04	95.70	95.70
	Compression Rate	6.16	6.14	6.11	6.05	5.99	5.93	5.88	5.82	5.77	5.72
	Flops	0.99	0.99	0.99	0.99	0.99	0.99	0.99	0.99	0.99	0.99

Table 12. Product quantization for all combinations of *clusters* and $s = 2, 4$ on fully connected layer only.

	Splitting Parameter	**s = 2**						**s = 4**			
	Clusters	**1**	**2**	**4**	**8**	**12**	**16**	**1**	**2**	**4**	**8**
Model 1	**Loss**	1.38	0.52	0.22	0.23	0.24	0.24	1.39	0.38	0.22	0.23
	Accuracy (%)	35.46	90.01	94.14	94.14	93.97	94.49	35.46	90.53	93.97	94.15
	Compression Rate	6.14	6.11	6.05	5.93	5.82	5.72	6.11	6.04	5.93	5.72
	Flops	0.99	0.99	0.99	0.99	0.99	0.99	0.99	0.99	0.99	0.99
Model 2	**Loss**	1.40	0.47	0.31	-	-	-	1.37	0.76	-	-
	Accuracy	16.87	81.41	91.05	-	-	-	16.87	76.25	-	-
	Compression Rate	5.27	5.25	5.20	-	-	-	5.25	5.20	-	-
	Flops	0.99	0.99	0.99	-	-	-	0.99	0.99	-	-
Model 3	**Loss**	1.43	0.64	0.95	-	-	-	1.48	0.37	-	-
	Accuracy (%)	16.87	79.69	94.84	-	-	-	16.87	90.71	-	-
	Compression Rate	9.69	9.59	9.41	-	-	-	9.59	9.41	-	-
	Flops	0.99	0.99	0.99	-	-	-	0.99	0.99	-	-
Model 4	**Loss**	1.37	0.57	**0.18**	0.20	0.20	0.20	1.38	0.46	**0.21**	0.20
	Accuracy (%)	37.00	85.54	**95.35**	95.70	95.52	95.70	36.60	88.12	**95.01**	95.52
	Compression Rate	13.03	12.86	**12.57**	12.06	11.59	11.15	12.86	12.57	**12.06**	11.15
	Flops	0.98	0.98	**0.98**	0.99	0.99	0.99	0.99	0.99	**0.99**	0.99
Model 5	**Loss**	1.36	0.35	0.19	0.21	0.21	0.21	1.37	0.35	0.21	0.20
	Accuracy (%)	35.46	91.74	95.18	95.87	95.52	96.04	35.46	89.84	94.66	95.70
	Compression Rate	6.14	6.11	6.05	5.93	5.82	5.72	6.11	6.05	5.93	5.72
	Flops	0.99	0.99	0.99	0.99	0.99	0.99	0.99	0.99	0.99	0.99

4.3.6. Combining Filter Pruning and Quantization

Finally, we investigate the combination of the aforementioned methods by applying filter pruning in the convolutional layers and quantization on fully connected layers. In this way, we are able to reduce the requirements of the classifier in both memory and computational power. The approach of the iterative training is selected for filter pruning, since it yields better results than the approach with no retraining. Firstly, we perform filter pruning in order to exploit the fact that the weights adjust to the changes and, then, quantization algorithm, either scalar or product, is executed. Below, we present the classification accuracy of the developed architecture, with respect to the amount of the pruned feature maps in convolutional layers and the number of clusters in fully connected layer.

Figure 7 shows how classification accuracy changes as the number of pruned feature maps or the clusters, produced with scalar quantization, increases. It is clear that the accuracy of all models, apart from model 2, depends mostly on the number of clusters in fully connected part of the classifier. When we perform k means on it, with clusters equal to 1, the classification accuracy drops to 35% (Models 1, 2, 4, 5) and 16% (Model 3). Model 2 reaches 91% or above when 14 out of 16 filters have been removed and for 8 clusters in each fully connected layer. However, when we prune 15 out of 16 filters its accuracy drops to 35% without improving when the number of clusters is increasing. On the other hand, the rest of the models keep their classification accuracy at high levels, even when their convolutional layers are left with only one filter. The best architectures seem to be Models 1 and 5, which achieve an accuracy over 90% with 8 clusters and even with 15 out of 16 filters removed.

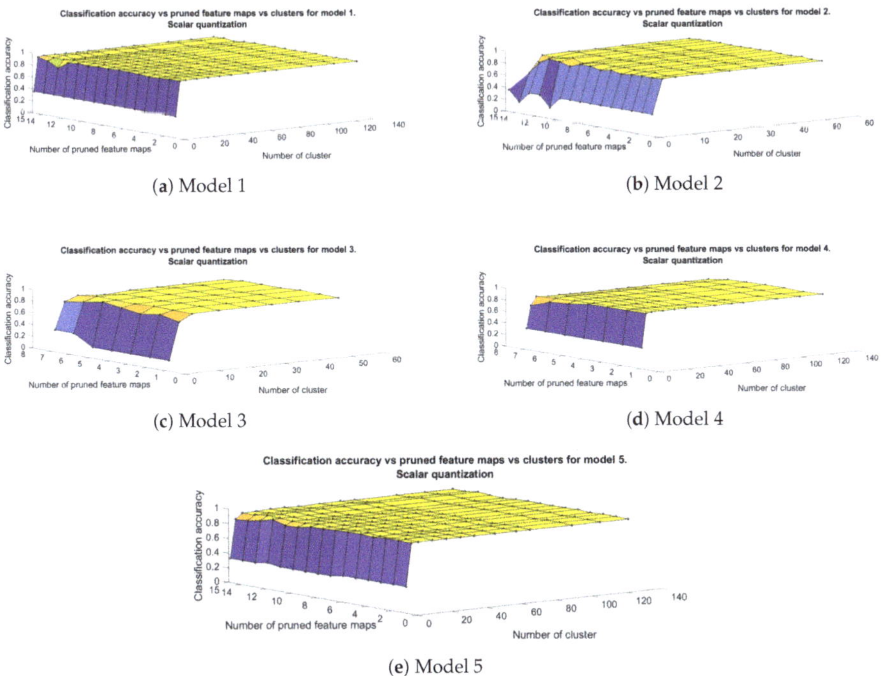

(**a**) Model 1 (**b**) Model 2

(**c**) Model 3 (**d**) Model 4

(**e**) Model 5

Figure 7. Classification accuracy of the different models in Table 1 that include filter pruning and scalar quantization. The horizontal axes represent the number of pruned feature map and number of clusters in fully connected layer, respectively.

Next, we proceed to the evaluation of combining filter pruning method with product quantization along y axis of the weight matrix of the fully connected layer. Figure 8 shows the classification accuracy of the developed models, versus the number of pruned feature maps and the number of clusters in each subspace, when the value of splitting parameter s is set equal to 1. For Models 1, 4, 5 the maximum amount of clusters is 32 whereas for Models 2, 3 is 12. Again, the parameter that affects mostly the performance of the classifier is the number of clusters produced by k-means algorithm. For $cluster = 1$, hence when we force all neurons to have the same output response, the classification accuracy drops dramatically to 35% (Models 1, 5) and 16% (Models 2, 3, 4). It is also clear that the highest classification accuracy can be achieved with intermediate values of the parameters. For example, model 5 reaches up to 95.52% accuracy, which is the highest among our architectures, after pruning 7 feature maps and for 8 clusters in fully connected layer achieving 8 times compression of the initial structure. For the same level of compression model 1 achieves 94.66% ($pruned\ feature\ maps = 7$, $clusters = 8$), model 2 93.97% ($pruned\ feature\ maps = 7$, $clusters = 8$), model 3 reaches up to 93.8% ($pruned\ feature\ maps = 2$, $clusters = 4$) and model 4 to 95.01% ($pruned\ feature\ maps = 3, clusters = 8$).

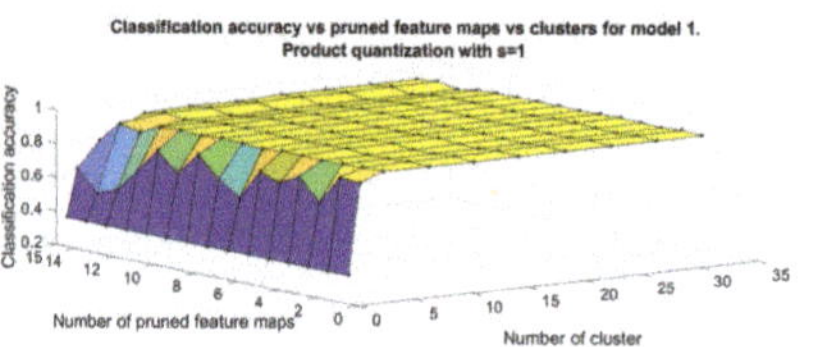

(**a**) Classification accuracy for model 1 with respect to the various combinations of pruned feature map and the clusters in fully connected layer.

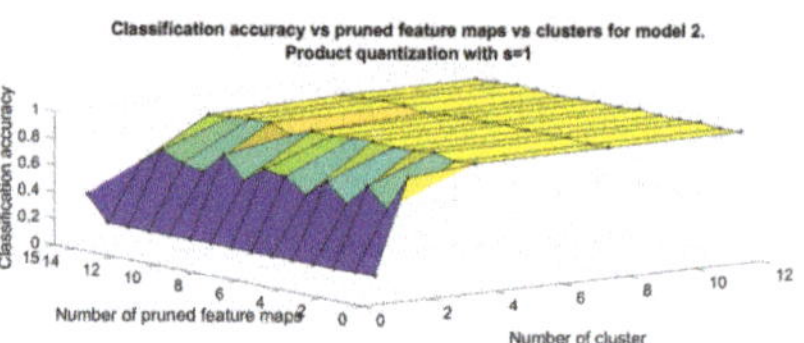

(**b**) Classification accuracy for model 2 with respect to the various combinations of pruned feature map and the clusters in fully connected layer.

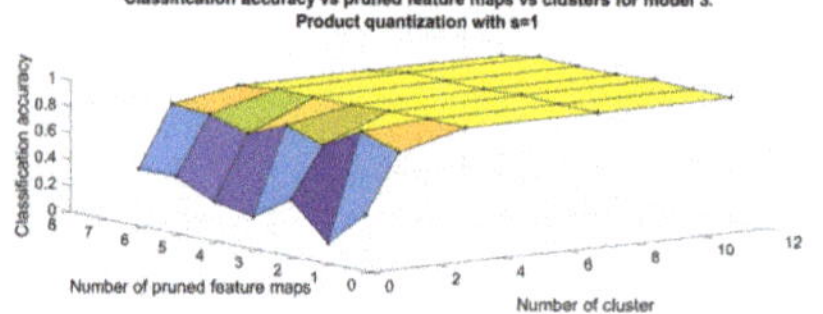

(**c**) Classification accuracy for model 3 with respect to the various combinations of pruned feature map and the clusters in fully connected layer.

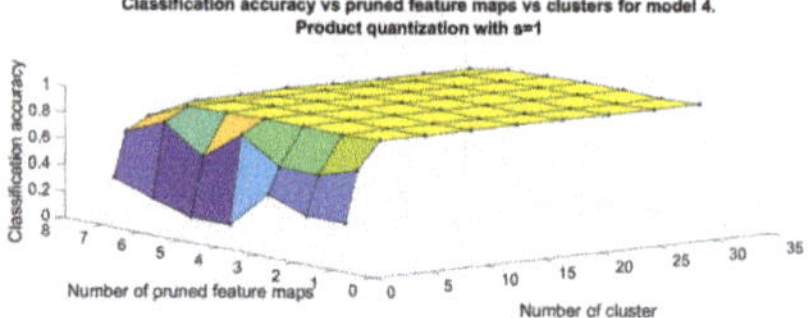

(**d**) Classification accuracy for model 4 with respect to the various combinations of pruned feature map and the clusters in fully connected layer.

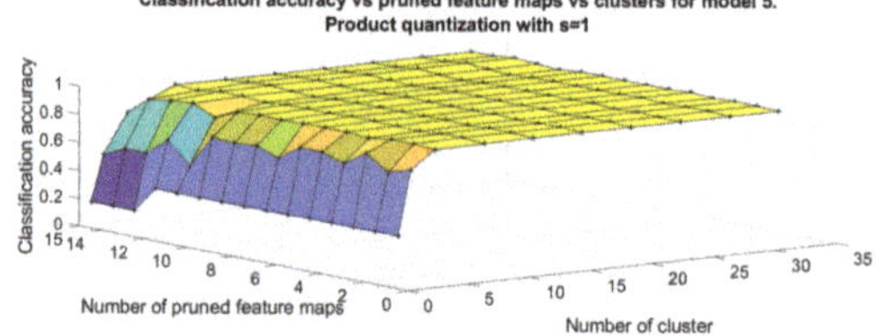

(**e**) Classification accuracy for model 5 with respect to the various combinations of pruned feature map and the clusters in fully connected layer.

Figure 8. Classification accuracy for models in Table 1 for the approach that include filter pruning and product quantization with $s = 1$.

Increasing the splitting parameter s to the value of 2 we take the results presented in Figure 9. Similarly to the results obtained from the previous experiments increasing the number of clusters in each subspace of the weight matrix we are able to improve the performance of the classifier. Again, the highest classification accuracy, 95.52%, is achieved by model 5 when we prune 3 filters from the convolutional layer and quantize the neural network at the back end, with 8 clusters in each subspace, leading to a compression factor of 3.5. However, we can compress it by a factor of

5 and at the same time achieve an accuracy up to 95.35%, which is an acceptable trade off between accuracy and compression, by quantizing it with 4 clusters instead of 8. For this compression rate model 1 yields 94.49% ($pruned\ feature\ maps = 3,\ clusters = 4$), model 2 reaches up to 93.11% accuracy ($pruned\ feature\ maps = 5, clusters = 4$), model 3 up to 92.6% ($pruned\ feature\ maps = 1,\ clusters = 4$) and model 4 achieves an accuracy of 94.66% ($pruned\ feature\ maps = 3, clusters = 8$). It is important to note that in order to achieve a compression rate of 8, we need, for model 5, to prune 7 filters and quantize with 4 clusters, but with a drop at classification accuracy of 2% achieving 93.97%. This result is consistent with those presented in previous sections, where it is shown that increasing the value of parameter s, the accuracy of the classifier at the same level of compression decreases.

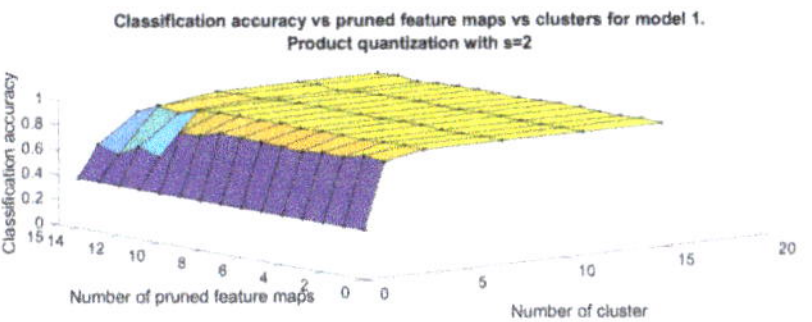

(**a**) Classification accuracy for model 1 with respect to the various combinations of pruned feature map and the clusters in fully connected layer.

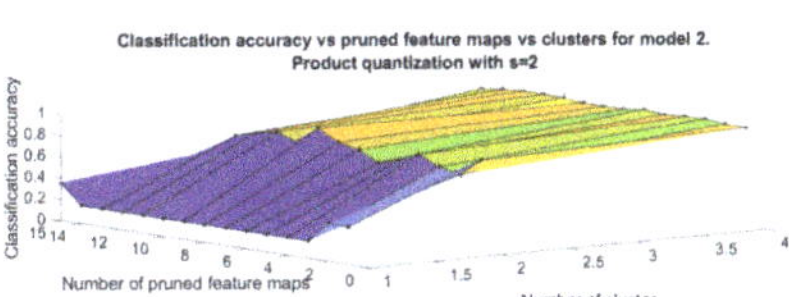

(**b**) Classification accuracy for model 2 with respect to the various combinations of pruned feature map and the clusters in fully connected layer.

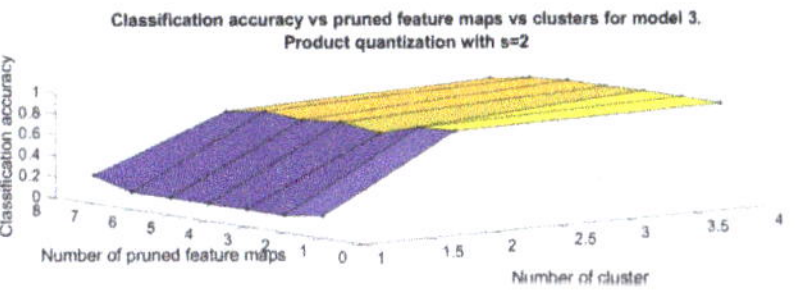

(**c**) Classification accuracy for model 3 with respect to the various combinations of pruned feature map and the clusters in fully connected layer.

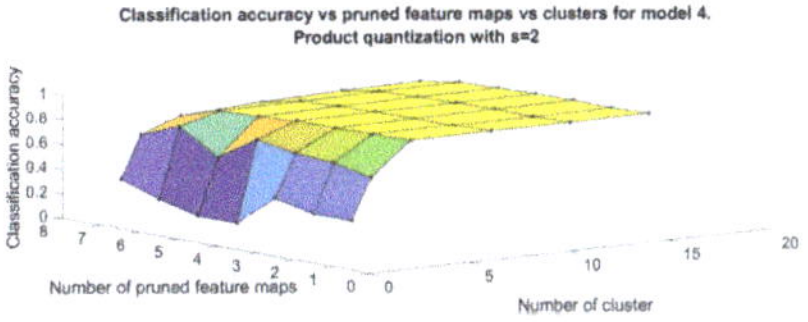

(**d**) Classification accuracy for model 4 with respect to the various combinations of pruned feature map and the clusters in fully connected layer.

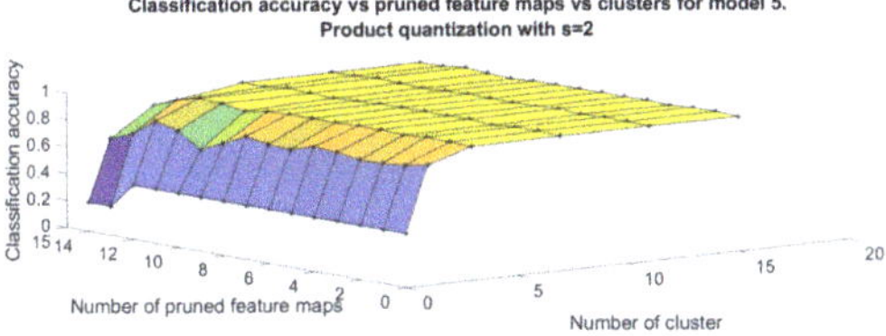

(**e**) Classification accuracy for model 5 with respect to the various combinations of pruned feature map and the clusters in fully connected layer.

Figure 9. Classification accuracy for models in Table 1 for the approach that include filter pruning and product quantization with $s = 2$.

Finally, Figure 10 presents the results for the classification accuracy with respect to the number of pruned feature maps and the number of clusters when the splitting parameter s is set to 4. For this approach the highest classification, 95.52%, is achieved by *Model 5* for the number of pruned filters set equal to five and for four clusters, which results in a compression rate of 4. The following table presents the number of clusters and pruned filters for each model, with a compression rate equal to 4.

Table 13 summarizes the results from the combination of filter pruning and scalar quantization in fully connected layers for the five proposed models and for the same compression rate. It can be seen that Models 1, 4, 5 yield similar classification accuracy despite the fact that Model 4 has fewer filters, while Models 2 and 3, which contain fewer parameters in the fully connected layer, fail to reach the same level of accuracy. In other words we can achieve high performance, even twith limited number of filters, as long as we retain enough parameters at the fully connected layer. Overall, when comparing the compression techniques, Model 5 seems to achieve the best performance in most of the *MultiSubj* evaluation experiments. This fact, along with its superiority in *SingleSubj* and *LOSO* validation schemes, indicates that Model 5 is the most preferable among the five architectures.

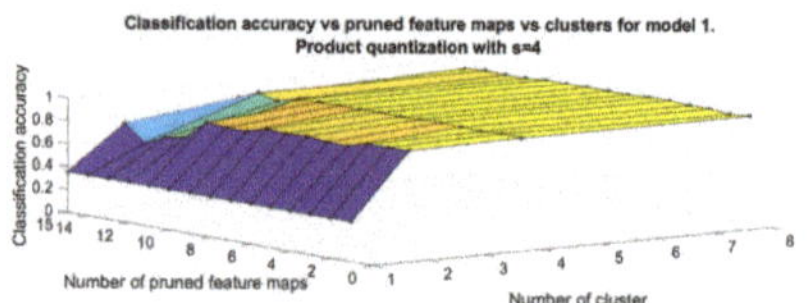

(**a**) Classification accuracy for model 1 with respect to the various combinations of pruned feature map and the clusters in fully connected layer.

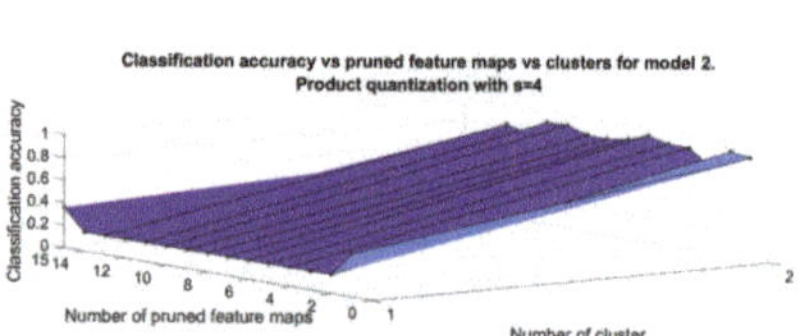

(**b**) Classification accuracy for model 2 with respect to the various combinations of pruned feature map and the clusters in fully connected layer.

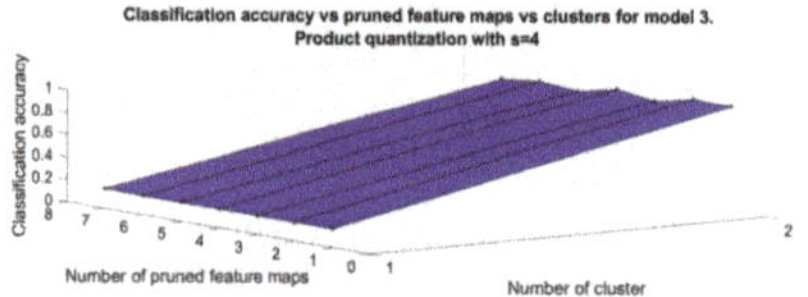

(**c**) Classification accuracy for model 3 with respect to the various combinations of pruned feature map and the clusters in fully connected layer.

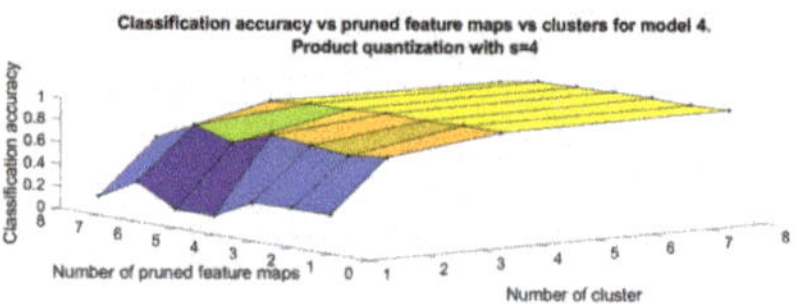

(**d**) Classification accuracy for model 4 with respect to the various combinations of pruned feature map and the clusters in fully connected layer.

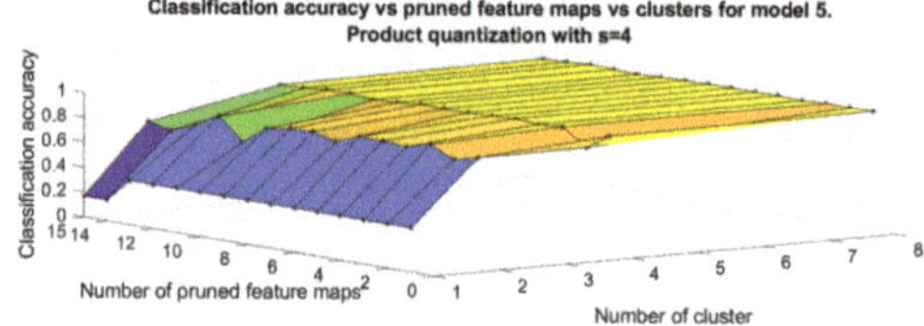

(**e**) Classification accuracy for model 5 with respect to the various combinations of pruned feature map and the clusters in fully connected layer.

Figure 10. Classification accuracy for models in Table 1 for the approach that include filter pruning and product quantization with $s = 4$.

Table 13. Number of clusters and pruned filters for each model with a compression rate equal to 4.

Model	Accuracy	Pruned Feature Maps	Clusters	Compression Rate
1	95.01%	5	4	4
2	92.43%	2	2	4
3	91.74%	1	1	4
4	95.35%	5	4	4
5	**95.52%**	5	4	4

5. Conclusions

Asthma adds an important socioeconomic burden both in terms of medication costs and disability adjusted life years. The accurate and timely assessment of asthma is the most significant factor towards preventive and efficient management of the disease. It outlines the need to examine the technological limitations for real time monitoring of pMDI usage, in order to create easy to use tools for safe and effective management. In this paper, we discussed on current medication adherence monitoring techniques, addressing related aspects that promote adherence with novel sensing capabilities. We, also, investigated acceleration approaches employing convolutional neural networks, trained to classify and identify respiration and medication adherence phases. Employing CNNs directly on the time domain, facilitated lower memory and processing power requirements. Evaluation studies demonstrate that the presented CNN-based approach results in faster execution time, requiring 0.4 s to perform a classification, whereas computationally expensive feature extraction approaches have a mean execution time of 7.5 s. Aiming at acceleration and compression we furthermore applied deep sparse coding strategies, namely filter pruning, scalar values pruning and vector quantization. Different CNN architectures were employed in order to assess the performance of deep sparse coding under various settings. The goal of such methodologies is to speed up the neural network outcome, allowing for real-time implementation, adaptable to IoT architectures and devices. Specifically, we achieve a compression rate of 6.0 in several cases, while maintaining a classification accuracy of 92%. The proposed work provides an experimental evaluation at a key area with renewed research interest, characterized by a high potential for novel improvements related to deep neural networks compression and acceleration. However, our approach is validated only through recordings of three healthy subjects resulting in a small dataset. More experiments with recordings from both healthy and subjects with respiratory disease should be carried out in order to thoroughly assess the presented approach and validate its potential in monitoring medication adherence.

Author Contributions: Conceptualization, A.S.L.; Data curation, S.N.; Formal analysis, V.N., S.N. and A.S.L.; Funding acquisition, A.S.L.; Investigation, V.N., N.D.F. and S.N.; Methodology, S.N. and A.S.L.; Project administration, A.S.L. and K.M.; Software, V.N. and S.N.; Supervision, A.S.L., M.B. and K.M.; Validation, V.N., N.D.F. and S.N.; Visualization, V.N.; Writing—original draft, V.N. and S.N.; Writing—review & editing, V.N., N.D.F., S.N., E.I.Z., A.S.L. and M.B. All authors have read and agreed to the published version of the manuscript.

Funding: This research has been co-financed by the European Regional Development Fund of the European Union and Greek national funds through the Operational Program Competitiveness, Entrepreneurship and Innovation, under the call RESEARCH - CREATE - INNOVATE (project code: T1EDK-03832).

Conflicts of Interest: The authors declare no conflict of interest. The funders had no role in the design of the study.

Abbreviations

The following abbreviations are used in this manuscript:

CNN	Convolutional Neural Network
pMDI	pressurized Metered Dose Inhaler
MFCC	Mel-Frequency Ceptral Coefficients
RF	Random Forest
ZCR	Zero Cross Rate

References

1. World Health Organization The Global Impact of Respiratory Disease. In *World Health Organization: Sheffield*; World Health Organization: Geneva, Switzerland, 2017.
2. Ngo, C.Q.; Phan, D.M.; Vu, G.V.; Dao, P.N.; Phan, P.T.; Chu, H.T.; Nguyen, L.H.; Vu, G.T.; Ha, G.H.; Tran, T.H.; et al. Inhaler Technique and Adherence to Inhaled Medications among Patients with Acute Exacerbation of Chronic Obstructive Pulmonary Disease in Vietnam. *Int. J. Environ. Res. Public Health* **2019**, *16*, 185. [CrossRef] [PubMed]
3. D'Arcy, S.; MacHale, E.; Seheult, J.; Holmes, M.S.; Hughes, C.; Sulaiman, I.; Hyland, D.; O'Reilly, C.; Glynn, S.; Al-Zaabi, T.; et al. A method to assess adherence in inhaler use through analysis of acoustic recordings of inhaler events. *PLoS ONE* **2014**, *9*, e98701. [CrossRef] [PubMed]
4. Jardim, J.R.; Nascimento, O.A. The importance of inhaler adherence to prevent COPD exacerbations. *Med. Sci.* **2019**, *7*, 54. [CrossRef] [PubMed]
5. Gupta, N.; Pinto, L.M.; Morogan, A.; Bourbeau, J. The COPD assessment test: A systematic review. *Eur. Respir. J.* **2014**, *44*, 873–884. [CrossRef]
6. Murphy, A. How to help patients optimise their inhaler technique. *Pharm. J.* **2019**, *297*.

7. Aldeer, M.; Javanmard, M.; Martin, R.P. A review of medication adherence monitoring technologies. *Appl. Syst. Innov.* **2018**, *1*, 14. [CrossRef]
8. Heath, G.; Cooke, R.; Cameron, E. A theory-based approach for developing interventions to change patient behaviours: A medication adherence example from paediatric secondary care. *Healthcare* **2015**, *3*, 1228–1242. [CrossRef]
9. Alquran, A.; Lambert, K.A.; Farouque, A.; Holland, A.; Davies, J.; Lampugnani, E.R.; Erbas, B. Smartphone applications for encouraging asthma self-management in adolescents: A systematic review. *Int. J. Environ. Res. Public Health* **2018**, *15*, 2403. [CrossRef]
10. Kikidis, D.; Konstantinos, V.; Tzovaras, D.; Usmani, O.S. The digital asthma patient: The history and future of inhaler based health monitoring devices. *J. Aerosol. Med. Pulm. Drug Deliv.* **2016**, *29*, 219–232. [CrossRef] [PubMed]
11. Khusial, R.; Honkoop, P.; Usmani, O.; Soares, M.; Biddiscombe, M.; Meah, S.; Bonini, M.; Lalas, A.; Koopmans, J.; Snoeck-Stroband, J.; et al. myAirCoach: MHealth assisted self-management in patients with uncontrolled asthma, a randomized control trial. *Eur. Respir. J.* **2019**, *54*, 745.
12. Polychronidou, E.; Lalas, A.; Tzovaras, D.; Votis, K. A systematic distributing sensor system prototype for respiratory diseases. In Proceedings of the 2019 International Conference on Wireless and Mobile Computing, Networking and Communications (WiMob), Barcelona, Spain, 21–23 October 2019; pp. 191–196.
13. Nousias, S.; Lalos, A.S.; Arvanitis, G.; Moustakas, K.; Tsirelis, T.; Kikidis, D.; Votis, K.; Tzovaras, D. An mHealth system for monitoring medication adherence in obstructive respiratory diseases using content based audio classification. *IEEE Access* **2018**, *6*, 11871–11882. [CrossRef]
14. Ntalianis, V.; Nousias, S.; Lalos, A.S.; Birbas, M.; Tsafas, N.; Moustakas, K. Assessment of medication adherence in respiratory diseases through deep sparse convolutional coding. In Proceedings of the 2019 24th IEEE International Conference on Emerging Technologies and Factory Automation (ETFA), Zaragoza, Spain, 10–13 September 2019; pp. 1657–1660.
15. Pettas, D.; Nousias, S.; Zacharaki, E.I.; Moustakas, K. Recognition of Breathing Activity and Medication Adherence using LSTM Neural Networks. In Proceedings of the 2019 IEEE 19th International Conference on Bioinformatics and Bioengineering (BIBE), Athens, Greece, 23–28 October 2019; pp. 941–946.
16. Howard, S.; Lang, A.; Patel, M.; Sharples, S.; Shaw, D. Electronic monitoring of adherence to inhaled medication in asthma. *Curr. Respir. Med. Rev.* **2014**, *10*, 50–63. [CrossRef]
17. Holmes, M.S.; Le Menn, M.; D'Arcy, S.; Rapcan, V.; MacHale, E.; Costello, R.W.; Reilly, R.B. Automatic identification and accurate temporal detection of inhalations in asthma inhaler recordings. In Proceedings of the 2012 Annual International Conference of the IEEE Engineering in Medicine and Biology Society, San Diego, CA, USA, 28 August 2012; pp. 2595–2598.
18. Holmes, M.S.; D'Arcy, S.; Costello, R.W.; Reilly, R.B. An acoustic method of automatically evaluating patient inhaler technique. In Proceedings of the 2013 35th Annual International Conference of the IEEE Engineering in Medicine and Biology Society (EMBC), Osaka, Japan, 3–7 July 2013; pp. 1322–1325.
19. Holmes, M.S.; D'arcy, S.; Costello, R.W.; Reilly, R.B. Acoustic analysis of inhaler sounds from community-dwelling asthmatic patients for automatic assessment of adherence. *IEEE J. Transl. Eng. Health Med.* **2014**, *2*, 1–10. [CrossRef] [PubMed]
20. Priya, K.; Mehra, R. Area Efficient Design of FIR filter using symmetric structure. *Int. J. Adv. Res. Comput. Commun. Eng.* **2012**, *1*, 842–845.
21. Proakis, J.G.; Manolakis, D.G. *Digital Signal Processing*; PHI Publication: New Delhi, India, 2004.
22. Salivahanan, S.; Vallavaraj, A.; Gnanapriya, C. *Digital Signal Processing*; McGraw-Hill: New York, NY, USA, 2001.
23. Ruinskiy, D.; Lavner, Y. An effective algorithm for automatic detection and exact demarcation of breath sounds in speech and song signals. *IEEE Trans. Audio Speech Lang. Process.* **2007**, *15*, 838–850. [CrossRef]
24. Taylor, T.E.; Holmes, M.S.; Sulaiman, I.; D'Arcy, S.; Costello, R.W.; Reilly, R.B. An acoustic method to automatically detect pressurized metered dose inhaler actuations. In Proceedings of the 2014 36th Annual International Conference of the IEEE Engineering in Medicine and Biology Society, Chicago, IL, USA, 26–30 August 2014; pp. 4611–4614.
25. Taylor, T.E.; Holmes, M.S.; Sulaiman, I.; Costello, R.W.; Reilly, R.B. Monitoring inhaler inhalations using an acoustic sensor proximal to inhaler devices. *J. Aerosol. Med. Pulm. Drug Deliv.* **2016**, *29*, 439–446. [CrossRef]
26. Taylor, T.E.; Zigel, Y.; De Looze, C.; Sulaiman, I.; Costello, R.W.; Reilly, R.B. Advances in audio-based systems to monitor patient adherence and inhaler drug delivery. *Chest* **2018**, *153*, 710–722. [CrossRef]

27. Nousias, S.; Lakoumentas, J.; Lalos, A.; Kikidis, D.; Moustakas, K.; Votis, K.; Tzovaras, D. Monitoring asthma medication adherence through content based audio classification. In *Computational Intelligence (SSCI)*; 2016 IEEE Symposium Series on IEEE; IEEE: New York, NY, USA, 2016; pp. 1–5.
28. Ganapathy, N.; Swaminathan, R.; Deserno, T.M. Deep learning on 1-D biosignals: A taxonomy-based survey. *Yearb. Med. Inf.* **2018**, *27*, 098–109. [CrossRef]
29. Sakhavi, S.; Guan, C.; Yan, S. Learning temporal information for brain-computer interface using convolutional neural networks. *IEEE Trans. Neural Netw. Learn. Syst.* **2018**, *29*, 5619–5629. [CrossRef]
30. Kalouris, G.; Zacharaki, E.I.; Megalooikonomou, V. Improving CNN-based activity recognition by data augmentation and transfer learning. In Proceedings of the 2019 IEEE 17th International Conference on Industrial Informatics (INDIN), Helsinki-Espoo, Finland, 23–25 July 2019; Volume 1, pp. 1387–1394.
31. Papagiannaki, A.; Zacharaki, E.I.; Kalouris, G.; Kalogiannis, S.; Deltouzos, K.; Ellul, J.; Megalooikonomou, V. Recognizing physical activity of older people from wearable sensors and inconsistent data. *Sensors* **2019**, *19*, 880. [CrossRef]
32. Angrick, M.; Herff, C.; Mugler, E.; Tate, M.C.; Slutzky, M.W.; Krusienski, D.J.; Schultz, T. Speech synthesis from ECoG using densely connected 3D convolutional neural networks. *J. Neural Eng.* **2019**, *16*, 036019. [CrossRef] [PubMed]
33. Wu, H.; Gu, X. Max-pooling dropout for regularization of convolutional neural networks. In *International Conference on Neural Information Processing*; Springer: Cham, Switzerland, 2015; pp. 46–54.
34. Li, H.; Kadav, A.; Durdanovic, I.; Samet, H.; Graf, H.P. Pruning filters for efficient convnets. In Proceedings of the 5th International Conference on Learning Representations (ICLR), Toulon, France, 14–26 April 2017.
35. Gong, Y.; Liu, L.; Yang, M.; Bourdev, L. Compressing Deep Convolutional Networks using Vector Quantization. *arXiv* **2014**, arXiv:1412.6115.
36. Wu, J.; Wang, Y.; Hu, Q.; Cheng, J. Quantized Convolutional Neural Networks for Mobile Devices. In Proceedings of the 2016 IEEE Conference on Computer Vision and Pattern Recognition (CVPR), Las Vegas, NV, USA, 27–30 June 2016.
37. Taylor, T.E.; Zigel, Y.; Egan, C.; Hughes, F.; Costello, R.W.; Reilly, R.B. Objective assessment of patient inhaler user technique using an audio-based classification approach. *Sci. Rep.* **2018**, *8*, 1–14. [CrossRef] [PubMed]

MDPI
St. Alban-Anlage 66
4052 Basel
Switzerland
Tel. +41 61 683 77 34
Fax +41 61 302 89 18
www.mdpi.com

Sensors Editorial Office
E-mail: sensors@mdpi.com
www.mdpi.com/journal/sensors

www.ingramcontent.com/pod-product-compliance
Lightning Source LLC
LaVergne TN
LVHW070104170726
843515LV00007B/1610
9783039434305